INTEGRATED BUFFER PLANNING

Integrated Buffer Planning

Towards Sustainable Development

JERZY KOZLOWSKI
University of Queensland, Australia

ANN PETERSON
University of Queensland, Australia

with

The Guide

ANN PETERSON
University of Queensland, Australia

Routledge
Taylor & Francis Group
LONDON AND NEW YORK

First published 2005 by Ashgate Publishing

Reissued 2018 by Routledge
2 Park Square, Milton Park, Abingdon, Oxon OX14 4RN
711 Third Avenue, New York, NY 10017, USA

Routledge is an imprint of the Taylor & Francis Group, an informa business

A Library of Congress record exists under LC control number: 2004057480

ISBN 13: 978-1-138-31570-9 (hbk)
ISBN 13: 978-0-429-45618-3 (ebk)

Contents

PART FOUR: INTEGRATED BUFFER PLANNING (IBP) MODEL

THE GUIDE:

Ann Peterson

The Guide is presented as a stand alone document following the index.

List of Figures

List of Tables

Acknowledgements

This book would not have been possible without the significant contribution of many people. Special acknowledgement must go to Professor Danuta Ptaszycka-Jackowska, who was substantially involved in originating and developing the buffer zone planning concept in Poland in the late 1970s.

In addition, the research output of students studying resource management and environmental planning courses within the School of Geography, Planning and Architecture (previously the Department of Geographical Sciences and Planning) at The University of Queensland has been included in several of the Australian case studies, which focus on applications of the buffer zone planning method to areas of both natural and cultural value. The wide scope of this research has enabled us to rethink several of the central principles underlying buffer planning and design.

Special thanks also go to Jenny Roughan, Kerry Hruza, Nick Vass-Bowen and Chris Izatt whose research, undertaken as partial completion of their degrees in planning at The University of Queensland, has been cited in several of the Australian examples of buffers. Their case studies have provided powerful insights into buffer planning and have stimulated the development of the innovative integrated buffer planning model approach that is the cornerstone of this book.

Thanks also go to Ian Mansergh for allowing us to use illustrations of mountain pygmy-possum habitat. Finally, we would like to acknowledge the support of the School of Geography, Planning and Architecture at The University of Queensland in all aspects related to the final stages of this book.

PART ONE

BACKGROUND

Part one sets the scene for the book by focusing on issues and principles related to planning for sustainable development. It identifies a simplified model planning process, which incorporates the main components of planning and the key players in the process. The concept of buffering significant natural areas is introduced, together with an examination of a range of buffer approaches. The role of protected areas in conserving biodiversity is also examined and some of the more significant problems surrounding their management are identified.

Chapter 1

Buffers: Their Role in Conservation

Introduction

Buffers are a valuable planning tool that increasingly are being used by planners and landscape managers to conserve the values of protected areas and other remnant habitat, as well as freshwater and aquatic ecosystems. Such environmentally sensitive areas (ESAs) are linked in many ways with their adjoining landscapes or regions and many have been experiencing increasing levels of ecosystem stress due to impacts from incompatible and often unsustainable land use practices that occur in their surrounds. Such pressures have been increasing in recent decades due to high levels of habitat fragmentation and isolation, which are often the result of development expanding into natural environments.

The Third World Congress on National Parks and Protected Areas called on governments to initiate sustainable development measures to reduce the external threats to protected areas (McNeely and Miller, 1984). There was concern that unwise land use in the surroundings of protected areas might seriously endanger the security of those areas, if not their very existence. It was stressed that effective resource management could not occur when conservation planning and development planning proceeded in isolation (McNeely and Miller, 1984). These views were reinforced at the Fourth World Congress on National Parks and Protected Areas, with the Caracas Declaration (IUCN, 1992) calling on governments to undertake urgent action to consolidate and enlarge national systems of well-managed protected areas with buffer zones and corridors. The World Resources Institute similarly stressed the importance of the management of resources surrounding protected areas and stated that '... the concept of "buffer zones" or "transition zones" is an essential complement to protected area design' (WRI et al., 1992: 129-133). The Fifth IUCN World Parks Congress continued this theme and emphasised that regional landscape and seascape planning should incorporate zoning and management planning process to assist in designing and enhancing comprehensive protected area networks that conserve wide-ranging and migratory species and sustain ecosystem services. It also recommended the adoption and promotion of protected area design principles that reflect those inherent in the world network of biosphere reserves, where core protected areas (and associated buffers and transition zones) are part of landscapes that are designed to enhance overall conservation values (IUCN, 2003).

Reference to the need for buffers occurs in legislation, non-statutory policy and agreements, management plans and the scientific literature, and many groups within our communities continue to press for their introduction. However, despite the popularity of buffers, and the broad agreement among practitioners that they provide many benefits, planners frequently have little to guide them in their development and implementation. Design has traditionally focused on estimating minimum buffer widths for application in a wide range of differing circumstances. Often these distances are fixed from an identified core area, or important natural area such as a waterway or wetland, with policies or controls being applied uniformly throughout the buffer. The main advantages of fixed buffer widths are their ease of application and enforcement. They can be designed and implemented in a range of circumstances, and may require little expert knowledge, including an understanding of the ecosystem process in operation within the core and its surrounding landscape. However, such an approach may only rarely achieve the desired objectives of protecting the core's values, as individual sites may have a range of physical and social elements, both within the core and in adjacent areas, which need to be examined, understood and incorporated into the buffer design.

The development and application of a rigorous, yet easy to apply methodology, which takes into account the values of the particular area needing to be buffered, as well as its specific features, may be a more appropriate way of designing effective buffers. Use of such a methodology may require a greater input of time and resources. However, if the process of design is easy to understand and implement, little training may be required and hence application can be universal.

It is the purpose of this book to propose such an innovative integrated buffer planning (IBP) methodology, which is based on sound ecological principles, is easy to develop and implement, takes into account site specific issues and threatening processes, and places a high priority on community involvement and understanding. Therefore, a workable approach and a user-friendly guide for the application of the IBP methodology in real life are the principle outcomes of this book. However, the development of the buffer needs to be based on appropriate criteria, and as a consequence the introductory chapters of this book examine these criteria, which are based on several important principles of sustainable development and good planning practice. A brief history of buffers is also outlined and includes a classification of existing buffer approaches.

As one of the main reasons for buffering important natural areas, including protected areas, is to conserve their values, it is first necessary to examine what is meant by the term biodiversity and to consider some of the more important causes of its loss and degradation in recent times.

Biodiversity Loss – What are the Causes?

A diversity of biological resources provides a foundation for all life on this planet. Biodiversity can be thought of as the variety of all life forms, including the different plants, animals and micro-organisms, the diversity of genes they contain, the different ecosystems of which they are a part and the combinations of these ecosystems across the landscape.

There are four main components or levels of organisation of biodiversity (Figure 1.1):

- *genetic diversity* is the total diversity of genetic information contained in individual species. It represents the variation within and between related genes present in different individuals or different species of organisms. It affects a species' physical characteristics, its productivity, viability, resilience to stress and adaptability to change and may enhance population fitness and provide the heritage basis for environmental adaptation and species evolution (Saunders et al., 1998). It may be maintained if the populations of a species are kept above a minimum critical size, which can vary from a few hundred to a few thousand individuals depending on the species;
- *species diversity* is the variety of living organisms on the earth. The protection of this diversity of species has been a priority of past approaches to biodiversity conservation;
- *ecosystem diversity* is the variety of the associations of biotic (plants, animals and micro-organisms) and abiotic (the physical environment) components of the biosphere. Put simply, it is the diversity of the different types of communities formed by living organisms and the relations between them. Ecosystems such as wetlands or grasslands may contain a variety of habitats, biological communities and ecological processes and may vary in species richness, structure,

composition and function. Ecosystem diversity provides critical life-support functions necessary for life on earth; and

- *regional diversity* includes the variety in kinds of ecosystems and their patterns and linkages across regions. It consists of the diversity of the landscape components of a region and the functional relationships that affect environmental conditions within ecosystems. The terms 'ecoregion' or 'bioregion' have been used to describe this diversity.

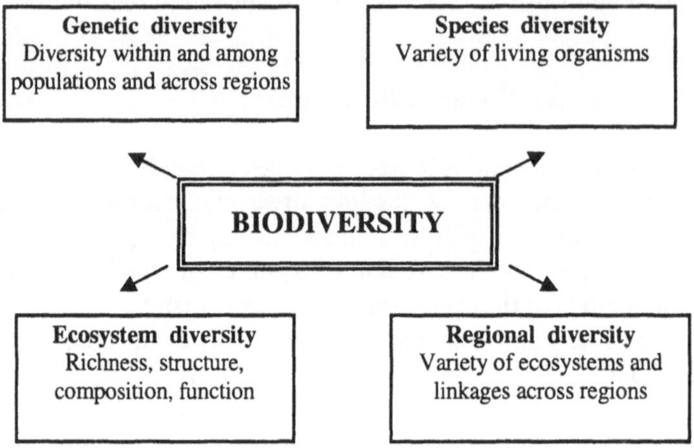

Figure 1.1 Main components of biodiversity

Biodiversity has many values (Figure 1.2). It provides renewable biological resources such as food, timber, shelter and pharmaceuticals, and contributes, through the use of wild relatives and genetic engineering, to the improved resistance of many crops to pests and disease. High levels of biodiversity support the proper functioning of ecological systems on which humans depend and provide the genetic variation necessary for continued evolution of species. For example, it provides environmental services such as the breakdown of pollutants and the recycling of nutrients, an important service for absorbing the waste stream from industrial and agricultural activities. It allows for the continuing exchange of carbon dioxide and oxygen, a critical cycle for life on this planet. Biodiversity is also important in aesthetic terms as people favour variety in their experiences, especially in the natural world. Maintaining and restoring biodiversity also helps to retain the community's options for the future. Ethical considerations further stress the right to life of all species.

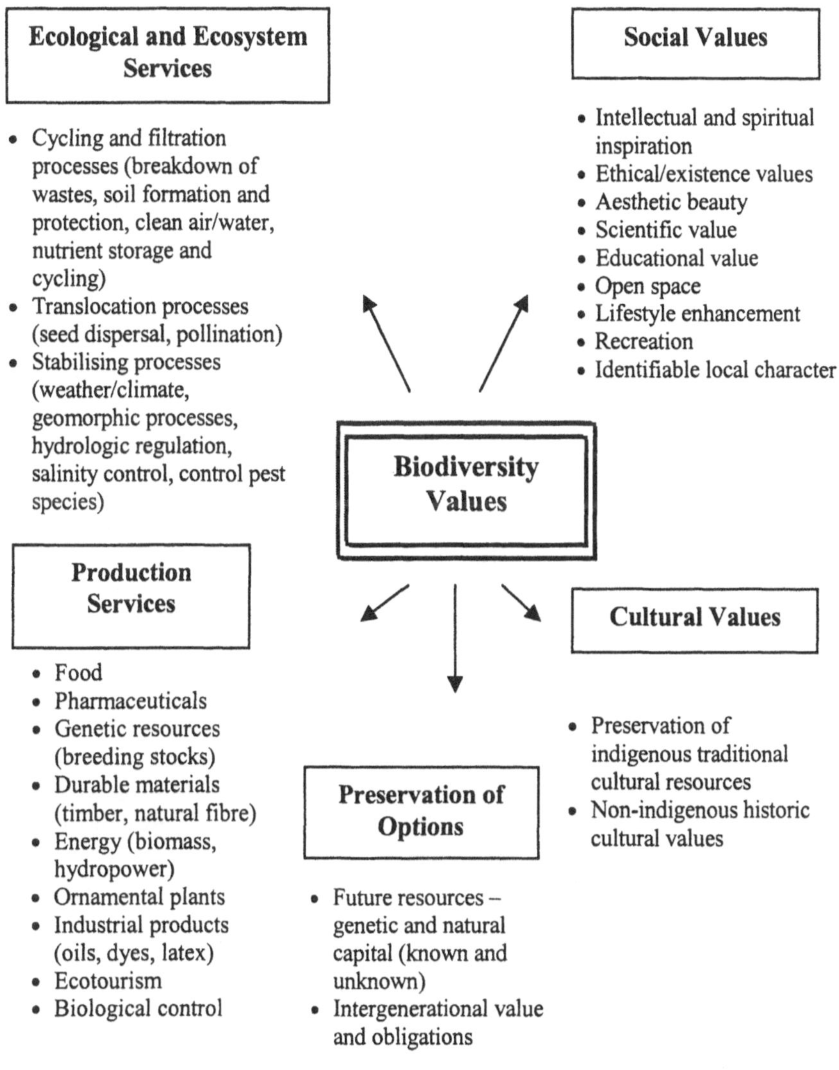

Figure 1.2 Biodiversity values

There are three primary attributes of biodiversity (Noss, 1990): composition, or the variety of elements, such as species in a collection; structure, or the pattern of a system across the landscape; and function, meaning the ecological and evolutionary processes. Biodiversity should be planned for, managed, monitored and evaluated at multiple levels of organisation (genetic, species, ecosystem and regional diversity) in relation to composition, structure and function, and at multiple temporal scales. Such a 'biodiversity hierarchy' (Noss, 1990; Sattler, 1993) illustrates the heterogeneous nature of biodiversity, involving many overlapping ecosystems and many interdependent elements.

Ecosystems and the species they contain are being lost at an accelerating rate. However, current extinction rates give little indication of the genetic impoverishment of many species that are on the verge of extinction due to fragmentation of their habitat, the progressive loss of sub-populations and their diminished capacity to adapt, thus ultimately threatening their survival status (Spellerberg and Hardes, 1992; IUCN Species Survival Commission, 2004). This loss of richness may be potentially dangerous because the environmental systems of the world support all life and little is known about the diversity and variability of plants, animals, micro-organisms and the key components of ecosystems that are necessary to ensure essential ecosystem functioning and the conservation of biodiversity.

Extinction of species is a feature of life on our planet, the average estimated 'natural' or background rate of extinction being about 25 species each year from a conservative estimated total of around 10 million species, or three species per year for the 1.3 million described species (Magin et al., 1994). Extinctions appear to be concentrated among certain taxa. Thus, although vertebrates constitute around 3.5 per cent of described animal species, they account for around 50 per cent of known extinctions and have experienced five times more extinctions than predicted from the estimated background extinction rate (Magin et al., 1994). While most past extinctions have occurred by natural processes, today human activities are overwhelmingly the main cause, both direct or indirect (Table 1.1) and the rate of extinction may be hundreds of times higher than in the past (World Resources Institute (WRI) et al., 1992; IUCN Species Survival Commission, 2004). 'Habitat change' is the major cause of species' extinction (Table 1.1), followed by hunting and species introductions (and associated links to competition, predation, parasitism and disease). Similarly, most population extinctions are the result of persistent changes in habitat or the local environment of a population (Noss and Murphy, 1995).

Habitat loss and degradation have resulted in fragmentation and isolation of habitat and resultant altered ecological processes and assemblages of species. This process is inextricably linked to human population size and growth rates and to increasing levels of consumption. No other agent of environmental change has been as devastating as humans, or so thorough.

Table 1.1 Causes of species extinctions (%)

Class	Human Predation*	Introduced Species	Indirect Effects	Habitat Change	Natural Causes
Invertebrates	16	33	1	49	1
Aquatic vertebrates	8	37	4	50	1
Terrestrial vertebrates	30	46	1	22	0

(* Includes species consumed as food, exploited for other biological products, and deliberately eliminated as pests. Where invertebrates (n=72); aquatic vertebrates [n=28]; and terrestrial vertebrates [n=111]). Source: Thomas (1994:374).

A study of 61 tropical countries (Primack, 1993) concluded that in 49 of these countries more than 50 per cent of the original wildlife habitat had been destroyed (Figure 1.3). In tropical Asia and sub-Saharan Africa approximately 65 per cent of the wildlife habitat has been lost. Relatively limited success has been achieved worldwide in stemming the loss of species. In most cases of extinction, species loss is due to a combination of factors. Thus, if habitat loss and alteration are the main causes of species

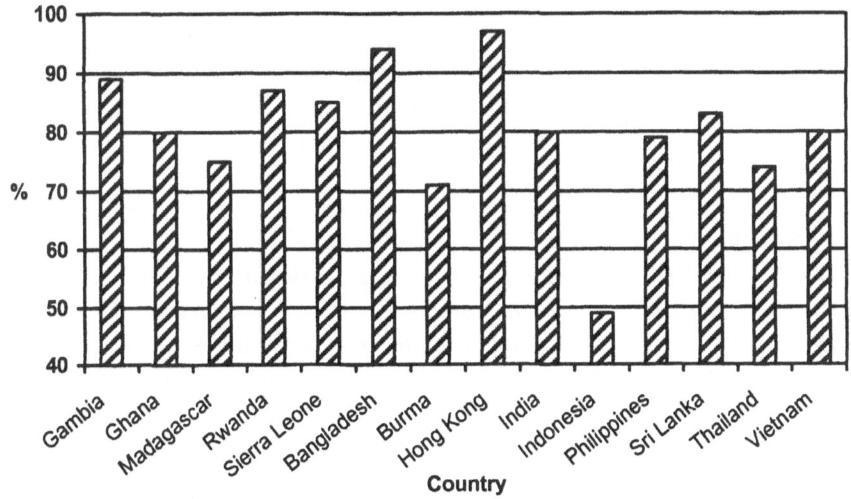

Figure 1.3 Wildlife habitat loss in selected tropical countries
Source: IUCN/UNEP (1986) (adapted from Primack 1993:119).

losses, habitat protection and management must be a central theme for any nature conservation strategy. These strategies should be based on the elimination or amelioration of major threatening processes and their causes before a particular species or ecosystem becomes endangered or before an endangered species reaches critically low levels. It is in this context that planners have an important role to play, and that integrated buffers may be a useful tool.

Planning, Conservation and Development

Planning frequently stems from the identification of problems and bridges the gap between the current situation and the desired future situation, with plans being produced to guide future development. Planning is thus a decision process incorporating the conscious determination of courses of action based on facts and considered estimates. Thousands of years ago it became clear that effective organisation of space and the allocation of land uses, while ensuring appropriate conservation of the natural environment, were necessities that determined the quality of everyday life, particularly its comfort and convenience. Dealing with these types of problems is the primary 'mission' and challenge, of what, over the past hundred years has become known as 'town planning'. However, accompanying the earth's unprecedented population expansion has been an increased demand for the use or development of land, for agriculture, forestry, industrial, residential or recreational purposes. As the total stock of land is fixed and limited in relation to the number of activities and the intensity of uses that it can support, there is inevitable conflict among competing uses. This conflict over the use of land arises on three levels as a result of the development process:

- *Change of use*
  Development that changes the use of a particular piece of land, for example from a forest to a residential estate or farmland, may have positive and/or negative influences on the community's social, cultural, economic and ecological interests. Planning should provide some level of control over the change of land use and aim to ensure that both the present and future community's interests are protected and a sustainable outcome is achieved.
- *Side effects of use*
  A development may produce unexpected side effects (both positive and negative) on its surroundings, including other developments, the

community and the environment. Typically many developments are encouraged for the employment and other economic benefits they may provide. However, they may also produce negative impacts such as the fragmentation of habitat or the introduction of pest species. Planning offers the possibility of a more rational location of specific activities based on an environmental assessment of the proposed development and its impacts. One of the prime responsibilities of planners is to assist in maximising the positive impacts and avoiding or minimising the negative impacts of development, through influencing the development decision-making process.

- *Form of use*
  Improper land use produces wide ranging losses to the community, economy and environment. Planning has an important role to play in ensuring an appropriate form of land use, one that will enable sustainable development and the conservation of biodiversity. Planning must, however, recognise the interests of the community at large and of key stakeholders, this being one of the hardest problems that local governments and state/provincial planning authorities have to face.

Planning is complicated by conflicting values, especially between those advocating development and those favouring conservation, and frequently the proponents do not realise that the one can not do without the other. A common form of conflict is that frequently the most favoured sites for protection of natural values are those that are sought after by agriculture, tourism and many other forms of development. Land suitable for nature conservation must compete with the demands of other uses for the diminishing and scarce land resource. Such conflict is part of the very essence of the planning problem, and planning should aim to deal in a rational manner with this conflict and develop a rationale for the organisation of space and land uses.

The ecological role and responsibility of the planning profession is clear. Planning should provide a base for guiding and controlling development and for addressing these conflicting interests in the use and development of resources. Planning has a very important role to play in ensuring the sustainable use and development of natural ecosystems and in particular of conserving biodiversity.

Protected areas, such as national parks, play an important part in the conservation of biodiversity. However, the existing network of protected areas will not afford effective long-term protection of biodiversity due to the increasing fragmentation and isolation of these areas, the wide range of external threats to individual protected areas and the high level of bias in representation of biodiversity within the network. Effective conservation of biodiversity requires the development of a comprehensive, adequate and representative (CAR) reserve system and the integration of these areas with other environmentally sensitive areas and with their surrounding landscapes. It is in this context that planning can contribute most, for planners deal with the spatial organisation of land uses and attempt to minimise the conflict between competing uses of land and water.

Planning aims to promote a use of land that will ensure the sustainable development of resources, and as such, it has a specific role in the conservation of land of high biodiversity value. The ecological role and responsibility of land use planning in biodiversity conservation, particularly at the local and regional level, is widely recognised, although it has not yet been effectively implemented by practicing planners. The development role in planning continues to take precedence over its role in the conservation of the environment, with the result that many past and present land use planning decisions, or indeed, the lack of them, are a key issue in the loss of biodiversity in many countries.

For many wildlife species, their habitat may be in competition with what are thought to be more productive land uses such as agriculture, mining or forestry. In the past ecosystems and their wildlife have suffered, particularly as a result of land use planning systems that have allocated use rights to landholders, enabling them to degrade or even clear habitat. Good planning should be comprehensive and ensure that all resources, including biodiversity, are considered when land and water are allocated to particular uses, and that biodiversity, including the diversity of genes, species, ecosystems and regions, is conserved. This will require protection of adequately sized natural areas, linkage of remnant patches, minimising the threats to these important core areas and corridors, and implementing land use strategies that are compatible with the retention of biodiversity, in particular, strategies that prevent the fragmentation and isolation of habitat.

A specific gap in current planning practice lies in addressing the impacts on protected areas and unprotected remnants of native vegetation, and freshwater and marine ecosystems, all of which could be termed environmentally sensitive areas (ESAs), from the land uses and activities taking place within and surrounding these areas. No longer is it possible to assume that their biodiversity will be conserved in perpetuity, for the most significant threats to these areas are likely to be from uses and activities originating in surrounding lands. Growing fragmentation and isolation of remnant patches is placing them under increasing levels of stress. Important roles for planning are to ensure an effective spatial arrangement and connectivity of remnant patches, based on an integrated planning approach across entire regions (or landscapes) and to ensure that the land uses and activities on lands within and surrounding ESAs contribute to the conservation of the biodiversity in these patches in the long term. This may require a fundamental shift in the types and forms of development that are permitted in these areas and the use of innovative tools or techniques to encourage community acceptance of such an approach.

Despite the rhetoric expressed by both conservation and planning agencies concerning the need to better manage ESAs and adjacent areas, few effective strategies are in place and the methodological platform for determining them are either missing, or at best, not convincingly established. This is definitely the case regarding the introduction of buffers around ESAs affected by external threats. Buffers can become an important means of giving added protection to ESAs. Previous buffer planning research has been mainly empirical, resulting in the analysis and development of individual case studies rather than the development of a general buffer planning theory and no practical planning tools in this field are available 'on the market'. Approaches to date have also largely failed to fully integrate the preparation of the buffer plan within the existing institutional framework. Lack of an appropriate methodology and planning framework hinders any attempts for wider introduction of buffer zones and ultimately their effectiveness.

Purpose of the Book

This book intends to address the evident gap in research and planning for buffer zones and it focuses on the development of a theoretically sound and user-friendly tool to guide the design and implementation of buffers to conserve areas of high nature conservation value. There are perhaps three basic approaches to solving this planning problem:

- *science-based approaches*, which look to science to derive principles and practices and ways to upgrade them continuously;
- *practice-based approaches*, which look to professionals and programs involved in the conservation process, in order to review state-of-the-art principles and practices, and identify ways to improve them; and
- *innovative approaches*, which look to improve principles and practices by putting something new into practice (Clark, 1995).

The following chapters will explore these three approaches. The general structure of these chapters is described below.

Part One: Generation of the Concept of Buffers

This part focuses on why there is a need for buffers. In Chapter 2 the concept of sustainable development is used as a basis for buffer planning. Several important principles of sustainable development are discussed, including the need to conserve biodiversity. The role of planning is outlined and the types of problems addressed by planning are described and evaluated. Important principles to guide plan development are also identified. These form an important component of the suggested 'model' planning process, which provides the framework for the integrated buffer planning model, which is the prime focus of this book.

Dedication of protected areas, the traditional approach to biodiversity conservation, is examined in Chapter 3 to assess its continuing effectiveness. The need for an ecological basis to the management of protected areas and other ESAs is advocated based on the growing recognition that biodiversity is not adequately represented in existing protected areas and that an approach that focuses on the lands and waters outside of protected areas may afford greater protection to biodiversity and a resultant increase in the long-term sustainability of species and their ecosystems. The nature of ecosystem management or integrated landscape management is examined to establish general principles that planners, who are responsible for the development of land use planning strategies, can draw upon to aid biodiversity conservation and management. Although science-based principles are important, the integrated landscape management approach also stresses the need to consider the wider planning framework including political, social and economic contexts.

Part Two: Critical Review of Buffers

Past and present approaches to buffer zone planning are described in Chapter 4. The benefits of buffers for ESAs are also described and an important set of criteria, which should underpin integrated buffer design, are developed. Chapter 5 presents a detailed evaluation of buffer approaches in the Australian context and an examination of the policy and legislative framework of these buffers.

Chapter 6 provides a critical review of a range of buffer approaches. The objective is to synthesise previous research and critically evaluate it, using the criteria outlined in Chapter 4, so that an effective integrated buffer planning (IBP) model can be developed (Chapter 11). Evaluation of a range of approaches to buffer planning for ESAs makes it possible to identify the adequacy of current and past approaches, and to then modify, or even abandon old practices and develop new, more environmentally sustainable strategies. In this way evaluation studies provide important information on which to improve the quality of decision making (Clark, 1995). This particular approach has been little used in biodiversity conservation. The actual buffer planning technique, the IBP model, which is recommended in this book, is embedded within a wider planning framework to ensure consideration of a range of important institutional, political, and socio-economic factors that affect plan preparation and implementation. Through this approach the IBP model will present a new approach to the problem of planning for the long-term conservation of biodiversity, one that integrates the principles of sustainable development, biodiversity conservation and good plan design.

Part Three: Applications of the Buffer Zone Planning Model

Part three provides more detailed descriptions of the Buffer Zone Planning (BZP) model (Kozlowski and Ptaszycka-Jackowska, 1987), which is identified as possessing several important and practical features upon which to base a 'best practice' buffer model. In Chapter 7 the application of BZP to protected area planning in Poland is described. Chapters 8 and 9 provide more detailed Australian case studies on the BZP approach, with Chapter 10 focussing on its application to cultural heritage conservation, with two Australian pilot applications described.

Part Four: The Integrated Buffer Planning Model

Part four is devoted to laying the foundation for a recommended integrated buffer planning (IBP) approach. Chapter 11 briefly describes the main features of the IBP method, while a 'ready to use' Guide that describes in detail the main phases, steps and actions to be undertaken in the IBP model is presented at the end of the book. The Guide also provides a wide range of examples of how the recommended actions have been applied. A partial application of the IBP model to significant natural areas with important koala habitat in Australia (Peterson, 2002) is included in these examples to indicate how the IBP could be developed and applied by planners. It illustrates that sufficient information frequently exists within planning agencies and local governments to adopt the IBP method. Chapter 12 provides a summary review of buffer planning and looks at new avenues for this innovative form of integrated buffer zone planning.

Conclusion

The overriding goal in biodiversity planning and management is achieving sustainable outcomes that must be able to be implemented effectively and efficiently on the ground. There is an abundance of scientific research and information relevant to the conservation of ESAs, yet this knowledge has had little, if any, effect on planning practice. It is our contention that buffer zones that are designed and implemented in an ecologically sound manner will help to ensure the conservation of species and genetic diversity and the proper functioning of ecosystems within and around ESAs. Further, it is argued that planning is necessary to effectively determine and establish effective buffers and, as a consequence, to aid in the protection of biodiversity. Such a process will provide advantages over the present, less systematic buffer strategies that lack a common agreed methodology.

The purpose of this book, therefore, is not to add more 'truth' to the body of existing scientific knowledge, but rather to develop an innovative integrated buffer planning methodology that will assist planners, in a practical sense, to arrive at sustainable land use policies for land within and around ESAs. The proposed approach is developed to incorporate wider socio-economic and political issues to enhance its applicability to a range of situations. Hence, the solution to the problem is in terms of the procedures developed. As Clark (1995:23) concludes:

> A combination of practice-based and innovation strategies, combined with the ongoing science-based strategy, are a proven means to accelerate progress in (biodiversity) conservation.

The value of this approach will be determined by the extent to which its methods and findings make possible improvements in practice.

References

Clark, T.W. (1995), *Developing Pragmatic Koala Conservation Management Policy,* Draft Koala Policy Paper, School of Forestry and Environmental Studies, Yale University and Northern Rockies Conservation Cooperative, np.

International Union for the Conservation of Nature (IUCN) (1992), Recommendations, IVth World Congress on National Parks and Protected Areas, Caracas, 10-21 Feb.

International Union for the Conservation of Nature (IUCN) (2003), Recommendations of the Vth IUCN World Parks Congress, 9-17 Sept. Available at: http://www.iucn.org/themes/wcpa/wpc2003/english/outputs/recommendations.htm

IUCN Species Survival Commission (2004), *The IUCN Red List of Threatened Species*, Available at: http://www.redlist.org/ [12/12/2004].

Kozlowski, J. and Ptaszycka-Jackowska, D. (1987), 'Planning for Buffer Zones' in P. Day (ed.), *Planning and Practice*, Department of Regional and Town Planning, The University of Queensland, Brisbane, pp. 200-15.

Magin, C.D., Johnson, T.H., Groombridge, B., Jenkins, M. and Smith, H. (1994), 'Species extinctions, endangerment and captive breeding', in P.J.S. Olney, G.M. Mace and A.T.C. Feistner (eds), *Creative Conservation. Interactive management of wild and captive animals*, Chapman and Hall, London, pp. 3-31.

McNeely, J. and Miller, K. (1984), *National Parks, Conservation and Development.* Proceedings of the World Congress on National Parks, Bali, 1982, Smithsonian Institution Press, Washington.

Noss, R.F. (1990), 'Indicators for Monitoring Biodiversity: A Hierarchical Approach', *Conservation Biology*, 4(4), pp. 355-64.

Noss, R.F. and Murphy, D.D. (1995), 'Endangered Species Left Homeless in Sweet Home', *Conservation Biology*, 9(2), pp. 229-31.

Primack, B. (1993), *Essentials of conservation biology*, Sinauer Associates, Sunderland.

Sattler, P.S. (1993), 'Towards a nationwide biodiversity strategy: the Queensland contribution', in C. Moritz and J. Kikkawa (eds), *Conservation Biology in Australia and Oceania*, Surrey Beatty and Sons, Sydney, pp. 313-25.

Saunders, D., Margules, C. and Hill, B. (1998), *Environmental indicators for national state of the environment reporting – Biodiversity*, Australia: State of the Environment (Environmental Indicator Reports), Department of the Environment, Canberra.

Spellerberg, I.F. and Hardes, S.R. (1992), *Biological Conservation*, Cambridge University Press, Cambridge.

Thomas, C.D. (1994), 'Extinction, Colonization, and Metapopulations: Environmental Tracking by Rare Species', *Conservation Biology*, 8(2), pp. 373-8.

World Resources Institute (WRI), IUCN and UNEP (1992), Global Biodiversity Strategy. Guidelines for Action to Save, Study, and Use Earth's Biotic Wealth Sustainably and Equitably, WRI/IUCN/UNEP, Gland.

Chapter 2

Planning for Sustainable Development

Introduction

This chapter critically assesses the role and responsibilities of planning for sustainable development, especially in relation to conserving biodiversity and managing ecologically sensitive areas (ESAs), as advocated in the Brundtland Report, Agenda 21, and at the Johannesburg summit. As a consequence, it highlights the current key problems and the most promising prospects of planning for sustainability, and emphasises the need for planning to continually enhance its interdisciplinary character, based on the effective co-operation and mutual understanding of the key players in the process. To facilitate the attainment of this end, the main forms of planning are briefly examined in this chapter and synthesised into an outline 'model' process to highlight how an 'integrated buffer planning' (IBP) methodology, which is the main topic of this book, can best be incorporated into this process.

Sustainable Development

The environmental crisis is certainly global, with human civilisation continuing to destroy the natural resources on which its existence is based. Scientists and politicians across the world have become aware of this new ecological challenge and are beginning to take preliminary steps towards devising potential strategies to control the threats involved.

A fundamental question is whether development can continue to achieve its ends while, at the same time, its negative impacts are reduced to a level at which they are no longer a major threat to our survival. This question was addressed by the World Conservation Strategy (WCS) in 1980, and a fresh approach to the problem was launched on the basis that 'Development and conservation are equally necessary for our survival and for the discharge of our responsibilities as trustees of natural resources for the generations to come' (IUCN, 1980: I). This statement promoted the idea of sustainable development and of the integration of development with conservation.

The WCS (IUCN, 1980: 1) defined conservation as 'the management of human use of the biosphere so that it may yield the greatest sustainable benefit to present generations while maintaining its potential to meet the needs of future generations'. The strategy indicated that 'for development to be sustainable it must take account of social and ecological factors, as well as economic ones; of the living and non-living base; and of the long term as well as the short term advantages and disadvantages of alternative actions'. (IUCN, 1980:1) The strategy set three fundamental goals for 'living resource conservation':

- to maintain essential ecological processes and life support systems;
- to preserve genetic diversity; and
- to ensure the sustainable utilisation of species and ecosystems.

The main, and most commonly recognised goal of sustainable development, formulated and widely promoted by the Brundtland Report, is to achieve a reasonable and equitably distributed level of economic well-being that can be perpetuated through 'development that meets the needs of the present without compromising the ability of future generations to meet their own needs' (WCED, 1987: 8). Clearly, if this goal is to be achieved, sustainable development must be based on both conservation and development. However, this goal has been criticised for advocating too much economic growth to achieve sustainable development, with limited attempts to redirect this growth. It failed, for instance, to indicate the importance of recognising the urgent need for a new approach to environmental accounting to address the proper pricing of free goods such as water or air.

Further milestones on the 'road to recovery' were, among others, such international agreements and conventions as the 1987 Montreal Protocol to reduce CFCs, the 1988 First World Conference on 'The Changing Atmosphere' and the 1992 UN Earth Summit in Rio de Janeiro, which directed its focus to treaties on biodiversity, climate change and the so-called Agenda 21, which was to address the problems of the twenty first century. This summit '...marked the beginning of a continuing dialogue between the rich and the poor nations over the management of the Earth…' (Pickering and Owen, 1994: 315), and culminated in 2002 in the Johannesburg summit.

Among many definitions of sustainable development, the one proposed by the Strategy for Sustainable Living (a follow-up to the World

Conservation Strategy) is particularly relevant for physical planning as it considers that the main aim of sustainable development is 'improving the quality of human life while living within the carrying capacity of supporting ecosystems' (IUCN et al., 1991:10). The concept of 'carrying capacity' is directly linked with that of the final limits of the earth's ecosystems and the impacts they can withstand, without irreversible damage, while the expected services of 'supporting ecosystems' clearly depend on conservation of biodiversity.

Traditional free market economists do not recognise any 'limits' to economic development and believe it can continue to grow exponentially forever. However, advocates of sustainability (Daly and Cobb, 1989; Pearce et al., 1989; Barrow, 1995) agree that there are final, or critical limits (constraints or thresholds) to what the natural environment can take or provide, that they determine a carrying capacity that cannot be continually violated without a threat to our survival, and that science and technology can never provide effective means for permitting the extension of these limits indefinitely. As a consequence, achieving sustainable development means ensuring that both its ecological and economic dimensions are achieved simultaneously.

However, social sustainability, commonly recognised as the third major pillar of sustainable development, must be briefly considered at this stage. Hawken et al. (1999) point to the similarities between environmental and social services, where in both cases, only the sale of 'monetised' goods and services is rewarded. The resulting destruction of natural capital and its impact on ecological and economic sustainability has already been discussed, but it is important to realize that the current approach is equally damaging to human capital and, in particular, the social services it provides. This usually destroys or diminishes social integrity at the expense of economic gains. Hawken et al. (1999) recommend that approaches, which aim to improve social sustainability, need to take into account our current situation, which is characterised by abundant labour and scarce resources. Several innovative approaches are suggested by these authors.

Several authors stress the need to address poverty, which has been 'one of the important enemies of sustainable development' (Streeton, 1992:129). Oodit and Simonis (1992) perceive the relationship between poverty, environment and development as a 'vicious circle'. Their extended discussion of this matter clearly implies that alleviating poverty and achieving socially sustainable development depends, primarily, on success in first achieving ecological and economic sustainability.

In summary, the concept of sustainable development has evolved from a narrow ecological focus to reflect a growing awareness of the complex interactions among the environment, economic, social and political systems, and to place emphasis on conserving biodiversity. This is evidenced by statements that conservation of 'species and their ecosystems ... is an indispensable prerequisite for sustainable development' (WCED, 1990:210), and that maintenance of biodiversity and ecosystem processes underpin ecologically sustainable development. While renewable resources are essential for human prosperity, without the environmental services provided by diverse ecosystems, many of which cannot be replaced, life on earth would not be able to continue in its present form, if at all (Munashinghe and McNeely, 1994; Hawken et al., 1999). As a consequence, achieving sustainable development means ensuring that ecological, social and economic components related to specific developments are integrated. Several core principles of sustainable development, many of which have particular relevance in biodiversity management and planning are briefly described in Table 2.1. These are the pillars of sustainable development and provide the main context within which the integrated buffer planning methodology will be examined and developed.

The Role of Planning

The problem of solving or easing pressures on natural resources caused by accelerating development, while at the same time retaining a relatively conflict free co-existence between people, is as old as our civilisation. It can be argued that development is commonly governed by three major factors:

- *social goals*, which reflect the physical and intellectual needs of a given community;
- *geographic environment*, which creates constraints and opportunities for development; and
- *development determinants* (circumstances), which may include the state of the economy, the level of technology, the social organisation, cultural traditions and political system etc.

A generic definition of any planning is that it is the process of defining goals and of indicating the ways and means by which these goals can be

Table 2.1 Principles to guide sustainable development

Principle	Meaning
Conserve biodiversity	⇨ plan to protect a variety of life forms and their genetic diversity, and maintain ecological process, landscape and regional diversity
Undertake values assessment	⇨ assess the value(s) of the natural/environmental resources within the planning area
Use resources sustainably	⇨ use renewable resources within the capacity of natural systems to regenerate and use non-renewable resources at a rate that enables substitution of other resources
Use an integrative approach	⇨ plans should integrate economic, socio-cultural, institutional, political and environmental goals in local, regional, state and national plans and policies
Plan within ecological boundaries	⇨ look at functioning ecosystems and avoid planning solely within a restricted administrative jurisdiction
Incorporate effective participation	⇨ include all relevant stakeholders in the planning process
Ensure equity (intergenerational and intragenerational)	⇨ look after the needs of future generations, while ensuring fairness and equity for the existing community
Use precaution	⇨ in the face of uncertainty, risk and irreversibility, plan cautiously to avoid damage to the environment
Take a global view	⇨ recognise that plans and policies may impact at the local, regional, state, national and global level
Incorporate effective implementation	⇨ use a variety of strategies – regulatory, non-regulatory and market based incentives
Plan within realistic time frames	⇨ adopt a long-term, rather than short-term view

attained. While the goals should be defined by the community and key stakeholder groups as part of the planning process, the socio-economic determinants are generally beyond the direct control of planning. As a consequence, the role of planning is:

> to indicate how, within a given geographical environment and socio-economic determinants, development can most efficiently be guided, controlled and co-ordinated to achieve the community's pre-determined goals and, at the same time, achieve ecological, social and economic sustainability.

Recognising this statement inevitably means accepting sustainable development as an integral part of the goal of planning. This has already been happening. For instance, a new Integrated Planning Act, gazetted in Queensland (Australia) in 1997, states that the 'purpose of this Act is to seek to achieve ecological sustainability' (s1.2.1). This is certainly not the only place in the world where planners have been moving in this direction although, regrettably, there are more places where they have not.

In this context, planners involved in generating and/or advancing various development proposals (e.g. policies, strategies or projects) and in determining their environmental and economic consequences have a responsibility to integrate the principles of sustainable development, formulated at global, national and local levels, into decision-making processes to ensure that the outcomes of development are sustainable and that biodiversity is conserved. To discharge this responsibility, planners should primarily concentrate upon the following:

- *Management of development*, with particular attention given to the sustainable use of land and associated resources. This should be carried out primarily through effectively establishing the preferred:

  - *location, scale, kind and timing of development*
    Development should be contained within the capacity of the natural and built environment, which is determined by territorial and quantitative constraints. For example, territorial constraints may include areas that are unsuitable for development, such as steep slopes, protected areas, riparian zones, flood prone areas, erosion prone areas and culturally/historically significant areas etc. Quantitative constraints may relate to the capacity of a river system to absorb runoff and effluent; and
  - *form of development* (structure, pattern and connecting networks)
    The form of development should facilitate the attainment of ecological, social and economic sustainability for the identified range of development options.

- *Conservation of nature*, with particular attention given to the conservation of biodiversity. This should be achieved primarily by:
  - *conservation and sustainable management* of the natural environment and its resources; and
  - *rehabilitation and restoration* of elements lost and/or degraded in the past.

Thus, in conclusion, the potential, primary role of contemporary planning in effectively contributing towards achieving sustainable development is to ensure that development is contained within the final limits of the earth's ecosystems and that conservation of biodiversity is treated as the top priority issue.

However, planning also must be accountable for providing a reliable base for day-to-day development decision making related to various aspects of the functioning and development of settlements and the conservation of nature and natural resources. The real 'value' of planning depends on how efficiently it is implemented, that is, on how successfully it intervenes into an ongoing process of development and decision making. In this regard the prime responsibility of planners in the field seems to be, at least:

- to examine all possible development proposals (alternatives or strategies) leading to the attainment of socio-economic goals;
- to indicate environmental and economic consequences of pursuing these proposals;
- to ensure that each proposal submitted for consideration is able to be implemented; and
- to ensure that decision makers (politicians and developers), key stakeholders and the community at large, are fully informed of the scope, magnitude and character of these consequences.

Planning – What's in a Name?

There are many terms describing planning for towns and settlements. The term 'Town and Regional Planning' has been commonly used in the British Commonwealth countries while 'Urban Planning' and 'Land Use Planning' have been more popular in the USA. Other frequently used terms are 'Physical Planning', 'Town and Country Planning', 'Spatial Planning' and 'Settlement Planning' (a favoured UN term). Increasing use of the term 'Environmental Planning', particularly in the 1980s and 1990s, has been a response to traditional planning's failure to adequately address environmental concerns.

In relation to environmental issues, there has been a gradual change in the focus of planning in recent decades. In many developed countries, environmental concerns in the post second world war period focussed on regulating development to enhance and safeguard amenities. Planners produced strategies for conservation and ecosystem enhancement and natural areas were reserved for their ability to provide 'refreshment' to humans. Planning focussed on economic and social issues and tended to regard the environment as a setting for human activities. A heightened awareness of resource scarcity in response to population growth in the 1960s saw planning focus on enabling economic growth and associated technological change and conservation strategies were based on the conservation of resources for exploitation. Growing community awareness in the 1970s of the negative effects of development on the environment saw planning begin to emphasise the need to reduce environmental degradation and enhance environmental outcomes. From the 1980s the concept of sustainable development began to be integrated into planning documents, which identified environmental parameters as constraints on development. Concepts of thresholds, carrying capacities and limits of acceptable change were developed, although little implemented during this period. Planning in the 1990s, due to the influence of global environmental concerns, began to take a more precautionary approach and moved towards a more ecologically based approach, to ensure long-term sustainability. Planning in the new millennium may take on the recommendations of the Johannesburg summit and begin to examine and find solutions to the human dimensions of planning, including ensuring safe, healthy and secure environments for human communities. Achieving improved human outcomes may be a precursor to improved biodiversity or nature conservation outcomes.

Various definitions and core components of Environmental Planning have been enunciated (Beale, 1980; IUCN, 1982; Grumbine, 1994a). Many of these definitions incorporate the broad principles of sustainable development and are based on the concepts of community participation, planning with nature, minimising long-term negative effects on existing environmental quality, rehabilitating ecosystems, adopting a precautionary and strategic approach and utilising a trans-sectoral, trans-boundary and trans-media approach. 'Landscape Planning', 'Ecological Planning' and 'Ecosystem Management Planning' are additional terms used to reflect the need for planning to be based on sound ecological principles. Environmental Planning was defined by the IUCN as a process whereby regional, national or sub-national resource conservation or development plans are created in ways that consciously seek to minimise long term negative effects on existing levels of environmental quality. There are, however, views that consider Environmental Planning as all planning, that is, including planning for regions, towns and settlements of various kind (see, for instance, Faludi, 1987; Evans, 1997; Queensland and New South Wales planning legislation).

In this book, the term 'planning' is used to reflect the need for all planning activity to be based on a thorough understanding of ecosystem processes and the principles of sustainable development. Planning encompasses the concepts of conservation and good management, and provides a framework for preventing or at least minimising the impacts of human activities on the natural environment, and ensuring that the outcomes of development are sustainable. Thus planners, who are involved in evaluating the implications of proposals for the use and development of land and resources on the environment, should have regard to the intrinsic values of land and water, of ecosystem functioning, natural processes and cultural values.

The Types of Problems Addressed by Planning

A problem is 'any question or matter involving doubt, uncertainty, or difficulty' (Macquarie Dictionary, 2000: 317). Therefore, the process of formulating questions is directly related to the process of defining problems. According to Mazur (1976) all problems can be classified into two major groups, which reflect that in relation to reality, or to any part of it, only two attitudes are possible. Either:

- reality is accepted and all efforts are directed to observation and examination of the real world in order to find out as much as possible about it. In this case various *cognitive problems* have to be faced; or
- reality is to be transformed, in which case *decision problems* have to be solved.

In other words, there is either intervention or non-intervention. It is worth noting, however, that where intervention does not occur, reality will not necessarily remain the same. Similarly, reality will not necessarily change because intervention takes place. For example, intervention may be needed to prevent change. However, change may occur in spite of the intervention aimed at stopping it. Fighting against senility is a good illustration. It is also important to realise that although cognitive and decision problems are essentially different, they are closely interrelated in most cases. To gain knowledge about something, an action is needed to achieve it. Thus a decision is required. Decision making therefore assists in learning, and learning assists in decision making.

Knowledge of these relations is essential for understanding the nature of planning and the planning problem, particularly if planning is to be recognised as a scientific discipline. This is often questioned mainly because planning, as a relatively young sphere of knowledge, has not yet formed its basic theory and because its prime field of interest is in solving decision problems, whereas science has concentrated traditionally on dealing with cognitive problems.

This dilemma is not new having been addressed by Batty (1979) in his excellent analysis of the planning process. He draws attention to the sui generis dualism resulting from knowledge representing the image of the perceived reality and action as the consequent behaviour based on those perceptions. Hence, according to Batty, there are two major and interrelated processes: one related to the gathering of knowledge about a system; and another that uses this knowledge to generate actions. This reflects a difference between science and design as presented by Friedman (1976), who wanted to see planning as neither concerned with knowledge, nor with action, but with the mediation between them. This view concurred with Batty's (1979) final conclusion that it is probably impossible to do science without design, or design without science.

The same conclusion can be derived from Mazur (1976), who makes a major philosophical and all-embracing statement that science as an activity concerned with problem solving is one whole. Then he points out that, as a tragic misunderstanding, for thousands of years only cognitive problems were considered part of science, while decision problems were left to those who had little understanding of the rational approach to solving these problems. Even today many decision makers do not know that each decision, as the solution to a decision problem, should be founded on a rational, scientifically sound base and that its correctness should be evaluated. The consequences of this are too well known to require further discussion and the relevance of this statement to the realm of planning is quite conspicuous.

Cognitive problems, which have to be addressed to provide the basis for dealing with decision problems, can be subdivided into the following three categories:

- *exploratory problems* are solved by fact finding (examination and survey) and answer the question: 'What is?';

- *classification problems* are solved by defining properties (measuring and systematising) and answer the question: 'What is it like?'; and
- *explicatory problems* are solved by identifying relationships (understanding mutual interdependence between things) and answer the question: 'What depends on what?'

Planning is involved in addressing cognitive problems but, by its nature, it is primarily concerned with decision problems. These problems are encountered when a specific aim cannot be achieved, and will reflect the following formula:

'Problem = Aims + difficulties in achieving it' (Chadwick, 1971).

This means that it is not possible to define any planning problem without first knowing, at least in broad terms, what the relevant aims are, and what the issues are, that affect their achievement.

Decision problems also fall into the following three categories:

- *postulation problems* are related to determining the ends for which a particular intervention is directed and look for answers to the question: 'What is to be achieved by this intervention?' In a rational decision making methodology this is the earliest phase, known as goal formulation. As far as planning is concerned this calls for wide stakeholder participation in formulating goals. There is an inherent difficulty in this process because when developing long-term strategies planners must be concerned with the needs and desires of future communities, as well as present communities. In brief, solving postulation problems requires an indication of goals;
- *optimisation problems* are related to answering how the goals are to be achieved. In these problems, emphasis is placed on the choice of effective ways (often called strategies) to meet the goals. There are various difficulties in dealing with these types of problems, the major ones arising when no way is known to achieve a given goal. Effective optimisation should scrutinise all possible ways (or strategies) and examine their effects. By this process an optimal way should be found. In brief, solving optimisation problems is achieved by identifying and evaluating the ways, or strategies for goal achievement; and
- *realisation (implementation) problems* are associated with the issue of how the ways (strategies) should be implemented. This requires an indication of the concrete means, or resources that are needed (energy, labour, materials, tools and so on) for the strategies to be implemented

and for the goals to be achieved. Realisation requires, in particular, a determination of whether specific means are available, and if not whether there are any other means that can be substituted. Realisation problems may be solved by defining a process for strategy implementation.

There is thus a logical sequence in the process of solving decision problems. From postulation the goals are derived, from optimisation the ways (strategies) are determined, while from realisation the available means by which the goals and strategies are to be implemented are identified. These are summarised in Table 2.2.

Table 2.2 Types of decision problems

Problem	Meaning
Postulation	⇒ What are the desired future goals, both objectively needed and/or preferred by the existing and future community?
Optimisation	⇒ What are the optimum strategies to achieve these goals?
Realisation	⇒ What means (resources) are available for implementation?

This entire classification of all problems is logically complete and there cannot be any other than cognitive or decision problems. Both are, however, strongly interrelated, and solving problems in the first group not only precedes, but usually provides the basis for solving those in the second.

The Nature of the Planning Problem

Planning deals primarily with decision problems related to postulation, optimisation and realisation. The framework in which plans are prepared presents a number of obstacles for achieving effective solutions to each class of decision problem. For example, postulation is hampered by the uncertain futures for which planners must develop plans, and by the difficulties with ensuring effective consultation and involvement in the goal setting phase. Optimisation is limited by the frequent lack of suitable data on which to base planning strategies and the sectoral nature of plan preparation. Realisation is affected by the institutional and organizational context concerning the roles and responsibilities of all the participants involved in the planning and decision-making processes. Hence the issues or problems confronting planning relate to:

- an uncertain future;
- addressing issues associated with conflicting land uses;
- the availability of suitable data on which to make decisions;
- the particular institutional framework;
- the political context; and
- the complexity of problems and the need for an interdisciplinary focus.

Each of these elements of the planning problem is briefly described below and is followed by an assessment of the existing and potential effectiveness of planning in overcoming these obstacles and ensuring sustainable development.

Uncertain Future

Planning is often initiated on the basis of an identified problem. Although planning must be involved in, and based on, answering cognitive problems, it is by its very nature a decision-making process. It incorporates the conscious determination of courses of action based on facts and considered estimates. Where possible, planners should attempt to reduce uncertainty by using prediction and forecasting strategies. However, achieving the desired future state is difficult as unforeseen changes and factors beyond the control of the planner may intervene to disrupt the planning process. For example, uncertainty may surround the following: the objectives and criteria to be used for approving particular developments; the socio-economic and bio-physical conditions in the project area; the possible current and future responses to the problem; the effects of the responses; the sources of funding for projects; and the political context in which plan making occurs.

Uncertainty in a planning context may be minimised by incorporating comprehensive consultative processes during plan preparation and implementation. This may involve representatives from different cultural, social and economic backgrounds, as well as relevant sectors/agencies, both government and non-government, and interest groups. Wide representation is necessary to ensure that precise problem definition is achieved, along with consensus concerning the plan's stated aims and objectives. Where possible, future end users should also be identified and consulted. It is also precisely because uncertainty surrounds the achievement of the desired future state that plans should be dynamic and responsive to changing circumstances. Although desired future states may be difficult to define, many countries endorse the principle of intergenerational equity and the need for planning to accommodate not only current human activities, but to keep future options open. Thus, where uncertainty prevails, plans should incorporate the precautionary principle.

Conflicting Land Uses

Accompanying the earth's unprecedented population expansion are increased demands for the use or development of land for agriculture, forestry, industrial, residential and other purposes. However, as the total stock of land is largely fixed and limited in relation to the number of activities and the intensity of uses that it can support, there is inevitable conflict among competing uses. A common form of conflict is that frequently the sites of high biodiversity value are also those that are in demand from other forms of development (e.g. residential and tourist uses) and the conservation use must compete with these other land uses for the diminishing and scarce land and water resources.

Planning provides a base for guiding and controlling development and thus it has a very important role to play in ensuring the sustainable development of natural ecosystems and in particular of conserving biodiversity. The ecological role and responsibility of the planning profession has been clearly enunciated. 'Altering economic and land use patterns seems to be the best long-term approach to ensuring the survival of wild species and their ecosystems' (WCED, 1990:201), and one way to achieve this is through planning:

> ...measures governments could take to confront the crisis of disappearing species, recognising that it constitutes a major resource and development challenge, include consideration of species conservation needs and opportunities in land use planning... (WCED, 1990: 208).

The integration of conservation and development, which is the very essence of sustainable development, can best be achieved at the point in the planning process where decisions on the location of development, its scale, type and timing are determined (Kozlowski, 1993a, 1993b). The planning process has both restrictive and promotional strands. The determination of territorial, quantitative, qualitative and temporal constraints, within the restrictive strand sets an effective ecological frame for ongoing planning. It is the identification of this 'solution space' that is perhaps planning's main responsibility in terms of ensuring sustainable development (see The Role of Planning above).

Availability of Suitable Data

The cognitive strand of science is essential for reliable planning, as a good information base is required so that decisions on land use are made with all available knowledge relating to the suitability and fragility of the available resources, and the consequences of development of these and other resources, both now and in the future. This includes scientific data on biological and physical resources, ecosystem processes and functioning, land suitability and constraints to land use, as well as socio-cultural data and a knowledge of the political and policy framework. Although there may be significant gaps in data in many countries, especially in relation to ecological knowledge of organisms and ecosystems, planning decisions still have to be made. This situation is a persistent and pervasive part of the context for considering environment problems and seeking solutions.

Coupled with limitations in data availability and access are difficulties in integrating often incompatible data that are dispersed in a range of departments, industries and local governments with no common register detailing the type, scale, reliability and form of the data. Other problems include the time it takes to gather data, to order and analyse it, the risk being that it will be inaccurate by the time it reaches a decision maker. This increases the risk that the planning action taken might be unnecessary, mistimed or even counterproductive.

In the absence of perfect information and uncertainty about future human needs and the ecological effects of human activities, the use of soundly based methodologies may help to improve decision making. The integrated buffer planning (IBP) methodology to be outlined in this book aims to fill such a gap in planning for the conservation of significant natural areas, where data on regional ecosystems, target species and the precise nature of their habitat, and the socio-cultural and political framework may be limited.

The Institutional Framework

Organisational matters also surround the planning problem. Institutional structure and the roles individuals perform within organisations may determine the conditions under which problems and opportunities are appraised and acted upon (Brewer and de Leon, 1983). Communication channels, complexity, uncertainty, the age of the institution and a reluctance of bureaucracies to adapt to change (Grumbine, 1994b) may influence how issues are viewed and treated by an organisation. Westrum (1994) identified that bureaucracies may compartmentalise information into disciplinary channels, discount those who are adept at sharing information widely, neglect bridge building and often consider new ideas as potential threats. These are all considerable impediments to effective biodiversity conservation and sustainable development. Planners need to be aware of these potential problems and to identify clearly the roles and responsibilities of individuals involved in the planning process.

Political Context of Decision Making

Planning generally occurs within a political context and it is politicians who usually make the final decisions or direct particular planning actions to be undertaken. However, information available to decision makers is not always complete and this may hinder the decision-making process. Planning can assist decision making by developing a range of options, evaluating these options (especially the environmental consequences of development policies or projects) and recommending preferred strategies for the consideration of decision makers. Through such a planning process, planners can make a significant contribution towards sustainable development.

As wildlife do not respect political boundaries, decision makers involved in biodiversity planning, and particularly, in planning for and implementing buffers, can be many, including any number of local government decision makers, as well as numerous state or provincial government sectors (e.g. environment, natural resources and transport), in which mandates may overlap. Successful planning frequently relies on joint support. Where a buffer plan is developed across a number of administrative areas, the failure of a particular jurisdiction to participate may have serious consequences for the long-term viability of significant natural areas and their wildlife. Hence good planning should be based on decision making that is integrated across political boundaries, becoming transboundary in nature.

Complexity and Interdisciplinary Focus

The issues and problems that planning must confront are usually complex and multi-faceted. For example, biodiversity, as a resource or value that is to be sustainably managed, comprises interactions at many levels within the biodiversity hierarchy of genes, species, ecosystems and regions. Many of these complex interactions are only beginning to be understood, with many species and even ecosystems disappearing before they have been adequately recorded and documented. Of particular concern for planners is the related problem that the loss of biodiversity may be irreversible, as evidenced by the extinction of species and the destruction of ecosystems.

Complexity and the possibility of irreversibility heighten the need for effective planning to incorporate the principles of sustainable development.

Issues or problems related to biodiversity are embedded in social, cultural, economic and political contexts. A difficulty for effective long-term planning occurs when the environment is managed on a sectoral basis. Planners must continually understand, as well as solve problems in a multi-dimensional framework, and hence they need to ensure that the desired functioning of one sector or system is conducive to the desired functioning of other sectors/systems. Integrated planning is needed to replace sectoral fragmentation of planning and decision making and to reduce the complexity associated with planning problems. Integration is an important principle to guide sustainable development (Table 2.1) and also an important principle to guide planners in plan preparation. In particular, biodiversity conservation requires planners to draw data from a number of disciplines, including the physical, biological and social sciences, landscape architecture, engineering, economics, law and administration and to examine the data within a political framework. A single disciplinary approach limits examination of the problem to only part of it and may cause other aspects of the problem to be under-estimated or even dismissed. This approach encourages 'partial blindness' about policy process and hence a clear empirical picture is not possible (Clark, 1995).

An important role for planners is that of synthesising and integrating the results of specialised research into a more manageable whole. This often requires a sound knowledge about other, affiliated disciplines and, as a consequence, a move towards a general, interdisciplinary approach to planning. Planners must play a functional role in the process, acting as integrators of policy, theory and practice and mediators of knowledge and power (Clark, 1995). Maximising interactions between specialists is required and planning teams should include a variety of specialists, with the head of the planning team acting as arbiter and coordinator (Rosier, 1992). Increasing specialisation within scientific disciplines places added responsibility on planners to draw from this knowledge and integrate it into plans, so that planning strategies are economically, ecologically, socially, culturally and technologically viable in the long term. During the process of developing buffers for environmentally sensitive areas (ESAs), which is the central topic of this book, planners must particularly rely on the biological sciences to aid the design of an environmentally sound physical buffer plan, on the disciplines of sociology and economics to help effectively apply the plan within a functioning community, and on understanding the administrative and political structures of the local area to enable successful implementation.

Response of Planning to its Problems

Planning deals with problems that are highly complex and interdisciplinary in nature. Specifically in relation to ESAs and their biodiversity, planning has to address the potential problem of conflicts between conservation and development values and options, and this may cause difficulty in negotiating agreed future states and in obtaining political support for biodiversity conservation. Planning also has to cope with a lack of suitable data on natural resources, particularly wildlife and their ecosystems. Some of the more common responses by planners to these renowned planning problems are briefly analysed below.

Failure to Adequately Address Environmental Issues

There are ongoing claims that a significant share of planning's 'outputs', that is, its strategic, development control, local and regional plans, planning studies and/or more scientifically oriented products in the form of papers, articles and research publications, has been of negligible use in the process of solving everyday and/or long-term problems facing communities and their environment. However, for many planning tasks, especially those related to planning for conservation of nature, knowledge of broad desired sustainable outcomes exists and the principles of sustainable development, including the need to integrate conservation and development, have been espoused since the early 1970s. Despite this, a significant outcome of the planning process has been the rapid loss of biodiversity due to the conversion of land to other uses and the degradation of land and water as a result of unsustainable practices. Graham and Pitts (1996) argue that this occurs because the philosophy of sustainable development is difficult to understand and that there are problems in translating broad and sometimes vague philosophical concepts such as ecological sustainability, biodiversity, social justice or sustainable development into meaningful real-life actions.

In a planning sense, the natural environment has frequently been seen as a resource to be developed, or a constraint on development, for example, in terms of its steep slopes and susceptibility to flooding. These constraints are frequently viewed as factors that increase development costs. Resultant plans are often 'patterns of land use', with zones used to prescribe permitted or prohibited uses, the determination being related to degrees of compatibility of various activities (Kozlowski, 1990; Kuiken, 1990; RPAG, 1993a,). Goudberg (1992:82) comments on planning in Australia up to the 1990s:

> In the past, environmental considerations have had little impact on land use decisions, leaving us with an inadequate system of protected areas, fragmentation of bushland, dustbowl paddocks, half-baked tourism developments ... and large tracts of prime agricultural land buried beneath the sprawling suburbs.

This failure to adequately address environmental issues also relates to the role of decision makers in the planning process. Problems related to nature conservation historically have had little real commitment to lasting solutions from all tiers of government. For example, in Australia, the protected area estate fails to fully encompass the nation's biodiversity, with the expansion and consolidation of the estate in several bioregions being in conflict with the interests of other industry sectors. At all levels of government, the Minister responsible for environment matters has traditionally been one of the least powerful members of government, and it is often only when the effects of overuse of natural resources influence the prevailing power base that action is taken, or more generally when public opposition to a particular issue becomes a dominant factor affecting decision makers.

Due to the frequent failure to effectively implement plans that emphasise sustainable resource use, the IUCN stated that there is an immediate need 'to adopt and implement an ecological approach to human settlement planning to ensure explicit embodiment of environmental concerns in the planning process and thus promote sustainability' (IUCN et al., 1991:106).

Recent global initiatives, which are being integrated into national, state/provincial and local level plans, may begin to reverse these past practices. Planners, in particular, have a responsibility to ensure that these principles of sustainability are incorporated into decision making to ensure that biodiversity is conserved and that the outcomes of development are sustainable.

Short-term Time Frames

Although planners usually work within short-term time frames that are set by the political process, environmental cycles may operate over periods of centuries and longer. A focus on day-to-day management at the expense of a long-term vision for the environment has resulted in a failure to produce effective long-term strategic plans that consider the environmental implications of development or to comprehensively develop proactive and anticipatory approaches for achieving long-term objectives.

Sustainable development requires that planning adopt a long-term view about utilising the environment and its resources. The integrated buffer planning model to be outlined in this book adopts a long-term planning framework to build cooperative structures that will help to ensure compliance in achieving the buffer plan's goals and objectives. However, planning must also take into account the fact that plans may never be 'complete', due to the short planning time frame, and that plans may need to be continually redesigned and also recorded in the form of short and long-term proposals. This has particular relevance to buffer planning, where it may take many years of education and negotiation to gain community acceptance of sustainable land use practices that may be implemented to minimise negative impacts on biodiversity.

Ignoring the Need to Link 'What is Produced' with 'How it is Produced'

The planning process should be based on developing the following:

- *a physical plan*, which may identify appropriate spatial relationships between land use activities, and include policies to achieve sustainable development, performance measures to indicate the effectiveness of the plan in achieving its objectives, and relevant monitoring and associated review procedures; and
- *an institutional plan*, which should specify how to put the necessary people, agencies and resources together to develop a range of options to solve the identified planning problem (IUCN, 1992).

As the developing physical plan must fit the institutional context of the planning area, planners need to understand institutional structures and their responses in order to create innovative approaches to deal with cross-sectoral problems and conflicts. Hence the physical plan should be developed within an institutional framework, which allocates responsibilities to agencies and people (Rosier, 1992). This has not always been the context in which actual planning occurs.

The integrated buffer planning model to be developed later in this book is based on developing a physical buffer plan. However, the model is structured to ensure that planners focus on the process to achieve the desired outcomes, requiring planners to prepare the legal, institutional, financial and social groundwork on which plan implementation is based.

Unwillingness to Develop and Evaluate a Range of Planning Options

Ideally, the planning process aims to maximise opportunities for the community to make choices across a range of options, these being based on an interpretation of the available data, with recommended courses of action being suggested on the basis of the short and long-term consequences of each option. As planning occurs largely in an environment of uncertainty, there is an even greater need to produce a range of alternatives and to effectively evaluate their ecological, economic and social consequences. Such a process is likely to produce plans that are flexible and adaptable. Planning should aim to develop strategies for establishing priorities and consistent guidelines for resource use based on criteria which address present and projected needs (Rosier, 1992). However, final plans are frequently implemented without effective evaluation of alternatives and their long-term consequences and in essence this demonstrates a major failure in the planning process. Frequently this results from limited time frames, resources and political will.

Underestimating the Importance of Interdisciplinary Planning

Historically planning was mainly concerned with the development of human settlements at various levels, together with their impact on the natural environment, state of the economy and the quality of human life. Development is a multi-complex process, and planning usually has been recognised as responsible for its coordination both in space and time. In a general sense planning evolved from architecture, but later, other disciplines became entangled in it because of their interest in various aspects of development. At present, the prime responsibilities of planning cannot be discharged without close affinity with, and knowledge of, economic, managerial, ecological, technological and social issues. This is, of course, a very large field and it has compelled planning to evolve primarily as a generalist activity that must examine development from a wide range of viewpoints and disciplines, and then try to integrate this very broad spectrum into planning decisions.

At the same time, science and many professions have become more specialised and polarised, with several disciplines showing a tendency to split, and thereby further narrow their sphere of concern. New disciplines have emerged, leading to the growth of a 'jungle' of terminology and axioms superimposed upon an array of disciplines, previously homogeneous and well established. Thus, as the number of new, highly specialised disciplines grew, the greater became the need for generalisation and integration. In planning, therefore, there was a need, not only to prevent the narrowing tendencies, but rather to promote those associated with generalisation.

As a consequence, it can be argued that planners should concentrate on:

- formulating questions directed to relevant disciplines to indicate specific problems that require interdisciplinary examination; and
- widening their knowledge of development processes and their implications, based on the results from, and the perspectives of, other disciplines.

This implies that an ability to formulate the right questions is critical for both planners and those operating within other related disciplines. Similarly important is the ability to listen to questions asked by others and, as a consequence, to make necessary adjustments. Therefore, the formulation of questions seems to be one of the prime requirements and skills of contemporary planning and is essential for its evolution. Achieving a comprehensive knowledge of development processes appears, in turn, unattainable by individual planners, and one of their basic skills must be a capacity to synthesise and integrate the results of research, retrieved primarily from other disciplines, into a coherent whole.

All that leads to another logical conclusion, which is that learning about the methods and findings of other disciplines becomes a prerequisite of almost any responsible planning research and practice. This may be best accomplished by interdisciplinary cooperation and understanding between various specialists or various scientific and/or professional disciplines.

The practice of planning, however, is constrained by the sectoral structure of government departments and also legislation. Creating a sector dedicated to the environment is a common response in most countries at all levels of planning. Also, many of the institutions tend to be independent and fragmented, and to work within relatively narrow mandates with closed decision processes. Such a structure may be adequate when there is no conflict between the goals of each agency. However, when each sector sets its own policy path, this may run counter to sound ecological management.

Although environmental concerns should be a responsibility of all sectors, governments often lack an institutional or technical framework for integrating environmental considerations into decisions on a continuous basis. Planning in relation to the environment usually takes place sporadically, in sectoral agencies that are often marginalised or isolated from mainstream decision-making processes. This structure excludes the environmental dimension from decision making in other sectors and hinders the translation of environmental policies into effective action. One of the challenges of promoting sustainable development and biodiversity conservation is to ensure that decision makers in all sectors take into account the biophysical world. This reorientation, which requires a change from after-the-fact (ex post) repair to preventative (ex ante) care, is a major challenge for planners.

Although cross-sectoral planning is crucial, where this cooperation occurs, it usually is on the basis of multidisciplinary teams, where each component produces a report that is coordinated by the plan leader. Thus, although the process considers other sectoral interests, the resultant plan may not coordinate important sectoral issues throughout the planning process and resultant plans may be ineffective. Even within departments, plans are frequently developed in isolation from other departmental sections, with responsible people (e.g. departmental heads) being given drafts for comment. The wide range and complex nature of environmental concerns dictates that planning must be interdisciplinary with all professions contributing early in the planning process to produce a wide array of possible planning solutions.

Also of concern is the development of legislation by individual departments to deal with particular issues, this being independent of legislation and policy in other departments. It is not uncommon in many countries for there to be hundreds of regulations relating to the environment and resource management. There is a need for an integrated statutory mechanism to allow policies and controls associated with these regulations to be coordinated.

In relation to establishing buffers for protected areas and ESAs, an important component of the integrated buffer planning model should be the selection of a planning team that will ensure that those who are responsible for developing the plan, work as an interdisciplinary team and that decision makers, the main stakeholders, members of the implementing agency, and all concerned sectors of the general community are fully informed about the progress and findings of the plan.

Lack of Effective Community Involvement

Plans affect people, and community participation is an essential and frequently mandatory element in the planning process, achieved mainly through consultations and negotiations. However, such processes are often tokenistic, being based on minimalist approaches to ensure compliance with relevant requirements (for instance, as identified in legislation or policy). Local Agenda 21 stresses the need for consensus building and partnerships at the local level. However, as land that has important biodiversity values is frequently in conflict with other uses, conflict is built into decision processes. Hence, in the process of developing planning goals and objectives, all major stakeholders should be effectively involved to better understand the broader community's views, and where conflict is evident, appropriate action should be taken by the planning team to minimise conflict and to help ensure sustainable development and the protection of ESAs and their biodiversity values. The planning process must also capitalise on local knowledge, including indigenous traditional knowledge, and develop a greater degree of stewardship and sense of community ownership of resources. Planning must also help communities to make decisions that foster sustainable development. An important aspect of the integrated buffer planing model is to ensure wide stakeholder involvement from the very beginning of the planning process through to final plan implementation and even review.

Poor Performance in Integrating Planning at All Levels

Planning at the local level has frequently occurred without integration of regional, state/provincial or national policy frameworks, while broader planning has often lacked the practical means to integrate decision making at lower levels. The policy framework within which a plan is being developed, should integrate national, state, regional and local policies to ensure that the policies are compatible, appropriate and endorsed by all levels of government. Communication and coordination, by planners, within this organisational or political hierarchy is critical to reduce duplication, ensure consistency among planning objectives and transparency in the process. The integrated buffer planning model should emphasise the need to understand the legislative framework within which the buffer plan is to be developed, and integrate wider social, environmental and economic considerations into the plan. This inter-related planning structure may help to encourage public acceptance of the final plan as well as ensure institutional planning consistency.

Reactive vs Proactive Approach

Last but not least, it should be recognised that the predominant approach of planning to its problems, in general, and to environmental problems, in particular, has been traditionally 'reactive', rather than 'proactive'. That is, it has concentrated on curing the symptoms, rather than finding how to prevent the causes. This means, for instance, that planning may be eager to go ahead with an ostensibly attractive development proposal before environmental and social effects are considered and/or action taken to first restore ecologically and socially sound conditions. Thus, instead of preventing the causes of problems, as ex ante strategies, planning is based on 'react and cure', ex post strategies, where frequently adverse changes are not even seen until it is too late and planners have little scope to put in place environmentally sound development proposals. Preservation of the earth's biodiversity requires the use of a proactive, 'anticipate and prevent' strategy that incorporates predicting the results of destructive policies/actions and their cumulative effects, and taking appropriate action to avoid or ameliorate potential negative impacts. Certainly, the main prerequisite of proactive planning is that development to day must not create problems in the future.

The Present 'State of the Art' of Planning

At this stage it may be useful to present a simple but astute synthesis of the present state of planning (Kozlowski, 2002). England (2001) convincingly argued that in the recent period, planning has consolidated into three main types, 'minimalist', 'instrumental' and 'incremental'.

The main aim of *minimalist planning* is to achieve an ordered development of land, and at the same time minimise negative environmental impacts and economic loss. Its only vision is to prevent chaos, and therefore, minimalist planning is concerned 'more with development control on a case-by-case basis than with formulating policies and strategies to guide development' (England, 2001: 2). As a consequence, the main instruments of minimalist planning are zoning and development control plans. This type of planning greatly facilitates urban development and has been well supported by the development community, whenever it has been applied. Minimalist planning is very pragmatic and concentrates on what is real and obtainable and not on often esoteric and endlessly debated goals. As such it definitely upholds the status quo. This is not necessarily wrong, but it may be a hindrance to any reforms that try to improve it.

Instrumental planning aims to identify socio-economic goals and to make sure that they are effectively implemented. It has developed in two primary forms. The first concentrates mainly on improving and protecting the physical environment as one of the major ways for improving the quality of human life. Evans (1997:56) has even recognised it as 'classical town planning', which would normally venture beyond that purely physical agenda to concern itself with the goals and aspirations of society in general. The second form addresses social problems and society's goals, seeking from planning its active assistance in their achievement. This moves planning away from its traditionally affiliated disciplines of architecture and engineering towards a whole array of social sciences. While accepting that differences between these two forms may vary Friedman (1996) is convinced that it is the degree of social orientation that determines what is good planning.

However, instrumental planning is criticised for being incapable of achieving its ambitious aims because it is totally dependant on, and constrained by, the dynamic mechanism of capitalist economies. This could imply that it is 'the urban social movements and not planning institutions which are the sources of change and innovation within the city' (Kirk, 1980:84). Secondly human behaviour and the complex links between urban form and societal well-being are unlikely to be effectively 'managed' and, in addition, establishing a clear consensus on goals is also unlikely. England (2001:4) comments that,

> ...although instrumental planning cannot alter society's fundamental structural problems it may, nevertheless, have a role to play in implementing more modest reform goals, in the short term or in specific situations. The preservation of a particular habitat ... may be an achievable goal of planning even if the sustainable management of whole species is beyond the grasp ...

and that,

> ...the absence of any vision is an invitation to preserve the status quo, however unsatisfactory that may be. Planning may actively obstruct reform if it fails to move with the times and reflect the dominant goals and aspirations of society ...

In conclusion England (2001) argues that instrumental planning, if sufficiently well integrated with all essential economic and social aspects of society, may become quite effective and that it may naturally develop into holistic, integrated and multi-faceted planning.

The *incremental planning* approach is best suited to addressing compound planning problems, in which the individual parts are understood in detail, but where the relationships among them and the potential impact of additional factors can not be anticipated. This model does not require the prior specification of goals and it recognises the limitations of time and other resources. It is essentially a linear approach, but has several sub-optimum solutions that have to be worked out separately and integrated into one overall decision. As it recognises complexity and the inability to predict the future, incremental planning suggests changes only at the margins. Hence mistakes can be kept tolerably small and more easily rectified. However, this incremental change may be sufficient only if societal values are stable and the results of present policy are satisfactory. If past policies have produced environmental degradation, incrementalism may not result in improved environmental outcomes and more sustainable futures.

The main responsibility of planners who use this approach is not to discuss how to change the world, but to use their knowledge and experience in the application of planning law. As a consequence 'the claim to expertise here is based upon a knowledge of the policy process in managerial and political terms and of procedures and case law, linked to a knowledge of the economic processes by which urban development is generated and shaped and a capacity to mediate' (Evans and Rydin, 1997:58). Incremental planning is based on recognition of existing, competing interests and the need for mediation. In this respect it is similar to minimalist planning as both are not interested in any major changes in the existing status quo and believe that planning goals should only be set incrementally and within a specific, not general context. Planning as social learning is advocated (Friedman, 1996) and planners are expected to learn from practical experience what good planning is. As a consequence, this form of planning has started to focus on community participation:

> ...Participatory incrementalism suggests state planning can adequately incorporate the views and goals of urban social movements if the right type of participatory mechanisms are established ... Accordingly, planners are facilitators trained in mediation and procedural processes rather than strategists attempting to operationalise any particular planning goal. Nevertheless, participatory planning does not deny the feasibility of establishing context specific goals...' (England, 2001:6).

All those forms of planning may be reinforced by incorporating in them elements of what can be called an *interactive planning process*, which focuses on defining the values and roles of individuals and organisations participating in the planning process and which aims to produce a plan that accommodates most groups, thus resulting in a 'win-win' solution. An interactive planning team or organisation aims to ensure effective interaction among groups to help resolve existing and potential conflict between groups. This may result in greater acceptance of limits derived through negotiation. The planning process is not linear, but iterative, where a planner re-evaluates phases of the process as new information is gained. A particular problem with this process is in identifying future groups who may be affected by the plan and the possibility that ecological concerns may be devalued by the interacting groups. It may also be difficult to assess the environmental impacts of solutions arrived at through interactive planning due to the unpredictability of the trade-offs between environmental quality and other societal goals. However, as environmental problems are 'meta' problems frequently involving conflicting groups, the proposed integrated buffer planning process should emphasise interactive or participatory planning to produce an optimal solution. In the process this may broaden the social basis of decision making, help reconcile divergent groups, better manage uncertainty, educate the public and stakeholder groups and produce solutions that may be easy to implement due to their wide acceptance.

'Model' Planning Process and its Guiding Principles

Planning needs to place primary emphasis on its environmental framework if it is to play an effective role in achieving sustainable development. In this section key principles that should underpin the planning process are outlined and a simplified 'model' planning process is described (based on Kozlowski, 2002). The place of the integrated buffer planning model in this 'model' process is also considered.

Planning Principles

From the preceding analysis of the nature of the planning problem and planners' responses to it, several important principles can be identified (Table 2.3), which should be incorporated into any plan designed to conserve biodiversity and ensure sustainable development. These principles will form an important foundation for the formulation of a simplified 'model' planning process, which is used to assist in the development of an integrated buffer planning method, to be outlined later in this book.

Table 2.3 Principles to guide plan development

Principles	Meaning
Identify restrictive and promotional aspects	⇨ identify the environmental constraints (restrictions) to development and plan within the remaining solution space (promotional)
Incorporate a multi-objective approach	⇨ the plan should reflect the varied outcomes desired by the community
Incorporate effective participation	⇨ effectively consult with all stakeholders early in the planning process and throughout all stages of plan development to identify and define the planning problem and alternative ways of solving the problem
Use an interdisciplinary focus	⇨ complex problems need to be solved by effective communication among specialists
Incorporate a trans-sectoral approach	⇨ all sectors should effectively consider the environmental consequences of their plans and thus minimise fragmentation of decision making
Incorporate a trans-boundary approach	⇨ plan within ecosystem boundaries (bioregions) rather than narrow administrative boundaries to ensure the conservation of natural systems and processes
Use an integrative approach	⇨ the plan is consistent with agreed local, regional, state, national and global policies
Be proactive	⇨ anticipate and prevent damage
Be strategic	⇨ develop a long-term plan for the environment
Be dynamic	⇨ allow the plan to be flexible and evolve as circumstances change
Incorporate monitoring	⇨ track the results of actions so that success or failure may be evaluated, thus creating an ongoing feedback loop for plan modification

Simplified Model Planning Process

The rational comprehensive planning process was one of the first approaches advocated and applied in the field of planning and it follows the traditional process that includes:

- *problem identification*, based on a diagnosis derived from extensive survey and analysis of data;
- *problem solving*, including identification and formulation of alternative solutions and evaluation and selection of the best option in a rational process that is as comprehensive as the available data permits. The process is based on a sequential, linear model and usually has a long-term focus. The rational planning approach is most suited to solving simple planning problems that are fully understood in both their scope and detail; and
- *implementation and monitoring*, which are the essence of successfully achieving the pre-established goals.

Efficient planning depends primarily on three groups of factors:

- *interrelations* between planners, decision makers and key stakeholders, including the community;
- *management* of development processes and the use of various incentives or sanctions to influence the behaviour of the main players in these processes; and
- *planning methodology* as reflected in the planning process.

The last group is primarily planning's domain. A clear process will make it easier for specialists from other disciplines, decision makers and members of the community to understand the main principles of planning and the thinking processes that planners employ in their development of a range of planning instruments, including local and strategic plans.

A simplified, model planning process is presented in Figure 2.1 and the key issues, planning questions and expected outcomes are summarised in Table 2.4. The model offers a basic structure of the planning process, as a mental framework for a rational approach to problem solving in the course of a planning exercise. It is a flexible guide for the preparation of any major local or strategic plan and also provides a platform for further discussion on how the planning process can be improved. If applied in practice it would need to be expanded and adapted to concrete circumstances (e.g. specific problems, the legal setting, local administration and so on).

The model (Figure 2.1) indicates that any planning process should start from a comprehensive survey and diagnosis of the existing situation and lead to the identification of the planning problem. It is essential to know exactly what is to be solved during the preparation and implementation of a particular plan. Once the problems are known, the main core of the planning process, that of problem solving, follows (refer to Table 2.4). This involves: defining the development program; identifying possibilities, through an examination of constraints and opportunities for development; formulating strategies; determining the means for achieving the plan's aims; and choosing which strategy will be most effective.

Ultimately, the real value of planning depends on how effectively it is implemented. In Component 3, that of implementation and monitoring, the initial problems are expected to be solved, although new ones may appear. Monitoring of the plan should be a major input into any subsequent cycles of the never ending process of planning. The main criteria for evaluation and monitoring should reflect the plan's aims (goals, objectives). To properly monitor the progress of the plan, the evaluation should be applied throughout the process and not only in its final stage. This points to the importance of feedback, intertwined with evaluation, as being equally essential and providing for continual review and adaptive planning.

The other key, though external, part of any planning process is forecasting, which is essential for identifying future problems and is a fundamental prerequisite of proactive planning. It can be subdivided into the following types:

- *demographic*, which deals with the size and structure of the future population as a function of expected natural growth and migration;
- *societal*, which addresses the most likely behavioural models of the future community;
- *economic*, which assesses the most plausible developments of the main economic activities;
- *technological*, which deals with possible significant changes and developments in such fields as energy supply, infrastructure, or industrial and agricultural technologies; and
- *environmental*, which deals with changes in key environmental indicators.

During the planning process the partial results of each of these components should be subjected to relevant forecasting techniques.

The last remaining aspect of the model planning process is that of community involvement, which is imperative for any planning process and

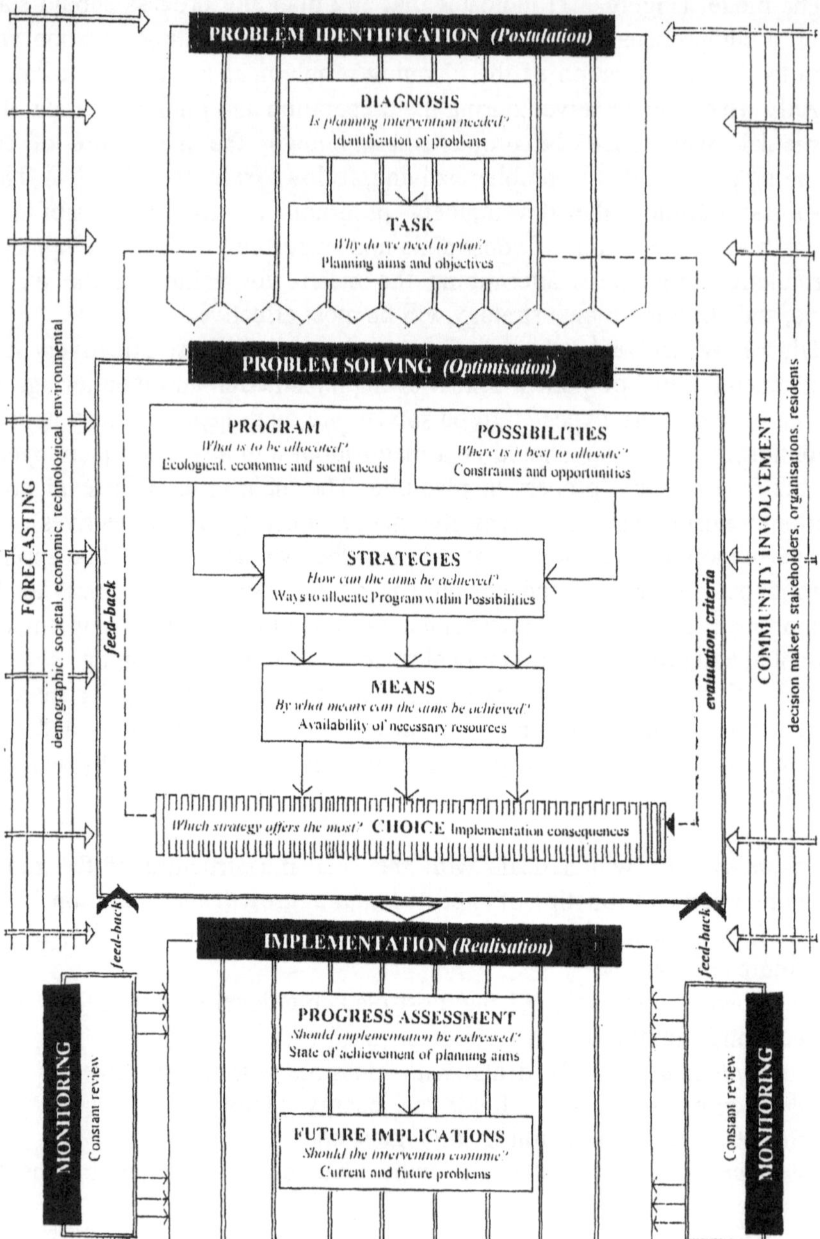

Figure 2.1 Model planning process

(Based on Kozlowski, 2002: 301)

Table 2.4 Model planning process: issues, planning questions and expected outcomes

Component	Issues to be Addressed	Planning Questions	Expected Outcomes
1. Problem Identification (Postulation)	***Diagnose the existing situation***	Is planning intervention needed?	Identification of planning problems
	Set the task	Why do we need to plan?	Determination of aims and objectives to be achieved through planning.
2. Problem Solving (Optimisation)	***Define the development program***	What is to be allocated?	Identification of ecological, economic and social needs (aspirations)
	Identify possibilities	Where is the optimal location?	Identification of development constraints and opportunities
	Formulate strategies	How can the aims be achieved?	Identification and evaluation of the potential ways or strategies for allocating the 'program' within the 'possibilities'
	Determine the means	Are the necessary resources available?	Identification of the necessary resources and confirmation of their availability
	Make a choice	Which of the strategies offers the most?	Evaluation of the consequences of implementation of each strategy
3. Implementation and Monitoring (Realisation)	***Assess progress in transforming the existing situation***	Is it necessary to redress and strengthen the implementation process?	Evaluation of the plan's progress in achieving its aims and objectives
	Implications for further planning	Was the intervention successful and should it continue?	Evaluation of the state of the identified problems and redefinition of the problem

(Based on Kozlowski, 2002: 300)

should occur throughout the entire process (refer to Figure 2.1) and include all relevant stakeholders (e.g. government and non-government agencies, community organisations and the like).

Integrated Buffer Planning and the 'Model' Planning Process

Planning for integrated buffers can be located within the model planning process. This is a significant improvement over many current approaches, as it places buffer planning in the wider field of interdisciplinary planning. In particular, planning for integrated buffers is very strongly entrenched in the *Problem Identification* and *Problem Solving* stages of the model process (Figure 2.1), and primarily addresses the 'Task' and 'Strategies' components of the processes. It also requires the early and ongoing involvement of surrounding communities and affected stakeholders.

Conclusion

Planning cannot successfully develop and become more reliable and efficient in addressing and solving the social, economic and ecological problems faced by the communities around the world, unless it becomes more interdisciplinary, and effectively incorporates other relevant disciplines. This, however, cannot be achieved without those disciplines becoming aware of planning, as seen by the planners themselves. Therefore, the main reason behind presenting this broad overview of the role and responsibilities of planning, including its shortcomings, and an outline of a 'model' planning process, has been to facilitate understanding of the basic principles upon which planning operates. These issues are of significance for the representatives of other disciplines, who are interested in and/or are required to be involved in planning, including professional discussions about the nature of planning and its processes. Further, as the prime purpose of this book is to develop a planning model to be used to identify integrated buffers for protected areas and other ESAs, now that the general planning process has been described, the more specific buffer planning process can be developed within this framework.

References

Barrow, C.J. (1995), *Developing the Environment*, Longman Scientific and Technical, London.

Batty, M. (1979), 'On Planning Process', in B. Goodall and A. Kirby (eds), *Resources and Planning*, Pergamon Press, Oxford, pp. 17-45.

Beale, J.G. (1980), *The Manager and the Environment: general theory and practice of environmental management*, Pergamon Press, Oxford.

Brewer, G.D and deLeon, P. (1983), *The Foundations of Policy Analysis*, The Dorsey Press, Illinois.

Chadwick, G. (1971), *A Systems View of Planning*, Pergamon Press, Oxford.

Clark, T.W. (1995), *Developing Pragmatic Koala Conservation Management Policy*, Draft Koala Policy Paper, School of Forestry and Environmental Studies, Yale University and Northern Rockies Conservation Cooperative, np.

Daly, H.E. and Cobb, J. (Jr.) (1989), *For the Common Good*, Beacon Press, Boston.

Delbridge, A. (ed.) (2001), *The Macquarie Dictionary*, Macquarie Library, North Ryde.

England, P. (2001), *Integrated Planning in Queensland*, The Federation Press, Sydney.

Evans, B. (1997), 'From Town Planning to Environmental Planning', in A. Blowers and B. Evans, *Town Planning into the 21st Century*, Routledge, London, pp. 1-14.

Evans, B. and Rydin, Y. (1997), 'Planning, Professionalism and Sustainability' in A. Blowers and B. Evans, *Town Planning into the 21st Century*, Routledge, London pp. 55-69.

Faludi, A. (1987), *A Decision-Centred View of Environmental Planning*, Pergamon Press, London.

Friedman, J. (1996), 'Two Centuries of Planning Theory: An Overview', in S. Mandelbaum and R. Burchell, *Explorations in Planning Theory*, Rutgers, pp. 10-29.

Goudberg, N. (1992), 'Local Government' in Australian Academy of Science (ed.), *Biological Diversity: Its future conservation in Australia*, Proceedings of the Fenner Environment Conference, 11-13 March 1992, Australian Academy of Science, Canberra, pp. 79-84.

Graham, B. and Pitts, D. (1996), *Draft Good Practice Guidelines for Integrated Coastal Planning*, Prepared for RAPI with assistance from the Federal Department of Environment, Sport and Territories, Canberra.

Grumbine, R.E (1994a), 'What is ecosystem management?', *Conservation Biology*, **8**(1), pp. 27-38.

Grumbine, R.E (1994b), *Environmental Policy and Biodiversity*, Island Press, Washington D.C.

Hawken, P., Lovins, A.B. and Lovins, L.H. (1999), *Natural Capitalism: The Next Industrial Revolution*, Earthscan, London.

International Union for Conservation of Nature and Natural Resources (IUCN) (1980), *World Conservation Strategy: Living Resource Conservation for Sustainable Development*, IUCN, Gland.

IUCN (1982), *The World National Parks Congress*, Bali 11-22 October 1982, IUCN, Gland.

IUCN (1992), 'Parks for Life: The Caracas Action Plan' Draft, IVth World Congress on National Parks and Protected Areas, Caracas, 10-21 Feb.

IUCN, UNEP, WWF (1991), *Caring for the Earth: A Strategy for Sustainable Living*, IUCN, Gland, Switzerland.

Kirk, G. (1980), *Urban Planning in a Capitalist Society*, Croom Helm Ltd., London.

Kozlowski, J. (1990), 'Queensland Strategic Planning: Pitfalls and Prospects', *Queensland Planner*, **30**(2), pp. 3-7.

Kozlowski, J. (1993a), 'Towards "ecological re-orientation" of professional planning', in J. Kozlowski and G. Hill (eds), *Towards Planning for Sustainable Development: A Guide for the Ultimate Environmental Threshold (UET) Method*, Avebury, Sydney, pp. 3-15.

Kozlowski, J. (1993b), 'Ultimate Environmental Threshold: An Alternative Tool for Planning Sustainable Development', *Sustainable Development*, **1**, no.1, pp. 51-63.

Kozlowski, J. (2002), 'To Sustainability through Interdisciplinary Planning: A Planner's Perspective', *Ekistics*, **69**, pp. 292-303.

Kuiken, M. (1990), 'Strategic Plans in Queensland: a Review of Gazetted Plans', *Queensland Planner*, **30**(1), pp. 31-6.

Mazur, M. (1986), *Cybernetyka i Charakter* (*Cybernetics and Character*), PIW, Warsaw.

Munashinghe, M. and McNeely, J. (1994), *Protected Areas Economics and Policy*, The World Bank and IUCN, Washington DC.

Oodit, D. and Simonis, U. (1992), 'Poverty and Sustainable Development', in F. Dietz, U. Simonis and J. van der Straaten (eds), *Sustainability and Environmental Policy*, Edition Sigma, Berlin, pp. 237-66.

Pearce, D., Markandya, A. and Barbier, E.B. (1989), *Blueprint for a Green Economy*, Earthscan Publications Ltd., London.

Pickering, T. and Owen, L.A. (1994), *An Introduction to Global Environmental Issues*, Routledge, London.

Regional Planning Advisory Group (RPAG) (1993), *Open Space and Recreation: A Policy Paper of the SEQ2001 Project* [Dept. Housing, Local Govt. and Planning] [Brisbane].

Rosier, J. (1992), *ESA-PLAN: An Ideal Planning Framework for Ecologically Sensitive Areas*. Thesis submitted for the degree of Doctor of Philosophy, Department of Geographical Sciences, The University of Queensland, Brisbane.

Streeten, P. (1992), 'Human Sustainable Development', F. Dietz, U. Simonis and J. van der Straaten (eds), *Sustainability and Environmental Policy*, Edition Sigma, Berlin, pp. 129-137.

Westrum, R. (1994), 'An organizational perspective: designing recovery teams from the inside out', in T.W. Clark, R.P. Reading and A.L. Clarke (eds), *Endangered Species Recovery*, Island Press, Washington, pp. 327-49.

World Commission on Environment and Development (WCED) (1987), *Our Common Future*, WCED, Washington DC.

World Commission on Environment and Development (WCED) (1990), *Our Common Future*, Australian Edition, Oxford University Press, Melbourne.

Chapter 3

Protected Areas: Problems and New Directions

Introduction

Chapter 3 focuses on protected areas as a key mechanism to conserve biodiversity. It examines the meaning of protected areas and briefly describes some of the more significant problems facing their establishment and management. These problems point to the need for additional strategies to conserve biodiversity. One such strategy that is recommended by the IUCN (2003) is the improved integration of protected areas with their surrounding landscapes and communities through buffering and linking natural habitats. A range of other strategies is also addressed in this chapter.

Protected Areas – What are the Issues?

A protected area is 'an area of land and/or sea especially dedicated to the protection and maintenance of biological diversity, and of natural and associated cultural resources, and managed through legal and/or other effective means' (IUCN, 1994). The Vth World Parks Congress (IUCN, 2003) highlighted that during the 21st century pressure on protected areas will increase as a result of the following changes:

- demographic shifts, population increases in urban areas, unsustainable consumption patterns and widespread poverty impacting on environmental services;
- greater demands for production of goods and services from protected areas;
- development of inappropriate infrastructure, climate change and invasion of exotic species;
- fragmentation of natural habitats;

- overfishing and collapse of marine fisheries and coral reefs, coastal and freshwater systems;
- increasing threats to the welfare and safety of protected area staff;
- consolidation and expansion of democratisation, decentralisation, 'deconcentration' and expanded public participation processes; and
- international assistance flows that focus primarily on the social needs of the impoverished.

Traditional approaches to biodiversity conservation, which have centred on declaring protected areas and similar reserves, supervising management within these areas, and on encouraging single species research, have failed to slow the loss in biodiversity and to satisfactorily preserve natural patterns and ecological processes. The main shortcomings of this approach are described below.

The Unsuccessful Percentage Area Approach

Specifying a desirable percentage area for the protected area estate in each country has been a common approach to achieving conservation goals (WCED, 1990; IUCN, 1992b). However, this has resulted in poor representation of biodiversity within the protected area network in several countries. The approach has been manipulated to include 'worthless lands' (Runte, 1973) at the expense of biologically diverse areas. Further, once the desired percentage area is achieved, opportunities to add new sites of high biological importance and to enhance the representativeness of the protected area network may be foreclosed. Percentage targets for terrestrial protected areas, which have been set at around 12 per cent, may be too small to prevent the extinction of many species (Margules et al., 2000) and may be totally unrealistic in situations where much of the remaining land is unprotected and subject to a range of threatening processes. The targets become more problematic as less of the surrounding matrix remains in a natural state. Hence a 12 per cent target may be appropriate if 30 per cent of the remaining land is cleared, but may be completely unsuitable if 90 per cent is developed.

Slow Growth in the Size of the Protected Area Estate

Although legal designation of protected areas worldwide has been increasing, there are signs of a decline in the growth rate of protected areas. Opportunities for establishing new protected areas, especially those offering a high level of protection to natural values, are diminishing as human populations expand into natural areas and the cost of acquisition becomes prohibitive, especially on land that is highly productive for other uses. Further, the traditional sources of reserves such as government owned land are diminishing (Pressey and Logan, 1997). This trend places at risk the objective of achieving a more comprehensive, adequate and representative reserve system and highlights the need for complementary approaches to ensure biodiversity conservation.

Diminishing Size of Protected Areas

There is a tendency, worldwide, for each new protected area to be smaller in size (Thorsell, 1992; IUCN, 2003) and thus less capable of conserving many rare or threatened species. Most parks, for example, may be too small to guarantee the long-term conservation of wildlife. This situation requires, in part, that planners integrate the management of environmentally sensitive areas (ESAs) with their surroundings to minimise the impacts of threatening processes and help to ensure that small habitat fragments have a higher chance of persistence. It is in this context that buffers for ESAs and corridors, as part of an integrated management strategy, will play an important role in maintaining biodiversity.

Inadequate Representation of Biodiversity Values

Protected areas in many countries, while failing to meet percentage targets, also fail to adequately represent the biodiversity within their national boundaries. The global protected area network is far from finished, with significant gaps in the coverage of protected area systems for threatened species and globally important sites, habitats and realms (IUCN, 2003). For example, the 'Interim Biogeographic Regionalisation for Australia' (IBRA) (Thackway and Cresswell, 1995) indicated moderate to high bias across almost 80 per cent of Australia in the extent to which the system of protected areas within each bioregion[1] represented the known biodiversity. In general, areas of high relief, low soil fertility, and steep rainfall gradients, the so called 'worthless lands', were more commonly sampled in protected areas than lands considered important for agriculture or other resource consumptive land uses typically occurring on lands of low relief or higher soil fertility. The consequences of this approach are that native wildlife may exist mainly on sub-optimal habitat, presenting difficulties for conservation and management of biodiversity and high probabilities of species' extinctions. The loss of many high quality environments since the arrival of Europeans in Australia also reduces the possibility of conserving or sampling much biodiversity in the future. In addition, the failure to link small and isolated reserves, further minimising the effectiveness of the reserve network in conserving biodiversity. As a consequence of these trends the IBRA report (Thackway and Cresswell, 1995) recommended the design of a more representative protected area system, and a focus on integrating conservation across the landscape.

Affected by Unrestrained External Threats

The legal boundaries of protected areas are those established by the legislative authority of a country or region and their location tends to be influenced by land tenure, property boundaries and administrative borders. In contrast, the biotic boundaries are 'hypothetical boundaries which would be necessary to maintain existing ecological processes and a given assemblage of species' (Newmark, 1985:197). As the legal and biotic boundaries rarely coincide due to the enormous potential size of the biotic boundaries (Lusigi, 1981; Garratt, 1982; Newmark, 1985), many protected areas are not fully functional ecosystems and thus protection and restoration of biodiversity is difficult. The interactions between protected areas and their adjacent lands at times produce ecosystem stress or threats, which can be defined as:

> ...those conditions of either human or natural origin that cause significant damage to park resources, or are in serious conflict with the objectives of park administration and management (Neumann and Machliss, 1989: 14).

Extensive research[2] in many countries indicates a wide range of external threats to ESAs, such that the ecosystem dynamics of these fragments are probably driven by external, rather than internal factors. The more prominent threats worldwide relate to unplanned colonisation (e.g. agricultural encroachment), unlawful subsistence hunting and plantation agriculture, fire, exotic plants, feral animals, chemical pollution, removal of vegetation (e.g. illegal logging and fuelwood collection) and illegal removal of animals.[3]

The effectiveness and viability of most, if not all reserves, are directly related to the impacts of surrounding land and water uses, and long-term protection can only be guaranteed by appropriate management of threatening processes within and outside of reserves. Hence, understanding the relationships between the elements of a reserve or habitat fragment and its surrounding environment is critical, not only for the conservation of wildlife, but also the proper functioning of ecosystems within a remnant natural area. For example, a protected area boundary may represent a barrier for wildlife, acting as a one-way filter, letting some individual members of population out, but enabling few to return. Due to the operation of internal and external processes, a generated ecological edge may be created, one that will probably not coincide with the protected area boundary. Where external pressures acting on the protected area are strong the generated edge may move inward towards the centre of the reserve. Such a situation may significantly reduce the effective size of the reserve for some species and encourage the invasion of generalist species.

Limited Funding for Effective Management

The World Commission on Protected Areas (WCPA) (IUCN, 2003) estimated that protected area budgets in the 1990s were about 20 per cent of the estimated US$20–30 billion required annually over the next 30 years to establish and maintain a comprehensive protected area system including terrestrial, wetland and marine ecosystems. Issues of concern raised by WCPA (IUCN, 2003) included:

- insufficient priority allocated to the conservation of natural and cultural values against other competing budget programs (e.g. mining and defence);
- revenues from protected areas not being earmarked for protected area management;
- institutional barriers restricting the flow of funding to protected areas; inappropriate management structures that fail to channel funding to protected area management;
- lack of mechanisms to encourage donor organisations to participate in supporting protected areas; and
- limited use of business planning at a protected area systems level and for specific protected areas.

Protected area managers are being required to devote resources to raise their own funding and as a result protected areas are facing greater degradation (IUCN, 2003), with many protected areas being no more than 'paper parks' (Stolton and Dudley 1999:11).

Other Issues

The Vth World Parks Congress (IUCN, 2003) provided a forum for over 3,000 delegates, representing many countries, interests and experience in protected areas to identify existing problems and issues and to provide direction for the future enhancement of protected areas. In addition to the issues raised above, other concerns related to the impacts of climate change on protected areas, the need to strengthen mountain protected areas and marine and coastal protected areas, the governance of protected areas, and issues related to poverty, indigenous peoples and the co-management of protected areas.

The Road to Recovery

The traditional approach to nature conservation, as described above, has focused on the declaration of protected areas. The main trends in protected area design and management are the declining growth rate of protected areas, the tendency for new areas to be smaller in size, the high level of bias in representation of biodiversity within protected area systems, the tendency for reserves to contain only a part of their fully functioning ecosystem, the increasing levels of external threats to which they are being subjected, the limited funding made available to manage protected areas and poor governance. Many reserves, being small in size and insular in character, are unable to support minimum viable populations of even small species and are increasingly subject to a range of edge effects.

Improved future prospects for protected area planning and management, and ultimately for biodiversity, will hinge on the extent to which change is initiated and embraced by governments, non-government organisations, local communities and civil society. Recommendations advocated by the World Parks Congress (IUCN, 2003) will provide guidance for the necessary changes. Important initiatives include the following:

Re-classification of Protected Areas

In 1994 the IUCN endorsed the new six-category, objectives-based classification of protected areas (IUCN, 2003), which encompass a variety of lands that may play a role in the conservation of natural resources and values. Category I (Strict Nature Reserve/Wilderness Areas), category II (National Park) and category III (National Monument) have high levels of protection and exclude incompatible extractive uses due to their natural, cultural and aesthetic richness. These protected areas are essential to the preservation of biodiversity and for providing the necessary benchmarks for monitoring change. Category IV (Habitat/ Species Management Area) protected areas are subject to active intervention for management purposes to ensure the maintenance of habitats and/or to meet the requirements of specific species, while categories V (Protected Landscape/Seascape) and VI (Managed Resource Protected Area) incorporate sustainable use of resources, including use by resident populations living within the protected areas (McNeely et al., 1994).

The concept of a protected area is now much wider than that of the traditional national park, encompassing a variety of lands that have been identified to play a role in the conservation of natural resources and values. Such areas usually serve several functions, with the overall goal being a progression towards sustainable living. The advantages to biodiversity conservation are significant in that the new classification, allows for many areas to be dedicated for conservation while allowing other compatible land use activities to continue, and it vests management responsibility in private owners as well as government and non-government organisations. This approach lends itself to the incorporation of buffers into the classification system as the buffered protected areas may include a variety of sustainable use practices, rather that exclusive protection of particular land and seascapes.

The Vth World Parks Congress (IUCN, 2003) identified that in addition to the conventional system of protected areas based on IUCN designated categories, a range of opportunities exist for enhancing the coverage of protected areas, including community conservation areas, community managed areas and private and indigenous reserves. Hence, in relation to the governance of protected areas, the Congress requested that the WCPA refine its protected area categorisation system to include a governance dimension that recognises the legitimacy and diversity of approaches to protected area establishment and management and makes explicit that a variety of governance types can be used to achieve conservation objectives and other goals. The recommended protected area categories for future inclusion are:

- government managed;
- co-managed (i.e. multi-stakeholder management);
- privately managed; and
- community managed (community conserved areas).

Such a range of protected areas will widen stakeholder participation in biodiversity conservation and result in an enhanced regional landscape approach to conservation.

Improved Representation of Biodiversity

A recent international review of conservation tools (James et al., 1999) demonstrated that nature reserves are cost effective in comparison with the cost of maintaining biodiversity in developed ecosystems, using a range of other tools. Hence, rather than aiming solely for percentage targets, which may not adequately conserve many species and ecosystems, planning is beginning to focus at the bioregional level to ensure the establishment of comprehensive, adequate and representative reserve systems. For example, the goal for Australia is for a comprehensive, adequate and representative National Reserve System, which is based on a bioregional assessment of biodiversity values (Thackway and Cresswell, 1995). In relation to coastal and marine protected areas the IUCN (2003) recommends a greatly expanded protected area system by 2012 to include at least 20 to 30 per cent of each marine/coastal habitat.

Integrated Planning Across Whole Landscapes

Protected areas, which cover 11.5 per cent of the Earth's land surface (IUCN, 2003), need to be complemented by off-reserve conservation strategies to avoid a continuing loss of biodiversity across the un-reserved landscape, and to fill the serious gaps in coverage of many important species and biomes. Merely taking land out of private ownership and adding it to the reserve system may not be a long-term solution, for even the most efficiently managed reserves are unlikely to halt the extinction process, although they may slow the loss of biodiversity (IUCN, 2003; WCED, 1990; WRI et al., 1992; IUCN, 1992b). Farrier (1995:15) terms this the 'ghetto approach' to conservation.

Although a comprehensive, adequate, representative, secure and well-managed protected area network remains an important basis for biodiversity conservation, the formally designated boundaries of protected areas may not coincide with the biological needs of their representative species and ecosystems and thus the management of protected areas and other ESAs should be meshed with planning considerations in their wider regions. Continuing discussion of regional landscapes (Noss, 1983; Noss and Harris, 1986), bioregions and regional ecosystems (IUCN, 2003, 1992a, 1992b; WRI et al., 1992; Thackway and Cresswell, 1995; Sattler and Williams, 1999) and ecosystem management (Shafer, 1990; Grumbine, 1994a, 1994b; Saunders, 1994; Stanley, 1995; Yaffee 1999) signifies a new way of looking at the conservation of biodiversity, for 'if we can't save nature outside protected areas, not much will survive inside' (Western, 1989: 159).

The World Parks Congress (IUCN, 2003) reaffirmed its position that while protected areas focus on biodiversity conservation, to be effective they must be planned and managed in the context of the broader land/seascape. Hence the Congress recommended the integration of conservation objectives into land/sea use and regional and sectoral planning at all levels. It also recommended that changes in biodiversity and key ecological processes affecting biodiversity in and around protected areas need to be identified and managed.

The Congress stated that the protected area system,

> … needs to comprise an ecologically representative and coherent network of land and sea areas that should include protected areas, corridors and buffer zones, and is characterized by interconnectivity with the landscape and existing socio-economic structures and institutions (IUCN, 2003:2)

In addition the Congress stressed that governments, non-government organisations, local communities and civil society should adopt and promote protected area design principles that reflect those inherent in the world network of biosphere reserves where core protected areas are part of landscapes designed to enhance the overall conservation value. Such reserves focus on combining conservation, development and research/ education objectives, by applying a zoning system, which includes a protected core, surrounding buffer and an outer transition area, which may be integrated into regional planning. These statements reaffirm the important role that buffers should play in an integrated approach to landscape planning and highlight planning's role in ensuring interdisciplinary processes, which co-ordinate and integrate the expertise of specialists in a range of related disciplines. This shift in planning methodology is slowly taking place.

Capacity Building and Awareness Raising

Protected areas need to be managed by effective institutions, within a supportive policy and legal framework, and by trained professionals with the necessary technical and management skills (IUCN, 2003). An improved future for protected areas requires the implementation of comprehensive capacity building programs. Integrated planning for protected areas also depends on the cooperative involvement of local communities. Hence strategies are needed to raise the awareness of the value of protected areas and the benefits they provide to society and to enhance the community's general commitment to support protected areas (IUCN, 2003).

Improved Financing of Protected Areas

Many countries with the highest levels of biodiversity are challenged by inadequate financial means and the imperative of poverty alleviation, and as a result compromise on creating and/or effectively managing a comprehensive and effective protected area system. There is thus a universal need to provide adequate funding to protected areas to ensure sustained conservation of biodiversity, natural and cultural heritage. Improved conservation outcomes will only come if protected areas receive adequate financial support (IUCN, 2003).

Principles of Integrated Landscape Management

As the model integrated buffer process to be detailed in this book relates primarily to the conservation of the values of significant natural areas, it is necessary to examine briefly some of the more important issues or principles related to understanding the ways in which an ESA may be linked to its surrounding landscape matrix. Many of the important concepts underlying ecosystem management or integrated landscape management mirror the important principles of sustainable development (Table 2.1) and of good planning (Table 2.3). For example, features of a landscape approach to biodiversity conservation include examining the whole system rather than individual parts and on understanding interrelationships among elements and levels of the biodiversity hierarchy (genes, species, ecosystems and landscapes), in terms of their composition, structure and function. This approach views ecosystems as dynamic structures, responding continually to change and having varying levels of resilience. It also recognises that particular environments may have certain carrying capacities and thresholds and hence limits to human activity. Further, it requires that the operational boundaries for planning and management encompass the functional landscape mosaic and importantly that the focus of management be based on ecological boundaries, rather than political or administrative units. This consequently requires that planners or managers should focus on gaining sectoral support and co-operation from all relevant levels of government.

Effective landscape planning and management may require extensive research and data collection (i.e. habitat or ecosystem inventory and classification, disturbance regime dynamics, baseline species and population assessment), better management and use of existing data, and monitoring to gauge the effectiveness of policy strategies so that current approaches may be adapted to changing circumstances (Grumbine, 1994a; Franklin, 1997; Yaffee, 1998).

Integrated landscape management brings with it awareness that humans are integral components of complex, interdependent ecosystems, and an acceptance that other living components of ecosystems, despite having use to humans, also have inherent value. Stanley (1995: 261) stresses that humanity must begin to view itself as 'part of nature rather than the master of nature'. However, he concedes this view may never occur and, if it does, will be a slow process that may come too late.

An integrated landscape approach requires the layering or nesting of conservation actions ideally beginning with continental-wide framework encompassing important natural areas and linkages, this being progressively implemented at the national, state/provincial, regional and catchment levels. Such an approach directs attention to protecting and managing remnants comprehensively within a wider regional context, rather than on an individual basis. There are several examples of this approach: United States (Noss, 1992; Florida Department of Environmental Protection, 1998; Wildland's Project, 1999); the Netherlands (vanZadelhoff et al., 1995); and Australia (EPA, 2002). Acceptance of an integrated landscape approach to biodiversity planning is essential to gain a broad perspective on the conservation priorities of specific bioregions and to extend the mechanisms for the protection of biodiversity beyond the reservation of land as a national park or similar reserve.

Use of the phrase 'landscape planning and management' is increasing and this shift from 'ecosystem management' is associated with the following:

- a growing recognition of the importance of ecological understanding for guiding human uses of environmental resources;
- an awareness that humans must see themselves as integral components of complex, interdependent ecosystems related to one another over a wide range of temporal and spatial scales; and
- acceptance that other living components of ecosystems have inherent value in and of themselves, regardless of their immediate usefulness to humans as resources.

Stanley (1995) examines this humanist context of ecosystem/landscape management, by questioning several assumptions, including: that science can determine how ecosystems function; that once function is known, the social/political system will be able to protect ecosystems to the extent needed for the survival of human society; that reality will take precedence over political expediency; and that humans can develop the technology needed to manipulate ecosystems.

Yaffee (1999) suggests that there are three faces of ecosystem management: environmentally sensitive, multiple use management; ecosystem based approaches to resource management; and ecoregional management. These three faces represent points on a five-stage continuum, beginning with dominant use and multiple use paradigms. Yaffee suggests that it is important to recognise that management agencies and individuals will be at various points on the continuum and that management strategies should be appropriately directed at progressively moving towards the ecosystem management end of the spectrum and that it is important to keep pushing behaviour at the margins. Thus policy instruments, incentives, information and training need to be targeted to the realities of specific settings.

Issues for Planners

Conservation biology theory has progressed from species-area curves, to a consideration of meta-population dynamics, environmental variation, nested subsets, and complicated habitat mosaics. However, these analyses have many inherent problems, as ecological situations do not generalise well and hence a species or system may not operate in the way envisioned by the theories applied to it (Doak and Mills, 1994:622). Further, these theories or generalisations may be of no practical value due to an inability to estimate the relevant parameters with the available data. However, ever increasing competition for a shrinking natural resource base, places extreme pressure on planners and other land managers to know the following:

- how much of a species' habitat must be protected;
- what should be the spatial arrangement of this habitat;
- what condition or quality of habitat is sufficient for the conservation of particular species;
- how can surrounding activities be managed to avoid or minimise their impacts on natural areas; and
- how should corridor systems be maintained and developed in order to conserve the species over the long term.

Planners frequently do not have all the data required to make the 'best' decisions. However, planning decisions are unable to be 'put off' until the data become available, for deciding not to address biodiversity conservation issues until reliable data are available, will frequently reduce the number and effectiveness of possible planning options, or in some cases, lead to ecological disasters. This problem becomes even more acute for land subject to high incoming earning potential from competing uses such as agriculture, residential and tourist uses.

Given the possibility that biodiversity may be affected by chance events, planning should also provide for the needs of species or ecosystems not only in 'normal' years, but also in exceptional circumstances. This can be likened to town plans that identify 1 in 50 year flood lines (i.e. abnormal rainfall rather than average rainfall figures) and restrict development in these flood prone areas. Such an approach, which recognises the likelihood of chance variations in demographic and environmental factors, as well as natural catastrophes and random genetic processes, is also needed in planning for the conservation of biodiversity. Planners must also ensure that their plans provide a high chance of long-term persistence for particular species and assemblages of species or ecosystems within a regional framework, and that these plans have minimal impact on the sustainable use of competing resources.

In the absence of scientifically validated relationships, the planner must be guided by currently accepted theory and inference from related disciplines. The broad ecological concepts discussed in this chapter point to a number of general planning principles that can be applied to biodiversity conservation. These include issues related to the size, shape and condition of ESAs and networks and corridors.

Size, Shape and Condition of ESA

Whether or not an ESA or habitat patch provides conservation benefit for species depends on the quality of the habitat and its relationship to its surrounding region (i.e. its landscape context) (Wiens, 1994; Haila, 2002). Perhaps the most important issues for planners are to stem the loss of habitat and to minimise the effects of fragmentation. However, fragmentation effects are difficult to translate into simple management rules to guide planners, because the effects tend to be species-specific, varying according to landscape structure, and its influence on species may be obscured by local effects such as declines in habitat condition (McAlpine, et al. 2002; Villard, 2002). Important principles for planners to consider include:

- retain larger habitat patches rather than smaller patches as they are likely to:
  - contain a greater sample of the original habitat than a smaller area;
  - contain a greater diversity of habitats for animals to occupy and consequently are likely to have sampled a greater diversity of species than a smaller area (Lynch, 1987; Bennett, 1990; Wilcove and Murphy, 1991);
  - sustain their component species, for local habitat deterioration will probably not affect the entire area simultaneously and the larger population size may help to prevent genetic drift. However, where small populations exist they may need to be managed as a meta-population, which may require natural migration or human directed migration of animals between sub-populations; and
  - experience fewer threatening processes and edge effects (Machliss, 1985; Coveney, 1993), and hence may experience decreased rates of extinction;
- establish appropriately shaped remnants, which decrease the edge to area ratios of ESA, so as to minimise edge effects;
- avoid abrupt edges to ESAs that interfere with fauna movement to outside habitat;
- the landscape context will have the greatest effect on habitat generalists, while habitat specialists will be more sensitive to abrupt edges and declines in habitat quality within the remnant patch; and
- small reserves or remnant patches are important, especially where they are interconnected and hence the aim should be to maintain natural habitat connectivity.

Networks and Corridors

Any guidelines on minimum viable populations, meta-population structure, corridor design and the like are, of necessity, general, as each situation is unique and will require specific research. However, in a highly fragmented region, where all that remains is a system of smaller reserves, cooperative management of ESAs and adjacent lands is very important. The small reserves or patches and their remnant populations should be comprehensively planned and managed as regional and inter-regional systems of interlinked reserves. Buffer zones, based on landscape management principles, may be an important tool to the planner to ensure a minimisation of threats to the ESA by integrating the management of the ESA with that of the surrounding region. Several principles include:

- manage the ESA within a bioregional context to facilitate gene flow and migration among populations and ensure adequate representation of species and habitats within bioregions;
- retain habitat in contiguous blocks rather than fragmented blocks of habitat (Wilcove and Murphy, 1991; Noss, 1992);
- retain closely spaced remnant patches of habitat rather than widely dispersed habitat patches. Close spacing of habitat may assist fauna migration thus decreasing the probability of local extinction due to random and other events. (*Note*: closeness may produce management difficulties, e.g. spread of disease and fire etc.);
- ensure habitat is well distributed across a species' range to reduce the susceptibility of individual species to extinction (Wilcove and Murphy, 1991);
- retain and restore interconnected blocks of habitat, rather than isolated blocks, particularly for species with effective dispersal mechanisms;
- maintain existing corridors, for example those along the existing water courses, rather than create new corridors across the landscape. As corridors may be biologically important habitats it may be important to include 'nodes' of habitat in corridors to provide stepping stones for movement;
- strive to retain and manage habitat outside an ESAs so that it closely resembles the habitat of the ESA, as this will enable dispersing species to more easily traverse the landscape mosaic (Wilcove and Murphy, 1991);
- avoid internal fragmentation of habitat, for example by roads, rail, fences, power lines, dams, and residential development. Where such structures are necessary, it is important that their design is in accord with the habitat and movement needs of the species in the ESA.

Buffering Ecologically Sensitive Areas

One important strategy for extending the generated edge of a reserve is to implement a buffer to effectively increase the size of the reserve and minimise the impact of externally originating disturbance regimes. As the reserve system itself becomes more comprehensive, adequate and representative, its effective buffering and integration with surrounding lands may significantly improve the conservation of biodiversity. Although this topic has been addressed in Chapter 1, a few extra comments may be useful.

Natural area managers have at their disposal a wide range of management strategies to deal with internal stresses to protected and reserved lands. However, strategies to effectively link the ESAs with their surroundings, to minimise external threats and to ensure compatible land uses outside of ESAs are generally lacking. A change in planning and management emphasis is required, for the viability of natural areas over the long term depends on how well they are ecologically, socially, and economically integrated into the surrounding region (WRI et al., 1992).

A buffer strategy is particularly important where habitat fragments are unrepresentative of the former ecosystems in a region and they consist of lower quality habitat. For example, removal of vegetation in the southeast corner of Queensland (Australia) has displaced the koala from its preferred habitat on the more fertile river flats along the eastern seaboard and it exists mainly in vegetation that is of marginal quality. This situation requires sensitive management to ensure the survival of the koala in this region. Buffer strategies have the potential to minimise the impacts of surrounding development on these remnants and enhance the conservation of biodiversity within the wider region.

Planning must ensure that conservation strategies are in place on the lands outside of protected areas and other ESAs and in particular that these areas are integrated with a range of 'off-reserve' management programs. Planners must attempt to define ecological boundaries and develop management strategies based on conserving ecological integrity (viable populations and effective ecosystem patterns and processes). Many native species can continue to live in unprotected areas, especially when those areas are planned and managed to maximise the benefits or contribution that the reserved lands can provide. It is in this context that buffering of ESAs from external threats through integrating their management with that of their surrounding lands may lead to a more effective way of maintaining biodiversity in the long term. Planning for integrated buffers to protected areas will also, by its nature, be part of a more general and long overdue move towards all planning becoming more interdisciplinary.

Conclusion

The acquisition of land for national parks and other protected areas has played an important role in nature conservation in the past and has served to slow the general pace of species' extinctions. However, protected areas represent only a small fraction of the total land area of any particular country, state or region. This role now needs to be re-examined, for the phase of large scale acquisition of protected areas is largely over. More steps need to be taken than just the reservation of land and the protection of individual species.

Natural area managers in the past have had to work with the fragments of habitat remaining after development has occurred. Opportunity to design an ecologically effective natural area system before development occurs has seldom been a reality. As a consequence, planners need a set of design criteria, guidelines or decision rules, to help conserve biodiversity and the survival chances of specific species in areas where development and consequent fragmentation of habitat already exist. Many species depend for their survival on resources outside the bounds of protected areas. There has been a reluctance to tackle the issue of external threats to protected areas, because they are considered more complex and difficult to deal with than internal threats. However, the mounting evidence of degradation to protected areas from a wide variety of external influences indicates a need to incorporate protected areas, as well as other environmentally sensitive areas with their surrounds in order to plan effectively for their long term sustainable development.

As natural areas become increasingly insular and colonization sources fewer and more distant, colonisation rates will diminish resulting in a decline of species over time. Conservation objectives need to adopt a bioregional approach that integrates planning and management of dispersed habitats with existing protected areas, and focuses on linking and buffering these systems. Such strategies have had some success and should remain as an important planning tool in fragmented landscapes. Specific land use policies should be devised for the lands surrounding or adjacent to the identified remnant patches or core areas, to ensure compatibility with conservation goals and acceptance and incorporation by local communities. This planning process will enable the ESA to form a core or integral part of a regional conservation strategy to ensure ecologically sustainable use of resources.

Incorporation of integrated landscape management into protected area planning requires an effective planning process. The simplified planning model outlined in Chapter 2 will form the basis for developing an 'Integrated Buffer Planning' (IBP) model, to facilitate its incorporation into planning processes and, as a consequence, to help ensure the improved integration of significant natural areas with their surrounding regions and communities.

Notes

1 A bioregion can be defined as 'a land and water territory whose limits are defined not by political boundaries, but by the geographical limits of human communities and ecological systems. Such an area must be large enough to maintain the integrity of the region's biological communities, habitats, and ecosystems; to support important ecological processes such as nutrient and

waste cycling, migration, and stream flow; to meet the habitat requirements of keystone and indicator species; and to include the human communities involved in the management, use, and understanding of biological resources. It must be small enough for local residents to consider it home' (WRI et al., 1992:99-100).

2 Refer to Lusigi, 1981; Newmark, 1985; Conservation Foundation, 1985; Machliss and Tichnell, 1985; Noss and Harris, 1986; Schonewald-Cox and Bayless, 1986; Wallace and Moore, 1987; Stottlemyer, 1987; Saunders et al., 1991; Department of Agriculture et al., 1991; Peterson, 1991; Coveney, 1991, 1993; Groombridge, 1992; Wells and Brandon, 1992; RPAG, 1993; Longmore, 1993; Thackway and Cresswell, 1995; Coates and Atkins, 1997.

3 There have been several major studies, which have examined the threats to protected areas and natural communities. These include the Forest innovations project (Dudley and Stolton, 2001); Brazil's protected areas (de Sa, 2000); Peru (Tovar, 2000); India (Singh, 2000); a report on 43 of the world's most threatened protected areas (IUCN, 1984, cited in Wells and Brandon, 1992:12); a survey of 135 parks in more than 50 countries (Machliss and Tichnell, 1985, cited in Wells and Brandon, 1992:12; Global Biodiversity Strategy (Groombridge, 1992); The State of the Parks -1980 (Conservation Foundation, 1985); a study of 23 protected areas from Africa, Asia and Latin America (Wells and Brandon, 1992); a study of parks in Victoria (Australia) (Department of Conservation and Natural Resources (Victoria) (Grace, 1990, cited in Coveney, 1993:209); an examination of threats to 20 parks in Victoria (Australia) from private land immediately outside their boundaries (Coveney, 1991); SEQ 2001 Project (Queensland, Australia – an examination of areas having regional conservation value within South East Queensland) (RPAG, 1993b); and an examination of threats to Australia's 12 biosphere reserves (Longmore, 1993).

References

Bennett, A.F. (1990), *Habitat Corridors: Their Role in Wildlife Management and Conservation*, Department of Conservation and Environment, Melbourne.

Conservation Foundation (1985), *National Parks for New Generation. Visions, Realities, Prospects*, Conservation Foundation, Washington, D.C.

Coveney, J. (1991), 'Threats to Parks from Adjacent Private Lands', *Park Watch*, June, pp. 22-24.

Coveney, J. (1993), 'Planning for Areas Adjacent to National Parks in Victoria', *Urban Policy and Research*, **11**(4), pp. 208-16.

De Sa, R.L. (2000), 'WWF Brazil's protected area effectiveness methodology and results', *Arborvitae, Management Effectiveness of Protected Areas.* An international workshop, Costa Rica, June 1999, p. 4.

Doak, D.F. and Mills, L.S. (1994), 'A Useful Role for Theory in Conservation', *Ecology,* **75**(3), pp. 615-26.

Dudley, J. and Stolton, S. (2001), 'Forest Innovations', *Arborvitae*, No. 17, pp. 8-9.

Environmental Protection Agency (EPA) (2003), Regional Nature Conservation Strategy for South East Queensland, EPA, Brisbane.

Farrier, D. (1995), Off-reserve management and the conservation of biodiversity, with particular reference to the management of land in private ownership, Consultant's Report to the Tasmanian Forests and Forestry Industry Council, np.

Florida Department of Environmental Protection (1998), *Plan for a statewide system of greenways: Five year Florida Greenways system implementation*, Florida Dept. Environmental Protection, Tallahassee.

Franklin, I.R. (1980), 'Evolutionary Change in Small Populations', in M.E. Soule and B.A. Wilcox (ed), *Conservation Biology. An Evolutionary – Ecological Perspective*, Sinauer Associates Inc., Sunderland, pp. 135-49.

Garratt, K. (1982), 'The Relationship between Adjacent Lands and Protected Areas: Issues of Concern for the Protected Area Manager', in A. McNeely and K. Miller (eds), *The Role of Protected Areas in Sustaining Society*. Proceedings of the World Congress on National Parks, Bali, 11-12 October 1982, IUCN, Gland, pp. 65-77.

Groombridge, B. (ed.) (1992), *Global Biodiversity. Status of the Earth's Living Resources*. A Report compiled by the World Conservation Monitoring Centre (WCMC), IUCN, United Nations Environment Program (UNEP), World Wide Fund for Nature (WWF), WRI, Chapman and Hall, London.

Grumbine, R.E (1994a), 'What is ecosystem management?', *Conservation Biology*, **8**(1), pp. 27-38.

Grumbine, R.E (1994b), *Environmental policy and biodiversity*, Island Press, Washington D.C.

Haila, Y. (2002), 'A conceptual genealogy of fragmentation research: From island biogeography to landscape ecology', *Ecological Applications,* **12**(2), pp. 321-34.

International Union for the Conservation of Nature (IUCN) (1992a), *Draft Recommendations*, IVth World Congress on National Parks and Protected Areas, Caracas, 10-21 Feb.

IUCN (1992b), *Parks for Life: The Caracas Action Plan,* Draft, IVth World Congress on National Parks and Protected Areas, Caracas, 10-21 Feb.

IUCN (1994), *Guidelines for protected area management categories*, IUCN and the World Conservation Monitoring Centre, Gland.

IUCN (2003), Recommendations of the Vth IUCN World Parks Congress, World Parks Congress, Benefits Beyond Boundaries, Durban, 9-17 Sept. Available at: http://www.iucn.org/themes/wcpa/wpc2003/english/outputs/recommendations.htm [20 October 2004].

IUCN, World Wide Fund (WWF) (2000), Forest Innovations project. Available at: www.iucn.org/themes/forests [3 March 2003].

Longmore, R. (ed.) (1993), *Biosphere Reserves in Australia: A Strategy for the Future,* Drawn from a report prepared for the Australian Nature Conservation Agency by P. Parker, Chicago Zoological Society for the Australian National Commission for UNESCO, ANCA, Canberra.

Lusigi, W.J. (1981), 'New Approaches to Wildlife Conservation in Kenya', *Ambio*, **10**(2-3), pp. 87-92.

Lynch, J.F. (1987), 'Responses of Breeding Bird Communities to Forest Fragmentation', in D.A. Saunders et al. (eds), *Nature Conservation: The Role of Remnants of Native Vegetation*, Surrey Beatty and Sons, Sydney, pp. 123-40.

McAlpine, C.A., Lindenmayer, D.B., Eyre, T.J. and Phinn, S.P. (2002), 'Landscape surrogates of forest fragmentation: synthesis of Australian Montreal Process case studies', *Pacific Conservation Biology*, **8**, pp. 108-20.

Machliss, G.E. and Ticknell, D.L. (1985), The State of the World's Parks: An International Assessment for Resource Management Policy, Westview Press, Boulder.

McNeely, J.A, Harrison, J. and Dingwall, P. (1994), 'Introduction Protected Areas in the Modern World', in IUCN (ed.), *Protecting Nature: Regional Reviews of Protected Areas*, IVth World Congress on National Parks and Protected Areas, Caracas, 10-21 Feb. pp. 5-28.

Margules, C.R. and Pressey, R.L (2000), 'Systematic conservation planning', *Nature*, **405**, pp. 243-253.

Neumann, R.P. and Machliss, G.E. (1989), 'Land-use Threats to Parks in the Neotropics', *Environmental Conservation*, **16**(1), pp. 13-18.

Newmark, W.D. (1985), 'Legal and Biotic Boundaries of Western North American National Parks: A Problem of Congruence', *Biological Conservation*, **33**, pp. 197-208.

Noss, R.F. (1983), 'A Regional Landscape Approach to Maintain Diversity', *BioScience,* **33**(11), pp. 700-6.

Noss, R.F. (1992), 'The Wildlands Project. A Conservation Strategy', *Wild Earth*, Special Issue, Cenozoice Society, Inc., Ann Arbor, pp. 10-25.

Noss, R.F. and Harris, L.D. (1986), 'Nodes, Networks and MUM's: Preserving Diversity at All Scales', *Environmental Management*, **10**(3), pp. 299-309.

Pressey, R.L. and Logan, V.S. (1997), 'Inside looking out: findings of research on reserve selection relevant to 'off-reserve' nature conservation', in P. Hale and D. Lamb (eds), *Conservation Outside Nature Reserves*, Centre for Conservation Biology, The University of Queensland, Brisbane, pp. 407-18.

Regional Planning Advisory Group (RPAG) (1993), *Open Space and Recreation: A Policy Paper of the SEQ2001 Project*, [Dept. Housing, Local Govt. and Planning], [Brisbane].

Runte, A. (1973), 'Worthless Lands - Our National Parks: The Enigmatic Past and Uncertain Future of America's Scenic Wonderlands', American West, 10, May, p. 11.

Sattler, P.S. and Williams, R. (ed.) (1999), *The Conservation Status of Queensland's Bioregional Ecosystems*, Environmental Protection Agency, Brisbane.

Saunders, D.A (1994), 'Habitat fragmentation: a symposium overview', in C. Moritz and J. Kikkawa (eds), *Conservation Biology in Australia and Oceania*, Surrey Beatty and Sons, Sydney, pp. 57-9.

Shafer, C.L. (1990), *Nature Reserves: Island Theory and Conservation Practice*, Smithsonian Institution Press, Washington D.C.

Singh, S. (2000), 'Protected area effectiveness in India', *Arborvitae, Management Effectiveness of Protected Areas*. An international workshop, Costa Rica, June 1999, p. 4.

Stanley, T.R.(Jr). (1995), 'Ecosystem Management and the Arrogance of Humanism', *Conservation Biology*, **9**(2), pp. 255-62.

Stolton, S,. and Dudley, N. 1999, 'Paper Parks', *Arborvitae*, Newsletter No.12, p. 11.

Thackway, R. and Cresswell, I.D. (eds) (1995), An Interim Biogeographic Regionalisation for Australia: A Framework for Setting Priorities in the National Reserves System Cooperative Program, Reserve System Unit, ANCA, Canberra.

Thorsell, J. (1992), 'The Road to Caracas: A Post Audit of the Implementation of the Bali Action Plan 1983-1992', in IUCN (ed.), *Parks for Life: Enhancing the Role of Protected Areas in Sustaining Society, Plenary and Symposium Papers*', IVth World Congress on National Parks and Protected Areas, Caracas 10-21 Feb., pp. 143-65.

Tovar, A. (2000), 'Protected area effectiveness in Peru', *Arborvitae, Management Effectiveness of Protected Areas*. An international workshop, Costa Rica, June 1999, p. 5.

van Zadelhoff, E. and Lammers, W. (1995), 'The Dutch ecological network', *Landschap*, **95**(3), pp. 77-83.

Villard, M. 2002, 'Habitat fragmentation: Major conservation issue or intellectual attractor?' in M. Villard (ed), *Invited Feature,* Ecological Society of America. Washington, D.C. pp. 819-20.

Wells, M.P. and Brandon, K.E. (1992), *People and Parks. Linking Protected Area Management with Local Communities*, World Bank, World Wildlife Fund, U.S. Agency for International Development, Washington D.C.

Western, D. (1989), 'Conservation Without Parks: Wildlife in the Rural Landscape', in D. Western and M. Pearl (eds), *Conservation for the Twenty-first Century*, Oxford University Press, Oxford, pp. 158-65.

Wiens, J.A. (1994), 'Habitat fragmentation: island v landscape perspectives on bird conservation', *Ibis*, **137**, pp. 97-104.

Wilcove, D.S. and Murphy, D. (1991), 'The Spotted Owl Controversy and Conservation Biology', *Conservation Biology*, **5**(3), pp. 261-2.

Wildlands Project (1999), *Summary of the Wildlands Project*, Available at: http://www.wildlandsproject.org /html/summary. htm [3 October 2000].

WCED (1990), *Our Common Future*, Australian Edition, Oxford University Press, Melbourne.

World Resources Institute (WRI), IUCN, UNEP (1992), Global Biodiversity Strategy. Guidelines for Action to Save, Study, and Use Earth's Biotic Wealth Sustainably and Equitably, WRI/IUCN/UNEP, np.

Yaffee, L. (1999), 'Three Faces of Ecosystem Management', *Conservation Biology*, **13**(4), pp. 713-25.

[illegible] (2000) [illegible] effectiveness [illegible]

[illegible] (1995) [illegible]

Robert, S. [illegible] (1998) [illegible]

Thackeray, R. and [illegible] National [illegible] Record [illegible]

[illegible] (1983) [illegible]

Tosun, A. (2000) [illegible]

[illegible] (1985) [illegible]

Villard [illegible]

[illegible] (1997) [illegible]

Wagner [illegible]

[illegible]

Williams [illegible]

[illegible]

PART TWO

CRITICAL REVIEW OF BUFFERS

Part two includes an historical examination of buffers and identifies several important criteria that should be considered in the design of buffers for environmentally sensitive areas. Buffers in a range of countries are critically evaluated and 'best practice' principles are identified. A broad classification of buffers is suggested, with an integrated buffer planning approach being identified as having several important attributes.

Chapter 4

Buffer Planning: Historical Overview

Ann Peterson

Introduction

The concept of buffering natural areas originated in the 1940s in the United States of America, with buffers being used to protect waterfowl habitat (Hilditch, 1992). Buffers were in place in the 1950s in Africa and Asia (Sayer, 1991). However, the term buffer zone, as applied to natural areas, came to prominence as a result of the United Nations Educational, Scientific and Cultural Organisation's (UNESCO) Man and the Biosphere program, launched in 1971. One of the program's themes centred on the conservation of natural areas to help counter the increasing loss of species and genetic diversity and to develop knowledge on how to conserve these areas. In response, UNESCO in 1974 endorsed the development of biosphere reserves incorporating buffer zones as a key component. Since the first biosphere reserve designation in 1976, 391 have been developed in 94 countries (Environment Australia, 2001).

Also in the 1970s Forster (1973) developed a buffer strategy based on 'internal zoning', while Lusigi (1981) applied the 'conservation unit' approach to wildlife management in Kenya. Since the mid 1970s there has been a proliferation of programs, particularly in developing countries, linking biodiversity conservation and protected areas with the development of local communities, through the creation of buffer type systems. These buffer programs have been termed multiple use conservation projects, buffer zone projects, integrated conservation-development programs (ICDP), support zones, and area of influence planning. In the 1970s Kozlowski and Ptaszycka-Jackowska (1981) developed the concept of protection or buffer zone planning (BZP) applying it to protected areas in Poland (refer to Chapter 7). Various applications and modifications of this method followed in Australia in the 1980s and 1990s (Roughan, 1986; Peterson, 1991; Hruza, 1993) (refer to Chapters 8-10).

Since the late 1980s many individuals and organisations have recommended buffer zones for environmentally sensitive areas (ESAs). Managers of important natural areas are well aware of the buffer zone concept, and management plans for numerous protected areas have made reference to such zones. For example, the IUCN (1992a, 1992b, 2003) recognises the importance of bioregions and the landscapes and seascapes that surround ESAs. As already indicated in Chapter 1, the need to buffer protected areas from external threats generated by their surroundings was a recommendation of the Third, Fourth and Fifth World Parks Congresses held in Bali (1982), Caracas (1992) and Durban (2003) respectively. The IUCN recommends innovative programs of integrated planning and cooperative management at the bioregional level to foster consistent and sustainable resource management practices and specifically calls on bioregional management to 'provide effective corridors and buffers.... Such a family of areas – strictly protected sites, multiple-use areas and the surrounding bio-regions under cooperative management – is essential' (IUCN, 1992b:3).

Overview of 'Practice-based' Buffer Approaches

This chapter is devoted to describing some of the more prominent practical approaches to buffer zone planning. By investigating several buffer methods, it will be possible to identify elements or features of these approaches that can be included directly into the recommended 'best practice' model, or which can be modified for inclusion. This approach will help to recognise shortcomings in current buffer designs, avoid unnecessary duplication and provide a sound basis for further discussion on, and development of, an effective buffer method for enhancing the protection of ESAs.

The range of approaches that are described below is broad, in an attempt to understand current thinking in relation to buffers and also to help to identify some of the more important key principles that should underpin any 'best practice' approach to buffering ESAs, and which can be applied by professional planners in a range of real-life contexts. While this chapter focuses on global approaches, Chapter 5 will examine buffer approaches in the Australian context, drawing upon the extensive research of Peterson (2002) in that country.

First, however, it is necessary to stress that the term buffer has been used in a wide range of situations and some of these applications are not consistent with the objective of conserving core ESAs. Hence the following categories are used in this chapter, namely 'separation zones', 'remnant habitat strips', and 'buffers to protect ESAs'. This grouping or classification of existing approaches is based on the 'buffer's' attributes in relation to the following:

- *purpose*: the role that the buffer is intended to play within the wider landscape;
- *composition*: the variety of elements (natural and cultural) within the buffer;
- *structure*: the pattern or physical organisation of the buffer. This relates to the buffer's complexity in terms of biological diversity and human related factors (e.g. size, boundaries, incorporation of threatening processes, heterogeneity, connectivity and spatial linkages); and
- *function*: the roles that the components of the buffer fulfil in driving the processes that sustain the functioning (ecological, social and economic) of the core.

Although the first two categories, namely separation zones and remnant habitat strips have been termed buffers by their proponents their purpose is not to aid the conservation of a core natural area and hence they are only briefly described in this chapter. It is the third category, the buffers to protect ESAs, which are its main focus.

'Separation Zones'

A separation zone is an area or feature that separates conflicting adjacent land uses. They are frequently advocated in planning documents, their purpose being to reduce or eliminate the impact of a particular noxious land use activity on the adjacent land, biota and people. They are often (and wrongly) termed buffers or protection zones. Their main known examples are:

Zones Separating Noxious Industry From Surrounding Land Uses

Separation zones are frequently applied in situations where a noxious industry needs to be separated from a surrounding community. For example, in Figure 4.1, a sawmill may be the source of noise and dust, which has the potential to impact negatively on the surrounding land uses and communities. A separation zone may be designed and implemented to eliminate or reduce the impact of specific threats from the mill on the

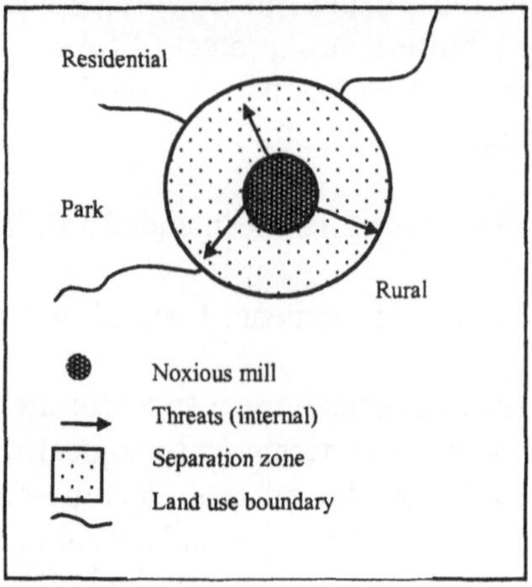

Figure 4.1 Separation zone

adjacent lands. The threats from the noxious source (i.e. the mill) are termed 'internal threats' as they originate within the central mill and their impact is outwards towards adjacent lands.

This concept of separation applies not only to isolated industrial activities or sites but also to general industrial zones that are to separate industrial activities from other land uses. Such separation zones are generally designated in planning documents as open space or non-development zones and are frequently and inappropriately termed buffer zones. Their design often relies on fixed distance measurements e.g. a separation zone extending 200 metres from the mill. They may also function as a visual barrier where existing vegetation communities are retained. Little emphasis is usually placed on maintaining the biodiversity values of the open space separation zone, thus restricting the purpose of the zone to minimising the impact of one type of land use on another. The purpose of these separation zones is not to eliminate the source of contamination within the industrial area, but to restrict its impact on surrounding land uses. In Lake County, Illinois 'bufferyards' or plantings are used to separate incompatible land uses, and bufferyard criteria may specify the width of the buffer and the type of plant materials to be used. Usually the number of plants required in the bufferyard varies inversely with the buffer width (Adams and Dove, 1987).

Zones Separating Agricultural and Residential Land Uses

In Australia, planning guidelines that have been implemented to separate agricultural and residential land uses, advocate 'buffer areas' or separation zones to minimise potential conflict between agricultural and urban uses (DNR and DLGP, 1997) (Figure 4.2). Agricultural uses may have impacts on adjacent urban land uses in terms of odour, noise, dust, spray drift and fire, while urban land uses may impact on the adjacent agricultural activities through dog attacks on stock, urban run-off, introduction of feral animals, vandalism and theft. A 'buffer area' is defined as an area of land separating

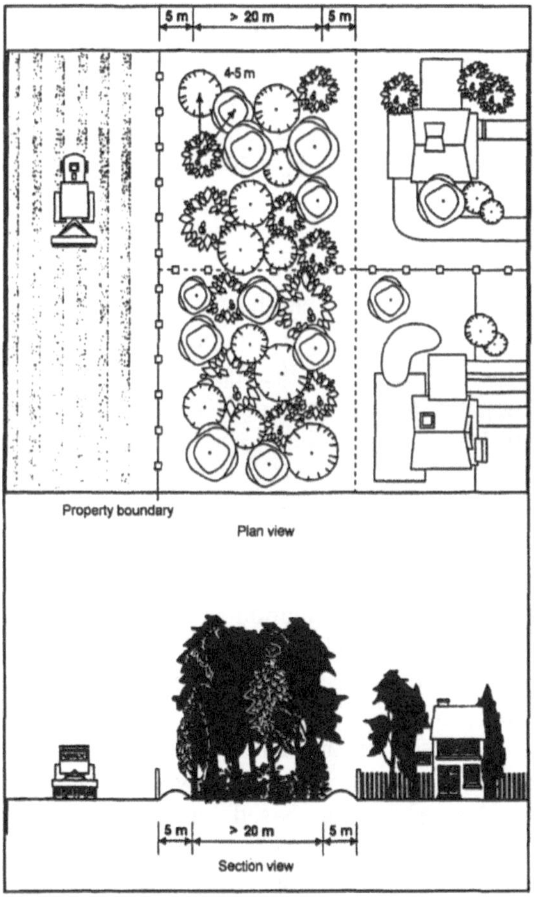

Figure 4.2 Land separating agricultural and residential land uses
Source: adapted from DNR and DLGP (1997: 28).

adjacent land uses, that is managed for the purpose of mitigating impacts of one use on another. A buffer area is said to consist of a '*separation*' distance and one or more '*buffer elements*' (or physical barriers) (DNR and DLGP, 1997:4). The planning guidelines state performance criteria to be achieved in relation to several elements (e.g. chemical spray drift, noise, dust, sediment and stormwater runoff), which may cause conflict with other land uses. To exemplify, for the element agricultural spray drift, performance criteria require that proposed residential development should be located, or incorporate measures, to prevent chemical spray drift from adversely affecting community public health and safety. A 'best estimate' separation distance of 300 metres is then recommended between existing agricultural land and proposed urban development, where open ground conditions apply, and a minimum width of 30 metres where a 'vegetated buffer' is satisfactorily implemented. The dimensions of separation distances and physical barriers vary according to the element causing conflict. Recommended noise separation distances are 120 metres for daytime agricultural activities and 500 metres for night-time agricultural activities. Zones of 150 metres are recommended to minimise the impact of dust. However, the distances are arbitrary and no separation distances are included to minimise the impact of the residential land uses on the agricultural land uses, the emphasis being on retaining agricultural activities and ensuring sufficient separation from encroaching residential areas. The guideline recognises the rights of the landholder to use the land forming the separation area, although within the confines of a range of planning controls (e.g. vegetation protection orders).

Zones Separating Urban Areas (greenbelts)

Green space separation zones, or greenbelts, are a long established planning tool to provide visual differentiation between large urban centres and to give form to a city. Although frequently called buffers, these areas more appropriately function as separation zones, being a means of confining urban areas and providing 'breathing spaces' for urban communities. The concept was central to the English new towns, the greenbelt towns in the United States and to most modern new towns throughout the world (Lynch, 1989). Their design is not based on ensuring effective ecosystem functioning or in minimising threats to important natural areas or even to the features within the separation zone. They may, therefore, become a static feature in the landscape, with surrounding urban development frequently impacting negatively on the natural values of the greenbelt.

Physical Barriers

The word buffer is also sometimes used to describe a physical barrier that separates two conflicting land uses and which limits the impact of a particular threat on adjacent land/water and people. For example, the 'Planning Guidelines: Separating Agricultural and Residential Land Uses' (DNR and DLGP, 1997:4) defines a 'buffer element' as 'a natural or artificial feature within a buffer area (separation area) that mitigates an adverse impact. A buffer element may include open ground, a vegetation buffer and/or an acoustic barrier'. 'Vegetated buffers' with a minimum width of 30 metres and containing trees of a certain leaf shape and density are recommended as effective barriers to prevent spray drift and noise from agricultural areas penetrating into residential areas. Buffers in this context function as physical barriers to a number of threats.

'Buffers', in the form of linear walls of varying heights and widths, are also used in transport corridors to minimise the impact of noise from vehicular traffic on adjacent communities. As such they function again as physical barriers to the movement of noise away from the transport line.

'Remnant Habitat Strips'

Remnant habitat strips are the second category of areas that have been (incorrectly) termed buffers. They frequently remain in the landscape after development has occurred in an area. For example, such strips or areas may be the result of forestry operations or more general development processes (e.g. urban or rural residential development).

Remnants in Production Forests

Forestry operations usually harvest timber in a patchwork across the landscape, by logging in coups and retaining unlogged remnant patches of vegetation to assist in maintaining native habitat for flora and fauna (Figure 4.3. For example, in Australia, the state of Victoria recommends, for the leadbeater's possum *Gymnobelideus leadbeateri*, 20 metre wide buffers consisting of unlogged remnant vegetation to separate 40 hectare coup conglomerates, thus allowing for trees to mature sufficiently for hollow development suitable for the possum. The primary function of such zones is not to separate conflicting land uses, but rather to provide suitable habitat for the remaining wildlife and a seed source to aid in forest regeneration. Typically such strips are interconnected to provide greater structural integrity.

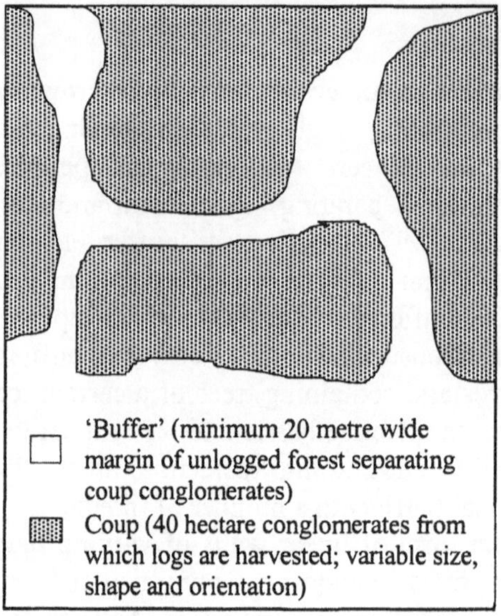

Figure 4.3 Remnant habitat strips designed to support leadbeater possums *Gymnobelideus leadbeateri*

Remnant Habitat in Urban Areas

In urban settings remnant habitat strips may be retained to enhance visual amenity and open space functions. For example, they may be retained on ridge crests to provide a scenic back drop or along watercourses to enhance recreational opportunities. Riparian strips in particular are frequently designed to achieve multiple objectives, including enhancing water quality, hydrologic functioning, flood mitigation, nature conservation and scenic landscape amenity. The primary purpose of such areas is not to better protect an identified core area. Also the size, configuration and distribution of riparian strips throughout a developed landscape may be haphazard.

Evaluation of Separation Zones and Remnant Habitat Strips

Table 4.1 summarises the main attributes of both separation zones and remnant habitat strips. Separation zones are implemented to separate conflicting land uses. They are typically utilised for open space purposes,

Table 4.1 Main attributes of separation zones and remnant habitat strips

	1. Separation Zone	2. Remnant Habitat Strip
Purpose*	• to reduce/eliminate the impact of one land use activity on an adjacent land use by separating the land uses • to provide a barrier (visual or aural) between conflicting land uses	• to provide a strip of habitat in an otherwise cleared matrix • to assist in maintaining / rehabilitating native habitats and their functioning
Composition*	• variable e.g. from cleared to well vegetated • usually open space land use allocation with limited development • may contain a physical barrier e.g. vegetated mounding or fence	• remnant habitat with a sample of the native species of the area • usually open space land use allocation with limited development
Structure*	• adjacent to core/source area • reflects the type of internal threats from a noxious source e.g. quarry • homogeneous • width usually based on prescriptive distances from the source area • no linkage to other zones	• frequently follow drainage lines or ridge crests • size usually based on prescriptive distances • frequently linear and may be interconnected
Function*	• zone of decreasing impact of threats from noxious source to surrounding matrix	• source of seed for regeneration in logged / cleared areas • conserve and maintain ecosystem function
Types / examples	(Refer Figure 4.1 and 4.2) • separating noxious industry from surrounding land uses • separating agriculture and residential land uses • urban greenbelts • riparian corridors e.g. in rural landscapes	(Refer Figure 4.3) • strips of habitat between logging coups • riparian strips in urban landscapes

(* Where: **Purpose**: the role that the buffer is intended to play within the wider landscape. **Composition**: the variety of elements in the buffer [e.g. natural and cultural features]. **Structure**: the pattern or physical organisation of the buffer. **Function**: the roles [ecological, social and economic] that the components of the buffer fulfil in driving the processes that sustain the functioning of the core).

both public and private, with little development being possible. A general methodology for defining separation type zones is lacking, with prescriptive distances being the norm. Such distances are unlikely to work in reality, may not be transferable from one situation to another and may give a false impression that they will achieve their objectives. They are certainly difficult to justify from an environmental perspective and can not, by their very nature, be supported by sound ecological concepts. There is frequently little or no examination of the nature of the threatening processes or the interactions between the conflicting land uses. Such zones tend to be homogeneous in structure and function. Other important values of the open space separation zones, such as their value as wildlife habitat and corridors, aesthetic, recreational and cultural values are often not examined in this approach and hence such values may be degraded over time. There is frequently no attempt to enhance regional landscape biodiversity or to implement compatible sustainable land use activities. Where arbitrary design principles are used, such separation zones have limited applicability to the long term conservation and restoration of ESAs. Hence, it is recommended that such areas be more appropriately termed 'separation zones' rather than buffer zones.

Remnant habitat strips may resemble separation zones in appearance and composition. However, they are functionally and frequently structurally different. Remnant habitat strips are usually designed to retain some of the existing vegetation within an otherwise cleared or altered landscape. Their purpose is not to give added protection to an existing ESA or protected area through a process of examining threatening processes. Their dimensions are usually determined arbitrarily and they may not represent fully functioning ecosystems capable of maintaining an area's biodiversity. Although frequently termed buffers, such strips may function more like corridor networks, especially where they follow drainage lines. However, their effectiveness as movement corridors for wildlife may be limited due to poor design, which is unrelated to the habitat and movement needs of an area's wildlife. As such it is appropriate that such areas be termed 'remnant habitat strips' rather than buffer zones.

As the focus of this chapter is on examining buffers that are used to give added protection to important natural or cultural areas, separation zones and remnant habitat strips are not discussed further.

Buffers for Environmentally Sensitive Areas – A General Overview

Many of the buffers that are currently in place to conserve ESAs typically represent zones designed to give an added layer of protection to areas of high environmental value such as protected areas, important wildlife habitat, remnant vegetation and water or wetland features. Such ESAs interact in many ways with their adjacent lands and communities and frequently these outside activities may threaten the viability of the ESA. For example, urban development that is adjacent to a national park may be the source of threats such as feral and domestic animals and exotic plant species (Figure 4.4). In response, buffers have been implemented to minimise the threats that originate in the surrounding landscape, external to the ESA, and which through a variety of mechanisms impact negatively on the composition, structure and function of the core natural area. For example, Pine Rivers Shire's (Queensland, Australia) *Green Plan* (Chenoweth, 1994: 4-17) recommends the implementation of buffers of 500 metres from reserve boundaries, 'within which tree cover and habitat should be retained and impacts minimised'. Similarly 20 metre wide buffers from building structures and hard surfaces are recommended 'to minimise edge effect deterioration, especially in residential and industrial areas' (Chenoweth, 1994: 5-4). In Luxembourg the Law of Luxembourg of 1982 on the Protection of Nature and Natural Resources prohibits any construction in buffers, which may extend up to 30 metres from a nature reserve (de Klemm, 1993). Similarly Tanzanian national parks are protected by buffers that are 100 metres in width and which prohibit most forms of development. The width of such buffers tends to be based on prescriptive distances from the ESA's boundary and the buffers generally do not have a methodology to guide their design. This ill-founded concept of fixed distance buffers can never be justified on sound ecologic grounds.

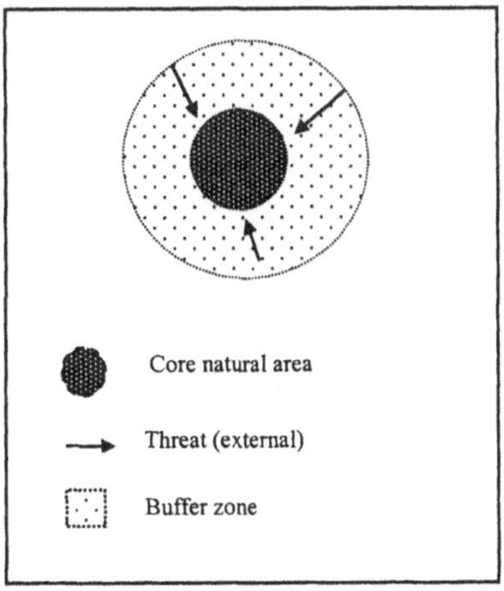

Figure 4.4 Buffer zone to minimise external threats

Buffers for natural areas in many developing countries are also implemented as 'socio-buffers' (Spellerberg, 1994), where specific development projects are targeted to local communities living on land adjacent to the ESA. The purpose of many of these programs is primarily to minimise the impact of people on nearby important natural environments and ensure land use activities are ecologically sustainable. Where buffer zones are implemented they may encompass human settlements and promote land use activities that are compatible with the goals of natural area management.

Some protected areas may have a buffer located within their bounds. In such situations a core area of high ecological significance within the protected area is surrounded by a buffer, which permits higher intensity uses than in the core. The buffer functions to minimise the impacts on the core and thus extends the protection of the core, usually to the park boundary. Some biosphere reserves typically have buffer zones located within the boundary of former protected areas.

From this broad overview, several types of buffer planning for ESAs that incorporate important design elements are briefly described below. These include: internal zoning; the wildlife conservation unit approach; buffer zone planning (BZP); biosphere reserves; nodes, networks and multiple-use modules; integrated conservation and development programs; and wildlife specific buffers.

Internal Zoning as a Form of Buffer Planning

Foster (1973) recommended the use of zones to reduce conflicts within protected areas and to provide effective internal management, particularly to meet the demands of recreational users of parks, while maintaining the natural values for which the parks had been dedicated. He suggested the use of internal buffer zoning to shield protected areas from negative external influences. The zones were defined on the basis of the intensity of use, and the size, types and quality of resource areas, especially in terms of their recreational potential. He suggested that natural features such as ridge lines could be used to identify the internal boundaries, although a formal process of delineating zone boundaries was not developed. Foster described three types of zones (Figure 4.5):

- *Concentric zoning*
  This is illustrated in Figure 4.5(A) by a protected area consisting of a core zone and surrounding buffer, or recreation zone. The core is to be

the least accessible and disturbed area, while the buffer represents an area of more intensive recreational use and greater accessibility.

- *Three zone configuration*
  This zonation consists of three zones (Figure 4.5[B]). The core contains strict natural areas or wilderness areas, the surrounding protective buffer restricts movement to authorized paths and the outer recreation zone allows free access and visitor facilities.
- *Nodes and linkages*
  For large protected areas Foster suggested a network of activity nodes connected by access corridors. In Figure 4.5(C) the outer boundary contains a large zone where no development is permitted, inside which is located a branching corridor zone, which may contain access roads and minor visitor services.

A Concentric Zoning

B Three zone configuration

C Nodes and Linkages

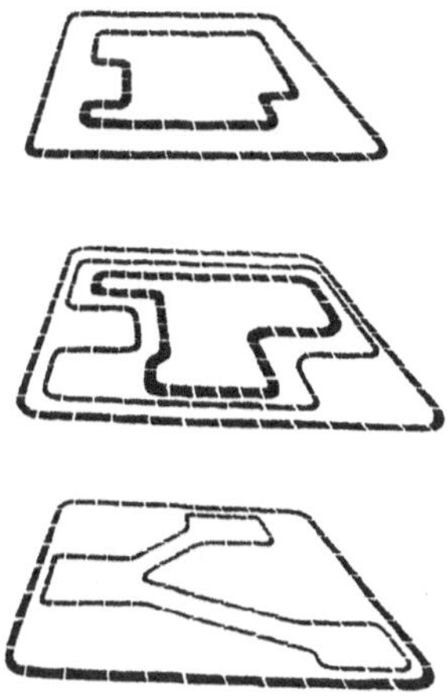

Figure 4.5 Internal zoning (Foster, 1973: 51-53)

In this buffer approach, the core of the protected area is identified as requiring a high level of protection, with the buffer functioning to reduce negative impacts on the core, particularly those threats arising from recreational activities. The model does not specifically deal with the impact of threats that originate outside the protected area. As the focus is on protection of the core, the model places less emphasis on the conservation and management of the land within the buffer. Such a strategy may be suited to very large protected areas that have core areas which are remote from their administrative boundaries.

Foster's model is not based on specific criteria that are used to define the zones and linkages. It requires intuitive application and may be of limited use to the practitioner. The implementation of zones that are located within the protected area may aid management, but may have limited effectiveness in minimising threats that arise outside of the protected area. The model may have limited suitability to the protection of areas of high ecological importance, particularly because it gives little guidance in terms of understanding the ecological and cultural interactions between the protected area and its surrounds and in developing a workable planning strategy.

Wildlife Conservation Unit Approach to Buffering

Lusigi (1981) developed the wildlife conservation unit approach based on his earlier experiences in park management in Kenya (Figure 4.6). The establishment of protected areas in many African countries interfered with the natural movement paths of wildlife and aroused the ire of local communities, who were not consulted and did not benefit from the establishment of the parks. Lusigi suggested that parks should be integrated with their surrounding landscape and he identified three land use categories:

- *National park*
  This includes the primary wildlife population or unique scenic features where there is to be minimal development and limited access. Management activities (e.g. prescribed burning and water development) are directed at retaining suitable ecosystems for the wildlife.
- *Protected areas*
  These areas surround the national park and include locations for intensively developed tourist lodges and associated maintenance and staff facilities. Restricted local grazing is allowed by permit and the area is to have good access facilities. Sustained harvesting of wildlife is permitted. Such areas represent an inner buffer zone surrounding the central core.
- *Multiple use areas*
  Multiple use areas surround the protected areas. The main purpose is related to wildlife management, coordinated with resident livestock management, tribal hunting, and tourism organized mainly by local residents. These areas represent an outer buffer zone surrounding the protected areas.

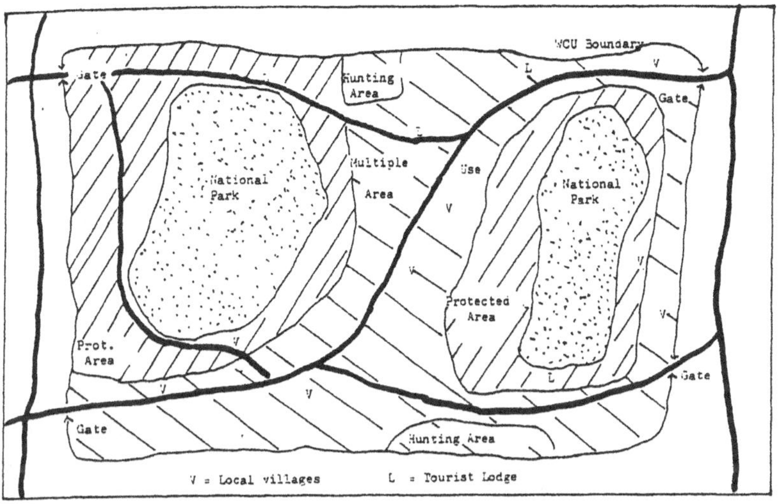

Figure 4.6 The wildlife conservation unit approach (Lusigi, 1981: 91)

The entire conservation unit is to be managed as a single entity, with marked and patrolled boundaries and entry only through gates on access roads. The unit could include one or more ecosystems and is to consider tribal and political boundaries, and thus coordinate ecological management across political boundaries.

The prime functions of Luisigi's model are to ensure the unimpeded migration of animal herds once they leave national parks, and to integrate the management of national parks with their surroundings. The model does not address more widespread external threats to the national parks and their component ecosystems. Lusigi aimed to integrate areas of high wildlife potential into the land use of whole areas, thus contributing to the social and economic development of the local areas. He stresses that cultural and socio-economic problems should be solved before ecological ones and hence the emphasis in this model is anthropocentric.

In summary, the conservation unit approach does not examine a wide range of external threats affecting the national park or core area. Its emphasis is minimising threats to migratory wildlife. As Lusigi does not precisely identify how the three zones are defined or implemented, the model has limited applicability to other situations.

Buffer Zone Planning (BZP) Method

The buffer zone or protection zone planning approach, developed and applied in the mid 1970s by Kozlowski and Ptaszycka-Jackowska (1981) recognizes that protected areas are part of the wider region in which they are located and that they may be subject to a variety of external threats, which over time will downgrade the ecosystems within the protected areas. The focus of this approach is on minimising external threats, both existing and forecast, to protected areas, or more precisely to the values to be protected. In this approach individual threats are identified. In Figure 4.7 a protected area is affected in a negative way by the threats of water pollution, pest animals and weed invasion. The specific source areas of each threat are identified and mapped and land use measures that could minimise or eliminate these threats defined. Such areas are termed elementary protection zones (EPZ). An EPZ is defined for each threat and the final buffer is a synthesis of all the delineated EPZ boundaries. The buffer thus reflects the heterogeneous nature of the surrounding environment and encompasses the area over which important threats to a core area's values may operate, indicating at the same time a range of desired counter measures. To manage the impacts of the threats, appropriate policies relevant to each EPZ area are identified. In such a way a heterogeneous buffer is designed that responds to specific external threats and is applied outside the bounds of the protected area, thus

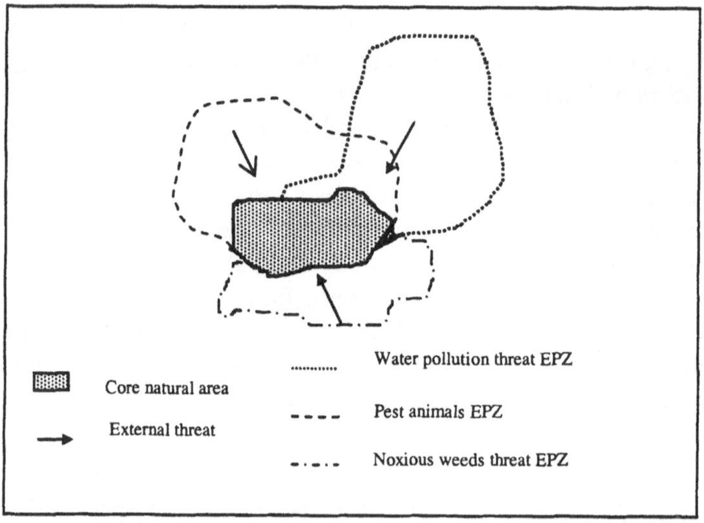

Figure 4.7 Buffer zone comprising areas of individual threat to the core natural area (EPZ: elementary protection zone)

extending its area of protection. A significant contribution of this approach is its seven step sequential planning process (Table 4.2), which guides implementation of the methodology.

Table 4.2 Steps in the buffer zone planning (BZP) method

Step	Process
1	Identification of particular values and characteristic features of the given area under protection.
2	Identification of the interrelations between the area and its surroundings to determine existing and potential threats.
3	Preliminary formulation of the criteria for demarcating and defining the principles of land use within analytical protection zones (APZ) to protect particular values.
4	A synthesis of the criteria and principles on the basis of the type of negative influence to determine elementary protection zones (EPZ).
5	Demarcation of elementary protection zones (EPZs) and the definition of the principles of land use within their boundaries.
6	Delineation of the buffer zone surrounding the area based on a synthesis of the EPZs identified in Step 5.
7	Formulation of guidelines or principles concerning different forms of use and activities within the buffer zone and introduction of these principles into development plans, which become legally binding after the plan's formal approval.

BZP was applied to Tatry and Gorce National Parks in southern Poland in the late 1970s (refer to more detailed descriptions in Chapter 7). Peterson (1991) refined the method and applied it to Cooloola National Park, Australia (refer to Chapter 8). Other applications include that of Hruza (1993) (Chapter 8) and Roughan (1986) (Chapter 9), while Vass-Bowen (1994) and Izatt (1995) applied the underlying concepts to the buffering of urban cultural heritage areas in Australia (Chapter 10). These and other applications and modifications to the BZP methodology are detailed in later chapters, as they provide an important foundation for the recommended integrated buffer planning (IBP) process detailed in Chapter 11.

Overall, the BZP model developed by Kozlowski and Ptaszycka-Jackowska (1981) is a significant advance on previous buffer approaches. It is based on understanding the important resources and values of the core natural area and its surroundings and places emphasis on identifying the range of processes that may threaten the core's values. However, its focus is directed at identifying and mitigating externally originating threats and does not cover internal threats, or those that originate within the core natural area. The model assumes that these threats should be dealt with by the direct management of the core area. The BZP method stresses the need for planning within ecological boundaries rather than administrative boundaries. Hence all major threatening processes are identified, with associated measures and policies designed to minimise their impact. Implementation is recommended through legally binding planning documents. Community participation is recognised as being important and the methodology places emphasis on the need for the buffer to be proactive, and to identify current trends and anticipate future developments within the plan area. However, perhaps the main strength of this strategy is that it is based on a logical model process which provides a strong basis for replication and adaptation.

Biosphere Reserves

Biosphere reserves integrate three areas – a core, buffer and transition area (Figure 4.8). The inner core consists of the best-preserved natural areas with the greatest genetic variability and ecosystems, where only non-destructive activities are permitted. Its size and shape are dependent on the type of landscape and the conservation objectives established for the biosphere reserve. It usually has legal protection (e.g. national park or reserve status), although its boundary may not be defined. Within the core, scientific data are collected over time providing a baseline area against which to compare management practices and measure long-term changes in the biosphere.

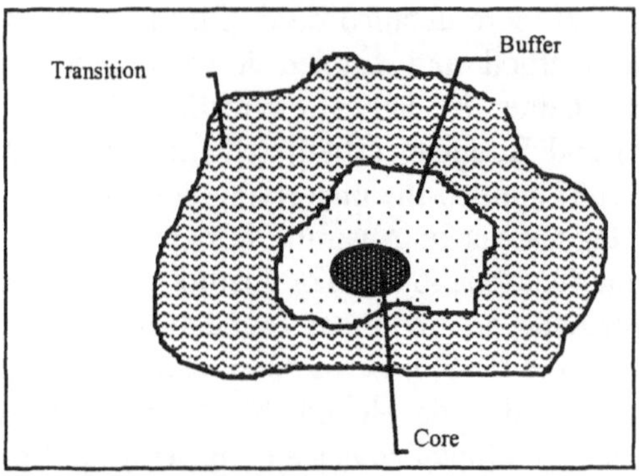

Figure 4.8 Biosphere reserve concept
Source: adapted from The Bookmark Biosphere Trust (1995:6).

The buffer zone which adjoins or surrounds the core allows limited human activity, including research, monitoring, education, training, traditional land uses, recreation and tourism. Its primary purpose is to decrease the direct and indirect impacts of development on the core and to ensure that any activities are compatible with conservation of the land and its uses. Its boundaries are usually legally defined and frequently correspond to the outer limits of the protected area.

The transition area represents a 'zone of cooperation' where the work of the biosphere reserve is applied to the needs of the local community in the region. It is 'not strictly delineated and corresponds more to biogeographic than administrative limits' (UNESCO 1986:73). This zone may contain settlements and a variety of economic activities that are ecologically, economically and socially sustainable. Within biosphere reserves people are an integral part of the environment and management programs aim to foster sustainable use of resources within the region. Co-operative activities are to be developed between 'researchers, managers, and the local population, with a view to ensure appropriate planning and sustainable resource development in the region' (UNESCO 1986:73)

Other zones may also be added to the bioshpere. For example, there may be a number of core areas surrounded by buffer and transition zones and linked by corridors.

Biosphere reserves contain land in a variety of tenures, private as well as public. The important issue is that there are agreed standards for managing the land in sustainable ways rather than a legal designation of the reserve boundaries.

One of the founding principles of biosphere reserves is to 'provide the information needed to solve practical problems of resource management' and to fill the gaps in the 'understanding of the structure and function of ecosystems and of the impact of different types of human intervention' (Longmore, 1993: 10). Each reserve should have value as a benchmark for measurements of long-term changes in the biosphere. Interdisciplinary research should be encouraged to develop models for sustainable use of ecosystem resources and the international network should provide a framework for comparative studies of similar problems (Longmore, 1993; UNESCO 1986). Biosphere reserves thus play a development role through problem oriented research and education and are linked into an international network.

Biosphere reserves, in theory, place importance on people as part of the biosphere. Human activities are fundamental to ensuring the long-term conservation of core areas and hence people are encouraged to participate in the reserve management. As a consequence, the implementation of the biosphere reserve concept in many countries has been slow, to give time for the local community to feel ownership of the reserve and to give support rather than cause conflict and antagonism. The March 1995 International Conference on Biosphere Reserves in Seville expressed the need to 'bring together all interest groups in a partnership approach to biosphere reserves both at site and network levels' (direction 9) (UNESCO, 1995).

The Conference however, noted a considerable gap between the concept of biosphere reserves and the reality. Many biosphere reserves had been superimposed on existing parks and reserves, with the agencies responsible for managing the areas lacking the resources to adapt their management approach to the new management philosophy. Further, the model in general did not provide detailed descriptions of tactics for its implementation, especially in relation to buffer delimitation. In response the Statutory Framework of the World Network of Biosphere Reserves (UNESCO, 1995) stated that 'biosphere reserves should strive to be sites of excellence to explore and demonstrate approaches to conservation and sustainable development at the regional scale (Article 3). Through a process of periodic review the objective is to ensure that the reality comes to match the concept (Price, 2002).

Multiple-Use Modules

Noss and Harris (1986) developed a multiple-use module (MUM) (Figure 4.9) to preserve large-scale ecological patterns and processes across the landscape. They see the MUM as a refinement of the biosphere reserve concept, with application at all levels in the biological hierarchy and in all landscapes. Its main components are nodes, corridors and networks, and buffer zones.

- *Nodes/Core areas*
  These are important natural areas, often part of the protected area estate, having variable size, distribution, density and dispersion in the landscape. They are selected on the basis of representing all ecosystems in each landscape and their ability to maintain viable populations of all native species and ecological and evolutionary processes. Some nodes are relatively permanent landscape features (e.g. national parks or caves), while others are dynamic (e.g. crocodile nesting site). The cores are to be managed as 'inviolate preserves'.

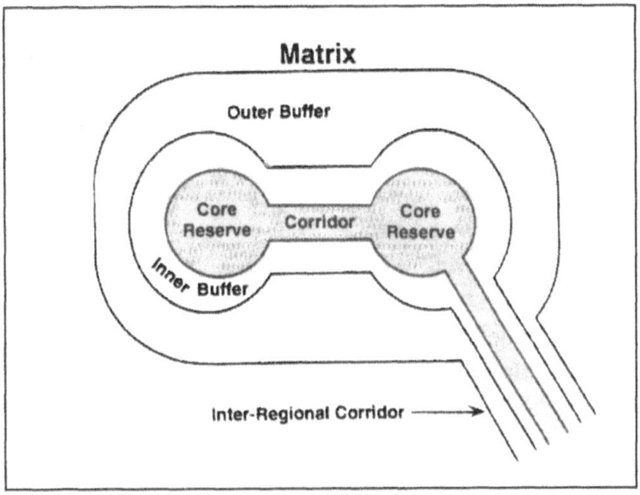

Figure 4.9 Multiple Use Module consisting of cores, corridors and buffers (Noss, 1992: 15)

- *Corridors and networks*
  Noss and Harris (1986) believed that the nodes would not persist over time if they existed as isolated fragments. They viewed each landscape as containing features that could either facilitate or inhibit dispersal processes. By examining the existing pattern of high quality nodes relative to potential travel corridors and dispersal barriers, a landscape conservation scheme could be devised to utilize the existing pattern. They recommended minimising artificial barriers and maximising connectivity and concluded that a network of corridors could also facilitate the shifting of habitat patches across the landscape in mosaic patterns of disturbance and recovery by providing colonisation sources and refugia. The pattern of high quality nodes should be determined relative to potential travel corridors and dispersal barriers.
- *Buffer zones*
  Buffer zones of centrifugally increasing utilisation surround cores and corridors. Within the buffers only human activity compatible with protection of the core reserves and corridors is allowed, the buffers being managed to restore ecological health, extirpated species and natural disturbance regimes. Intensive human activities such as agriculture, industrial production and urban centres may continue outside the buffers. Buffers are important to mitigate edge effects, especially for small reserves, to ameliorate external threats and to

provide supplementary habitat to native species inhabiting a core reserve, thus helping to increase population size and viability.

The MUM model is being applied in several locations. The Mesoamerican Biological Corridor is a strategy consisting of a number of local and regional projects to develop and implement MUMs across North America, linking through Central America to South America. Several projects have been proposed, including 'Passe Pantera' or the 'Path of the Panther', a project to reconnect, restore and better manage the fragments of the biotic corridor along the Central American isthmus, stretching from Belise and Guatemala to Panama, by developing a chain of parks connected with wildlife corridors and surrounded with buffer zones (Marynowski, 1992).

Evaluation of the MUM concept of buffers is difficult due to its limited application. The MUM concept is theoretically based on the need to recognize ecological rather than political boundaries for protected areas and wildlife habitat. It is based on buffer zones that are defined in relation to the impact of external threats operating on a core wilderness area. However, buffers that represent concentric type zones of increasing human use may not respond to the spatial location of specific external threats to the core area and such a structure may not effectively minimise all external threats. The model also does not specifically address the existing and potential impact of core areas on surrounding communities.

Integrated Conservation and Development Programs (ICDP)

Since the first biosphere reserve was dedicated in 1976 there has been a proliferation of programs, particularly in developing countries, linking biodiversity conservation and protected areas with the development of local communities, through the creation of buffer type systems. These have variously been termed multiple use conservation projects, buffer zone projects, integrated conservation-development programs (ICDP), support zones, and area of influence planning.

The term ICDP was coined by Wells and Brandon (1992:1) to describe protected areas in which management 'attempts to address the needs of nearby communities by emphasising local participation and combining conservation with development'. The central concern of ICDPs is to aid conservation of biodiversity while promoting social and economic development in adjacent communities. Such programs became popular in the 1980s due to public resistance to large development projects, many of which had devastating environmental consequences. Where ICDPs were targeted to lands located near protected areas and important wildlife habitat, the programs usually incorporated protected area management plans, buffer zones around the protected areas, and local social and economic development projects (Dang, 1991; Wells and Brandon, 1992; Spellerberg, 1994; Metcalfe, 1994). The purpose of many of these programs is primarily to minimise the impact of people on nearby important natural environments and ensure land use activities are ecologically sustainable. Where buffer zones are implemented they may encompass human settlements and promote land use activities that are compatible with the goals established for the area.

There are perhaps two main types of ICDPs, including those which provide:

- direct incentives for conservation of biodiversity by allowing the harvest of plant and animal resources, which are dependent on the protected area (e.g. 'Admade' and 'Campfire' programs and the 'Core-buffer Multiple Use Zone'); and
- access to alternative resources outside the protected area (e.g. social forestry projects).

A full analysis of ICDPs is beyond the scope of this chapter. Rather, attention will focus mainly on the game management areas in southern Africa, with selected reference to other, largely successful ICDPs.

Game Management Areas of Southern Africa

In Namibia, Botswana, Zambia and Zimbabwe, colonial governments gave ownership of wildlife to the State. Poaching escalated as local communities no longer felt they had a stake in sustainably utilising wildlife. The people also saw their resource rights to wildlife being taken away. In Zambia, the Lupanda Development project and later Administrative Design for Game Management Areas (Admade) project were developed (Lewis et al., 1990), while in Zimbabwe the Communal Areas Management Program for Indigenous Resources (Campfire) was launched in 1988 by the Department of National Parks and Wildlife. The game management/communal management areas in reality function as large buffer zones surrounding the national parks, working to extend the area of protection for species within the national parks, while also improving the standard of living of village people. Wildlife is strictly protected in the national parks, while in the game management areas, the wildlife is managed as a commercial resource, with revenues from hunting being used to improve the standard of living of local people and assist the conservation of many wildlife species. In 1995, 26 districts were implementing the program, in over half of the area of Zimbabwe. In Zambia 34 game management areas are established near parks in areas where agricultural potential is low (Physick, 1996). Levels of poaching were reported to be relatively low due to the cooperation of villagers (Wells and Brandon, 1992). However, political instability, may impact significantly on the levels of poaching in the future.

Core-Buffer-Multiple Use Zone

This concept was recommended in a report of a Task Force to the Indian Board for Wildlife in 1983. It suggested the zonation of Indian protected areas into inviolate cores, protective buffers and multiple use zones, to reduce park-people conflicts. The core is given a high level of protection. The buffer allows controlled use, while rural development projects are directed to the multiple-use zone, to assist in rehabilitation of these largely degraded lands. Buffer zones aim to protect the core from disturbance and extend wildlife habitat to the protected area boundary. This zonation occurs in 38 per cent of national parks and nine per cent of sanctuaries (Dang, 1991).

Buffer zoning of this type is an important objective of Project Tiger. The project established nine tiger reserves each with a core area free from all human use and a buffer area in which conservation land uses are allowed. The reserves are chosen to include as many biogeographic habitat types as possible and on the amenability of the specific area to a concerted conservation action. They are to be breeding nuclei from which surplus tigers emigrate to adjacent forests (Panwar, 1987). Critical to the success of the program is the development of compatible land uses in the multi-use surrounds, uses that both meet the needs of the rural people and habitat requirements of the tiger. Singh (cited in Dang, 1991) states that the tiger population of India grew at about 10 per cent during the late 1980s, leading Panwar (1987: 116) to conclude, 'Core-buffer units with compatible multi-use surrounds, are the only way wildlife reserves can be sustainably managed in India'.

ICDP Conclusions

It is difficult to assess the wide variety of ICDP programs. The general comments that follow place emphasis on the Wildlife Management Areas approach (Campfire and Admade) and Core-buffer Multiple-use model.

ICDPs are usually developed in response to increasing levels of external threat experienced by protected areas. They recognise the heterogeneity of the surrounding matrix and the wide variety of threatening processes, and they also attempt to minimise the impact of core areas on their surroundings. In Africa's game management areas, protected area managers realise that wildlife has frequently damaged property and even killed local people, and they work with local communities to set sustainable wildlife harvest levels and to reduce the impact of threats that a core may have on its surroundings.

The strategies implemented within ICDPs place emphasis on maintaining ecosystem processes and functions. However, they cannot address many of the underlying threats to biodiversity, in particular those that originate far from park boundaries:

> Among them are public ownership of extensive areas of land unmatched by the capacity of government agencies to manage these lands; powerful financial incentives encouraging overexploitation of timber, wildlife, grazing land, and crop fields; an absence of linkages between the needs of conservation and the factors encouraging development; and laws, policies, social changes, and economic forces over which poor people in remote rural areas have no influence (Wells and Brandon, 1992: xi).

In both Zimbabwe and Zambia, game management areas, which act as *de facto* buffers, usually correspond with tribal or community boundaries, and hence are not delineated entirely on the basis of ecological principles. However, these 'buffers' are usually significantly larger than the protected areas they adjoin and afford a high level of protection to wildlife. Wildlife management strategies in the buffer are based on sustainable utilisation.

Similarly in several Indian national parks, although external threats, particularly from human interference, were recognised in the design of the buffers, which were planned to 'extend' the protected area, a thorough analysis of the operation of all threatening processes was often not the basis of the buffer zones. For example, in Dudhwa National Park the buffer excludes threats operating along the southern boundary of the park, an area where approximately 60,000 villagers and as many cattle are found. There is no buffer for this southern section of the park and large scale poaching of prey species such as swamp deer has occurred, as well as hunting of so called 'man eating' tigers. Massive local resentment against tigers and the national park has developed (Dang, 1991). This situation exemplifies the need for buffers to address all major threatening processes. Partial solutions in restricted areas are unlikely to be successful in the long term.

One of the goals of ICDPs is to integrate local communities into the management of protected areas and their buffers in the hope of changing human behaviour and reducing pressure on protected areas. However, the extent of successful integration is more myth than reality (Wells and Brandon, 1993; Metcalfe, 1994), with local people frequently being treated as passive beneficiaries of project activities. Of 36 ICDPs reviewed by Kremen et al. (cited in Metcalfe, 1994), only five showed a positive relationship between development efforts and conservation of endangered biological resources and hence few projects achieved their conservation goals largely due to a lack of community driven incentives for protection. It is difficult to convince local people that restricted buffer zone access constitutes a valuable benefit (Wells and Brandon, 1993). Oldfield (1988:1) found few buffer zone management programs that 'succeeded in establishing stable and compatible land use systems around a protected area in such a way that local people are genuinely reconciled to the conservation function of the area'. Some of the difficulties in establishing formal buffer zones relate to the lack of legal authority to establish or manage buffer zones outside of park boundaries. Boundaries are thus related more to available suitable land, logical boundaries and prescriptive distance measures.

Where a clear link between conservation and development is evident (e.g. wildlife ranching and ecotourism) ICDPs have had some success. The Mountain Gorilla Project in Volcanoes National Park, Rwanda, illustrates a clear link between conservation and benefits to local people (Table 4.3). The proportion of respondents identifying personal benefits from the park doubled between 1979 and 1984. However, establishing such a link may be difficult where areas lack tourism potential or other avenues for income generation.

The Campfire program is theoretically based on co-management, incorporating a parks-and-people approach, with significant integration in decision making among local land authorities, national parks and district councils. Implementation of Campfire has varied throughout Zimbabwe. In Omay communal area (Sibanda, 1995), although the program aimed to involve local Tonga people in wildlife and conservation by transferring the wildlife ownership, use and management to 'appropriate authorities', these being the district councils, the Tonga people were not involved in decision making, the program having been decided elsewhere, with the Tonga being expected to participate without fully understanding why. The link between conservation efforts and monetary gains were not evident, with Sibanda criticising the Omay program especially as the system rewarded even those who were not located in wildlife areas, and further that it denied local

Table 4.3 Changing views on the value of Mountain Gorilla Project, Volcanoes National Park, Rwanda, 1979-1984

Subject	1979	1984
	%	%
Tourism		
national benefits	65	85
regional benefits	39	81
personal benefits	26	49
Forest values		
no value	17	22
wildlife refuge	19	28
tourism	0	6
Wildlife values		
no value	14	24
tourism	39	52
Open park to exploitation		
yes	51	29

people access to and utilisation of wildlife as a resource. Metcalfe (1994: 176) similarly criticises the program in Nyaminyami District:

> The communities themselves are not actively participating in the planning and management process and appear alienated from both the trust (Nyaminyami Wildlife Management Trust) and the wildlife on which they depend for their existence.

In contrast, the implementation of Campfire by the Kanyurira community in Guruve district (Metcalfe et al., 1995) was based on greater village/community based decision making, greater accountability (of council to community and community leaders to membership) and much closer links between conservation and development. Metcalfe (1994) comments that the majority of Campfire programs fall somewhere between the two extremes of the Nyaminyami and Kanyurira communities. Sibanda (1995) stresses the need for institutional capacity building to ensure more effective local participation and ownership of the program, while Metcalfe et al. (1995) recognised the challenge of including more women in decision making to ensure greater gender equity. Zimbabwe's Land Tenure Commission (cited in Metcalfe et al., 1995: 16) concluded that communities could benefit from Campfire, especially if they were empowered through local level institutions.

For ICDP programs that encourage alternative economic development outside of reserves, establishing a link between conservation of biodiversity and enhanced production in buffers has been difficult. Added problems have arisen in areas where profitable enterprises have been developed in buffers. Such areas have attracted large numbers of additional people, placing further stress on both the buffer and the protected area. Many ICDP buffer projects have also failed in situations where there is a rapidly increasing human population. Metcalfe (1994:184) comments in relation to the Campfire program that 'the political dimension of land tenure and settlement remains contentious', and that 'uncontrolled resource access risks rapid loss of wildlife diversity through overexploitation', this being a threat to ecosystem stability.

It has been technically difficult to achieve sustainable and productive agriculture in frontier situations where institutions are weak and immediate human survival is of more importance than conservation (Sayer, 1991b). However, where conservation organisations have made long-term commitments and developed close working relationships with protected area staff and the local communities the development of sustainable buffer projects has been more effective. Wells and Brandon (1992) believe that it probably takes more than a decade to change community attitudes and hence many short-term ICDPs are doomed to failure.

A shortcoming of many ICDP programs has been their failure to establish ecological monitoring programs (Wells and Brandon, 1992). Without knowing the total effects of ICDPs on biodiversity and overall ecosystem health, it is difficult for such programs to be responsive to change. Of the ICDPs reviewed by Wells and Brandon (1992) only Boscosa (Cost Rica) and Khao Yai (Thailand) surveyed surrounding communities to establish baseline data against which to measure subsequent project effects. Changes over time in local perceptions of wildlife are an important monitoring device in various districts implementing the Campfire program (Metcalfe, 1994), which rests heavily on being able to adapt to changing circumstances. Fundamental indicators (environmental, socio-economic and demographic) are also monitored (Metcalfe, 1994) to ensure the sustainability of the project's outcomes.

Management plans for traditional parks and multiple-use areas within ICDPs frequently refer to buffer zones, and several national conservation strategies have promoted the idea. A main difficulty with many ICDP buffers is the lack of a methodology to assist in the design and implementation of appropriate strategies within the buffer.

There has also been a lack of consensus in relation to buffer zone objectives, their location, whether they should be inside or outside parks, and what criteria should determine their area, shape and permitted uses. In reality buffer zones within ICDPs have usually been conceived as 'narrow strips of land on park boundaries, within which the "sustainable" use of natural resources will be permitted' (Wells and Brandon, 1992:26). Wells and Brandon (1992) comment that the scale of most ICDPs is too small in relation to the protected areas they are associated with, or that the surrounding populations have too great an effect to produce satisfactory conservation benefits for the protected area. However, despite their shortcomings, Wells and Brandon (1992:27) state that buffers remain 'a high priority for many conservation programs, a key component of traditional-park management plans, and a potentially important ICDP component'. In contrast, a 'Buffer Zone Consultation' in Thailand in 1993 (Gilmour, 1993:4) concluded that the concept of buffer zones, and the attainment of conservation goals through development in such zones, is clouded in uncertainty and ambiguity.

Wildlife Specific Buffers

Buffers are a commonly used management tool to protect individual species of wildlife, particularly birds, and their habitat (including nesting, roosting and foraging areas). Their spatial delimitation is usually specific to the species being protected and the individual characteristics of the site. The buffers are also usually based on a prescribed distance from say the nest site, and define an area where human activities are limited. Determining the size of these spatial buffers is problematic as disturbance regimes may depend on many inter-related factors. The common purpose of most wildlife habitat buffers however, is to extend the area of protection for the particular species and to minimise externally originating threats, particularly human activities, to the species and its habitat. Buffers have been implemented for a range of fauna species. In this section several approaches are briefly described.

Fish Habitat Buffer

Within Australia, the Department of Primary Industries (Queensland) has prepared 'Fisheries Guidelines for Fish Habitat Buffer Zones' (Bavins et al., 2000). Departmental policy recommends the implementation of buffer zones and/or other protective measures for every land use change or development adjacent to marine or freshwater fish habitat. Buffers with a minimum width of 100 metres, incorporating natural vegetation and other buffer elements, are recommended to be set back from the level of highest astronomical tide in tidal areas, especially if adjacent to a Declared Fish Habitat Area, and in freshwater areas a minimum 50 metres setback, also incorporating natural vegetation and other buffer elements, is recommended for freshwater fish habitat. These generic buffer widths are considered a 'starting point' from which site specific requirements can be negotiated to assist in the protection of marine plants and fish habitat areas.

The guidelines identify four steps to aid the selection, design and implementation of an effective buffer zone width including:

- confirmation of the need for the buffer zone;
- determination of the minimum buffer zone width to reduce impacts between fish habitats and adjacent land uses;
- provision of appropriate vegetation or structures (buffer elements) to fulfil the required functions of the buffer zone; and
- implementation of a management plan to ensure the integrity of the buffer zone is monitored and maintained.

In assessing the site-specific nature of the buffer, the landholders and stakeholders must be identified and involved in the site assessment, buffer design and implementation. Tenure and ownership details are confirmed, as responsibility for implementation and maintenance of the buffer rests with the landholder. The costs and benefits of the establishment of the buffer are also considered, including the promotion of the benefits of the buffer zone to the landholders and other stakeholders.

Several criteria which influence buffer width are identified, including:

- the value, functions and sensitivity of the wetland (e.g. these will determine the level of protection required);
- the intensity of the adjacent land use or land use impacts;
- the characteristics of the buffer (e.g. slope and vegetation cover);
- the specific buffer functions required; and
- the location of the buffer in terms of climate and rainfall.

An assessment proforma to guide the design of the buffer includes consideration of ecological issues, water quality issues, threats from nutrients, pesticides and heavy metals, and access issues. For each buffer function, minimum buffer widths are suggested, and are to be modified upon negotiation with the landholder and in relation to the local conditions.

Thus for spray drift, minimum buffer distances of 40 metres to 300 metres are suggested and for the protection of wildlife corridors, buffer widths of 15 metres to 45 metres are suggested. The final minimum buffer width is based on the threat that requires the greatest buffer width to minimise its impacts on the fisheries habitat. Although the guidelines recommend the application of performance criteria (as listed above) there is little information provided on the application of these criteria. The final buffer is to consist of natural vegetation including grasses, trees and shrubs, each of which will contribute to providing a range of buffer functions including bank stabilisation, filtering of sediments and nutrients, flood protection and aesthetic benefits.

In summary, the main purpose of the buffer zone design is to 'ensure the separation of conflicting land uses, whilst avoiding the need for any special maintenance requirements' (Bavins et al., 2000: 14). Its focus is however on ensuring the protection of fisheries habitat and associated values. While a consideration of appropriate separation distances is a useful component of this approach, the focus on a narrow zone, the main purpose of which is to filter pollutants, does not address wider landscape issues. Sustainable development requires that land uses which may be causing environmental harm should begin to address these threatening processes and reduce their impact on surrounding ecosystems and communities. This approach also does not effectively address future threatening processes that may arise from new or more intensive developments within the surrounding landscape. Hence a fish habitat buffer designed at one point in time and which is based on prescriptive distances may not afford long-term protection to fisheries resources.

Buffers for Bird Species

The buffer designed for the bald eagle *Haliaeetus eucocephalus* (McGarigal et al., 1991) was based on minimising the effects of human activities, both stationary and moving. The strategy used to establish the buffer width depended on:

- the predominant type of human activity (e.g. boats, pedestrians and aircraft);
- whether the predominant form of human activity is moving or stationary;
- the degree of eagle protection sought (i.e. protection from agitation or flushing); and

- the desired management level (i.e. management to ensure protection of a certain percentage of the population).

Temporal restrictions were also recommended at times when the bald eagles were using their critical resources, this being during the nesting and postfledgling stages (May to August). Restrictions involved prohibiting temporary human activities such as most recreational activities. Further, as eagle feeding activity was greatest during the early morning hours, buffers were recommended to be in place before 1000 hours, a time of high feeding activity (McGarigal, 1991).

Nest sites of the superb parrot *Dasyurus maculatus* are protected by 100m buffers or 'exclusion' zones (Figure 4.10), to reduce external threats from human disturbance, particularly noise.

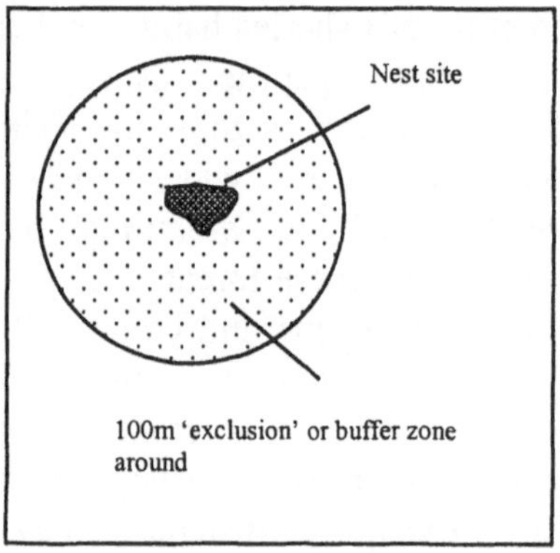

Figure 4.10 Buffer zone to minimise external threats to the superb parrot

Wildlife Specific Buffers – Conclusions

Buffers designed to protect wildlife species are usually designed to curtail the effects of external threats. However, they generally do not attempt to minimise all threats over the area in which they occur. Emphasis is frequently placed on minimising human disturbance threats and hence such strategies may not respond effectively to the area of impact of particular threatening processes.

The impacts of a particular wildlife species on its surroundings have not always been a consideration in buffer design, despite the frequent occurrences of animal wildlife moving from core habitat to surrounding landscapes and causing conflict with local communities. Protected area management strategies have been slow to recognise the negative impacts of wildlife on people, and to implement effective strategies.

Wildlife specific buffers typically nominate a set distance from a nest, roost or foraging area in which use is restricted. Although the buffer for the bald eagle was strongly based on a good knowledge of the biology and ecology of the species, minimum buffer widths of 400 metres and 800 metres were recommended by varying the management level for the bald eagle from 50 per cent to 90 per cent of the population to be protected. The buffer did, however, include both spatial and temporal considerations. The strategy for the northern spotted owl utilised the latest data (fecundity, adult survivorship, home range areas) on the species to determine the buffer areas. Seventy acres of suitable habitat closest to the nest or activity centre of the owl or within the radius of a circle with area equal to an estimate of the median-sized home range for the particular environmental province was to be preserved (Bart, 1995). In relation to other species, Voss et al. (1985) recommended buffers of 250 metres on land and 150 metres on water for great blue herons, and Anderson (1988) suggested a minimum 600 metres buffer for brown pelicans nesting on an island off the west coast of Mexico (see also Erwin, 1989; Rodgers and Smith, 1995).

Wildlife specific buffers usually incorporate monitoring of species and hence the management program and incorporated buffer would theoretically be responsive to change. Buffers of this type are largely 'one-off', with no particular methodology to aid implementation for a variety of wildlife species.

Evaluation of Buffers for Environmentally Sensitive Areas

This chapter has assessed several approaches to the design of buffers for ESAs. They include Internal Zoning (Foster, 1973), the Wildlife Conservation Unit Approach (Lusigi, 1981), Buffer Zone Planning (Kozlowski and Ptaszycka-Jackowska, 1981), Biosphere Reserves (UNESCO, 1974), Multiple-Use Modules (Noss and Harris, 1986), Integrated Conservation and Development Programs, and wildlife specific buffers. Key characteristics of buffers for ESAs are summarised in Table 4.4, their main common characteristic being that they are designed to protect sensitive natural areas from externally originating threats and thus extend the effective area of the core. Several of the approaches aim to integrate communities into the management of the buffer, although the reality is that this has been difficult to achieve. Typically the approaches also do not have a methodology which allows for easy application by practitioners.

Table 4.4 Key attributes of existing buffers for Environmentally Sensitive Areas

Attribute	Meaning
Purpose*	• to minimise the impact of threatening processes (mainly external threats) on an ESA • to extend the effective size of an ESA
Composition*	• variable, although usually contain natural environments • usually contain land uses compatible with ESA • may contain human communities
Structure*	• adjacent to core • usually based on examining external threats to an ESA • usually homogeneous • width usually based on prescriptive distances from ESA • usually not linked to other ESAs
Function*	• usually a zone of increasing intensity of land uses from an ESA (except for the BZP model)
Types / examples	(Refer Figures 4.4 – 4.10) • usually surround ESAs e.g. protected areas, biosphere reserves, wildlife habitat, habitat fragments

(* Where: **Purpose**: the role that the buffer is intended to play within the wider landscape. **Composition**: the variety of elements in the buffer [e.g. natural and cultural features]. **Structure**: the pattern or physical organisation of the buffer. **Function**: the roles [ecological, social and economic] that the components of the buffer fulfil in driving the processes that sustain the functioning of the core).

Conclusion

Well designed buffers are likely to result in many environmental, social and economic benefits. In this chapter, several existing approaches (e.g. separation zones, remnant habitat strips and buffers for ESAs) were presented and several short-comings relating to each type were identified.

The evaluation of the approaches revealed the lack of suitability of separation zones for developing integrated buffers for significant natural areas. The purpose of separation zones was to reduce or eliminate the impact of one land use type on adjacent areas. Similarly the remnant habitat strips have application mainly in forestry operations and have limited suitability as a general basis for buffering sensitive habitats. They are designed to retain tracts of vegetation, usually in a linear pattern in an otherwise disturbed landscape.

The buffers for ESAs, are designed to give an added layer of protection to sensitive natural areas and vary widely in their underlying structure and application. However, several of these approaches are based on criteria that will provide a useful foundation for the recommended best practice integrated buffer planning approach. In particular they are based on conserving the values of a core area, frequently a protected area, by examining external threats. Many involve effective community participation strategies and stress the need for integration of core areas across the landscape. The BZP's heterogeneous approach (Kozlowski and Ptaszycka-Jackowska, 1981) is particularly useful as it incorporates a seven step planning methodology and stresses the need to examine a full range of individual threats to protected areas and to develop specific policies to eliminate or minimise the impact of the threats. Biosphere reserves, MUMS and ICDPs were also identified as displaying several important aspects of good buffer design. A summary evaluation of buffers for ESAs is presented in Chapter 6, and key features that may be incorporated into a recommended buffer process are also outlined.

Chapter 5 will evaluate how buffers are designed and implemented in Australia. This investigation of practice-based approaches will provide further criteria to assist in the development of the IBP methodology.

References

Adams, L.W. and Dove, L.E. (1987), Wildlife reserves and corridors in the urban environment: a guide to ecological landscape planning and resource conservation, National Institute for Urban Wildlife, Columbia, MD.

Anderson, D.W. (1988), 'Dose-response relationship between human disturbance and Brown Pelican breeding success', *Wildlife Society Bulletin*, **16**, pp. 339-45.

Bart, J. (1994), 'Amount of Suitable Habitat and Viability of Northern Spotted Owls', *Conservation Biology*, **9**(4), pp. 943-6.

Batisse, M. (1986), 'Developing and focusing the biosphere reserve concept', *Nature and Resources*, **22**(3), pp. 2-11.

Bavins, M., Couchman, D. and Beumer, J. (2000), *Fisheries Guidelines for Fish Habitat Buffer Zones*, Queensland Fisheries Service, Brisbane.

Castelle, A.J., Johnson, A.W. and Conolly, C (1994), 'Wetland and Stream Buffer Size Requirements – A Review', *Journal of Environmental Quality*, **23**, pp. 878-882.

Chenoweth and Assoc. P/L (1994), *Pine Rivers Green Plan*, Report to Pine Rivers Shire Council, np.

Dang, H. (1991), *Human Conflict in Conservation. Protected Areas: The Indian Experience*, Har-Anand Publications and Vikas Publishing House Ptv Ltd, New Delhi.

de Klemm, C. (1993), 'Biological Diversity Conservation and the Law. Legal Mechanisms for Conserving Species and Ecosystems', *IUCN Environmental Policy Paper and Law Paper,* No.29, IUCN, Gland.

Department of Natural Resources (DNR) and Department of Local Government and Planning (DLGP) (1997), *Planning Guidelines: Separating Agricultural and Residential Land Uses*, DNR and DLGP, Brisbane.

Environment Australia (2001), *Biosphere Reserves*, Available at: http://www.ea.gov.
au/parks/biosphere/index.html [27 February 2001].

Erwin, R.M. (1989), 'Responses to human intruders by birds nesting in colonies: Experimental results and management guidelines', *Colonial Waterbirds*, **12**, pp. 104-8.

Forster, R. (1973), 'Planning for Man and Nature in National Parks: Reconciling Perpetuation and Use', IUCN Publication No. 26, IUCN, Gland, Switzerland.

Gilmour, D. (1993), 'Buffer Zone Consultation in Thailand', *IUCN Forest Conservation Programme Newsletter*, No. 16, pp. 4-5.

Grumbine, R.E (1994), 'What is ecosystem management?', *Conservation Biology*, **8**(1), pp. 27-38.

Hilditch, T.W. (1992), 'Buffers for the Protection of Wetland Ecological Integrity', Paper presented to INTERCOL's IVth International Wetlands Conference, Columbus, Ohio, 13-18 Sept.

Hruza, K.A. (1993), *Buffer Zone Planning. A Possible Management Tool for Fraser Island*, Bachelor of Regional and Town Planning thesis, Department of Geographical Sciences and Planning, The University of Queensland, Brisbane.

International Union for the Conservation of Nature (IUCN) (1992a), *Draft Recommendations*, IVth World Congress on National Parks and Protected Areas, Caracas, 10-21 Feb.

IUCN (1992b), *Parks for Life: The Caracas Action Plan Draft*, IVth World Congress on National Parks and Protected Areas, Caracas, 10-21 Feb.

IUCN (2003), Recommendations of the Vth IUCN World Parks Congress, World Parks Congress, Benefits Beyond Boundaries, Durban, 9-17 Sept. Available at: http://www.iucn.org/themes/wcpa/wpc2003/english/outputs/recommendations.
htm.

Izatt, C.S. (1995), *Planning for Protection Cultural. Heritage Precincts*, Bachelor of Regional and Town Planning Thesis, The University of Queensland, Brisbane.

Kozlowski, J. and Ptaszycka-Jackowska, D. (1981), 'Planning for Buffer Zones' in P. Day (ed), *Queensland Planning Papers*, University of Queensland, Brisbane, pp. 244-38.

Lewis, D.M., Kaweche, G.B. and Mwenya, A. (1990), 'Wildlife Conservation: An Experiment in Zambia', *Conservation Biology*, **4**(.2), pp. 171-80.

Longmore, R. (ed.) (1993), *Biosphere Reserves in Australia: A Strategy for the Future*, Drawn from a report prepared for the Australian Nature Conservation Agency by P.Parker, Chicago Zoological Society for the Australian National Commission for UNESCO, Australian Nature Conservation Agency, Canberra.

Lusigi, W.J. (1981), 'New Approaches to Wildlife Conservation in Kenya' *Ambio*, **10**(2-3), pp. 87-92.

Lynch, K. (1989), *Good City Form*, The MIT Press, London.

McGarigal, K., Anthony, R.G. and Isaacs, F.B. (1991), 'Interactions of humans and bald eagles on the Columbia River (Washington and Oregon, USA) estuary', *Wildlife Monographs*, pp. 5-47.

Marynowski, S. (1992), 'Paseo Pantera. The Great American biotic Interchange', *Wild Earth, Special Issue*, Cenozoice Society, Inc., Ann Arbor, pp. 71-74.

Metcalfe, S. (1994), 'The Zimbabwe Communal Areas Management Programme for Indigenous Resources (CAMPFIRE)', in D. Western and M. Wright (eds.), *Natural Connections Perspectives in Community-based Conservation*, Island Press, Washington, pp. 161-92.

Metcalfe, S., Chitsike, L., Maveneke, T. and Madzudzo, E. (1995), 'Managing the Arid and Semi-arid Rangelands of Southern Africa: The Relevance of the CAMPFIRE Programme to Biodiversity Conservation', Paper presented at the Global Biodiversity Forum on Decentralisation of Governance and Biodiversity Conservation, Jakarta.

Noss, R.F. (1992), 'The Wildlands Project. A Conservation Strategy', *Wild Earth*, Special Issue, Cenozoice Society, Inc., Ann Arbor, pp. 10-25.

Noss, R.F. and Harris, L.D. (1986), 'Nodes, Networks and MUM's: Preserving Diversity at All Scales', *Environmental Management*, **10**(3), pp. 299-309.

Panwar, H.S. (1987), 'Project Tiger: The Reserves, the Tigers and Their Future', in R.L.Tilson and U.S.Seal (eds.), *Tigers of the World*, Noyes Publications, Park Ridge, pp. 110-17.

Peterson, A. (1991), *Buffer Zone Planning for Protected Areas: Cooloola National Park*, Master of Urban and Regional Planning thesis, Department of Geographical Sciences and Planning, The University of Queensland, Brisbane.

Peterson, A. (2002), *Integrated Landscape Buffer Planning Model*, PhD thesis, the School of Geography, Planning and Architecture, The University of Queensland, Brisbane.

Physick, R. (1996), 'Conservation practices outside of protected areas: what meets the needs of today in southern Africa', Paper presented at Conference on Conservation Outside Nature Reserves, The University of Queensland, February 5-8, Brisbane.

Price, M.F. (2002), 'The periodic review of biosphere reserves: a mechanism to foster sites of excellence for conservation and sustainable development', *Environmental Science & Policy*, **5**, pp. 13-18.

Rodgers, J.A. and Smith, J.T. (1995), 'Set-Back Distances to Protect Nesting Bird Colonies from Human Disturbance in Florida', *Conservation Biology*, **9**(1), pp. 89-99.

Roughan,J. (1986), *Planning for Buffer Zones – An Application of Protection Zone Planning to Nicoll Rainforest*, Bachelor of Regional and Town Planning Thesis, The University of Queensland, Brisbane.

Sayer, J. (1991), *Rainforest Buffer Zones: Guidelines for Protected Area Managers*, IUCN, Gland.

Schonewald-Cox, C., Buechner, M., Sauvajot, R., Wilcox, B. (1992), 'Environmental Auditing. Cross-Boundary Management Between National Parks and Surrounding Lands: A Review and Discussion', *Environmental Management*, **16**(2), pp. 273-282.

Sibanda, B.M.C. (1995), 'Wildlife, conservation and the Tonga in Omay', *Land Use Policy*, **12**(1), pp. 69-85.

Spellerberg, I.F. (1994), *Evaluation and Assessment for Conservation. Ecological guidelines for determining priorities for nature conservation*, Chapman and Hall, London.

UNESCO (1974), Final Report, Task Force on Criteria and Guidelines for the Choice and Establishment of Biosphere Reserves. MAB Report Series No. 22, UNESCO, Paris.

UNESCO (1986), Report of the Scientific Advisory Committee on Biosphere Reserves, in UNESCO *Final Report, Ninth Session, International Co-ordinating Council of the Programme on Man and the Biosphere*, MAB Series No. 60, Paris, pp. 66-79.

UNESCO (1995), *The Seville Strategy and the Statutory Framework of the World Network of Biosphere Reserves*, UNESCO, Paris.

Vass-Bowen, N. (1994), *A Role for Buffer Zone Planning in Urban Heritage Conservation?*, Bachelor of Regional and Town Planning thesis, The University of Queensland, Brisbane.

Wells, M.P. and Brandon, K.E. (1992), *People and Parks. Linking Protected Area Management with Local Communities*, World Bank, World Wildlife Fund, U.S. Agency for International Development, Washington D.C.

Wells, M.P. and Brandon, K.E. (1993), 'The Principles and Practice of Buffer Zones and Local Participation in Biodiversity Conservation', *Ambio*, **22**(2-3), pp. 157-162.

Whitehouse, J.F. (1990), 'Conserving What? – The basis for nature conservation reserves in New South Wales 1967-1989', *Australian Zoologist*, **26**(1), pp. 11-21.

World Resources Institute (WRI), IUCN, UNEP (1992), *Global Biodiversity Strategy. Guidelines for Action to Save, Study, and Use Earth's Biotic Wealth Sustainably and Equitably*, WRI/IUCN/UNEP, np.

Chapter 5

Buffer Design and Implementation in Australia

Ann Peterson

Introduction

This chapter extends the review of buffer planning approaches outlined in the previous chapter, by examining how buffers have been implemented by Australian natural resource management and planning agencies. The research was initiated by Roughan (1986), continued by Peterson (1991) and Hruza (1993) and later comprehensively addressed by Peterson (2002), whose conclusions are based on the results of a questionnaire that was sent to 72 agencies or organisations involved in park management, forestry and natural area planning. The questionnaire aimed to obtain information on buffer design, with particular emphasis on whether buffers were used as a planning tool, the circumstances in which buffers were implemented, the criteria used to define them, their general design features or characteristics, the types of restrictions or controls in place within the buffer, their legal status and importance in planning. The questionnaire had a high response rate of 75 per cent, providing a sound basis on which to draw conclusions regarding buffer planning throughout Australia.

Implementation and Purpose of Buffers

Forty-six per cent of the respondents to the questionnaire had introduced what they called 'buffer zones' (Figure 5.1). However, there was considerable spatial variation in the application of buffers across Australia. In state-wide terms, there were high levels of buffer implementation in the Northern Territory (NT) and Western Australia (WA) and low levels of implementation (<30 per cent of agencies) in Queensland (Qld) and South Australia (SA). Approximately half of the respondents in New South Wales

(NSW), Victoria (Vic) and Tasmania (Tas) had implemented buffers, while the Commonwealth (C'wlth) had implemented none. It is important to note that where agencies responded positively to the existence of a buffer this did not necessarily mean that the buffer was formally identified in plans, policies or legislation. For example, one respondent stated that the 'buffer concept' was implicit in their park management techniques and strategies, although a buffer area was 'not overtly recognized as a "buffer" zone'.

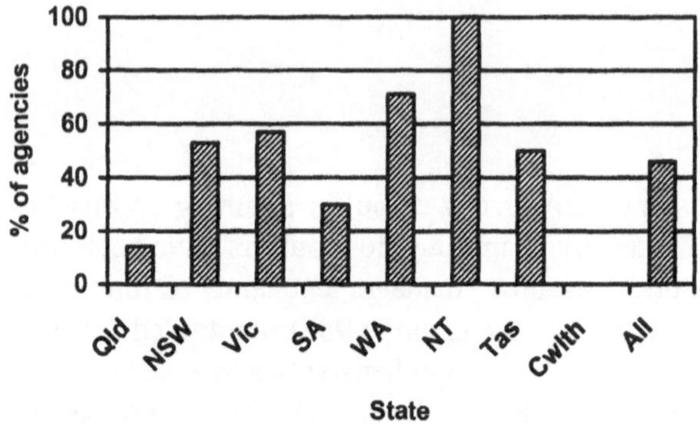

Figure 5.1 Per cent of Australian agencies, by state, to have implemented buffer zones

Fifty-six per cent of the agencies that had implemented buffer zones indicated their importance for wildlife protection (Figure 5.2). Other areas for which buffers were designed included:

- protected areas (36 per cent), and other ecologically sensitive areas (ESAs) (40 per cent), such as wilderness areas, lands with high aesthetic values, rainforests (especially littoral rainforests) and other remnant vegetation;
- water features (20 per cent), including water catchments, streams and wetlands;
- coastal areas (4 per cent);
- high resource use areas (4 per cent);
- areas subject to conflicting land uses (4 per cent), including buffers between agricultural, rural residential and urban developments; and
- hunting areas (4 per cent).

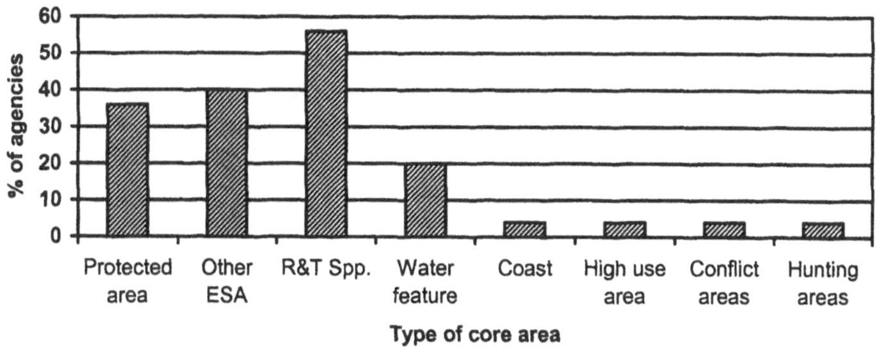

Figure 5.2 The types of 'core area' that are buffered in Australia (Where: ESA - environmentally sensitive area; R&T Spp - rare and threatened species; number of respondents = 25)

Sixty-nine per cent of the respondents regarded buffers as an important management tool and only three of these respondents qualified their support for buffers. Two indicated that buffers should be limited to the areas adjacent to urban development and saw no need for buffers 'where extensive agriculture such as grazing is the adjacent land use and there is no pressure for urban expansion or rural residential development'. Another stated that the area of application should be limited to minimising specific threats, such as invasive weeds and feral animals.

Twenty-four per cent of agencies did not see a need for buffers within their administrative areas. Some of the reasons offered by the respondents included:

- buffers give 'a false image of protection';
- planners should implement a conservation strategy across the whole landscape rather than rely on buffers;
- buffers are not a realistic planning strategy on privately owned land, and it would be difficult to implement them in such areas;
- buffers should be included within the protected area system;
- management practices in production forests ensured that areas surrounding important natural areas maintained forest canopy and high conservation values, and there was 'usually no need for additional buffers around reserves'; and
- buffers are not needed in areas that are well vegetated, or where topographic features, mainly lakes, protect parks from threats such as fire.

One respondent stated,

> I dispute the efficacy of 'buffer zones'. Energy and money is dissipated into buffer zones, which, in the end, give way to 'development'. Habitat must be fully protected, legally and on the ground, or not at all. The money is better spent fencing protected areas to make them fully protected, and permanently so.

This issue of the potential development that may flow towards effectively buffered areas is significant and highlights the need for effective policies to be in place to protect the values of the buffer and its core.

Buffer Design in Australia

In this section Australian buffers are evaluated using the 'important buffer criteria' detailed in Chapter 4. These include the consideration of both external and internal threatening processes, the need to be based on sound ecological principles, to effectively include local communities and to incorporate a workable planning framework or methodology.

Consideration of External Threats

All agencies that had implemented buffers viewed them as being important in reducing external threats. However, in identifying the nature and extent of the threats, 40 per cent of the agencies identified the external threats in relation to the core as a whole (i.e. comprehensively), with no attempt to examine how the individual elements of the core (e.g. fauna or water quality) were affected by a range of threatening processes. In contrast, eight per cent of agencies examined the individual elements or values of the core (e.g. its vegetation, fauna, hydrology, soil, cultural and aesthetic values) as a basis for understanding the interrelationships between the core and its surrounds (Figure 5.3). Twelve per cent of agencies considered both aspects, usually a comprehensive analysis followed by more specific consideration of the impact of particular threats on specific core elements. Thirty-two per cent of agencies did not consider this issue at all, the buffer usually being based on a set distance from the core.

Respondents were also asked to comment on whether the buffer was designed to reduce or eliminate a specific harmful influence (e.g. fire or weeds), or comprehensively to reduce or eliminate a wide range of harmful influences. Agencies were almost evenly divided in the way in which they

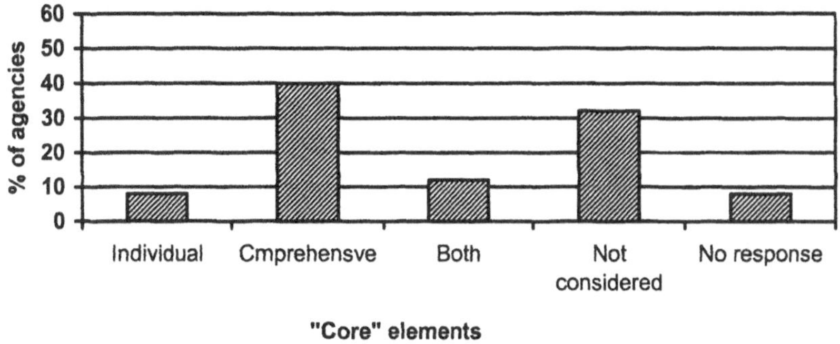

Figure 5.3 Process for examining external threats to core areas (Where 'individual' – examination includes assessment of individual elements of the core; 'comprehensive' – examination undertaken comprehensively with no individual assessment of each element of the core; 'both' – examination includes both comprehensive and individual assessments; and number of respondents = 25)

identified external threats (Figure 5.4). Fifty-two per cent of the agencies with buffer zones stated that their primary purpose was to reduce a specific threat, while 48 per cent stated that they were designed comprehensively to minimise a range of individual threats to a particular core natural area. For example, one respondent designed buffers comprehensively 'as a sponge around the biosphere core'. It is important to note that no agencies indicated that specific policies were applied to the area of operation of the

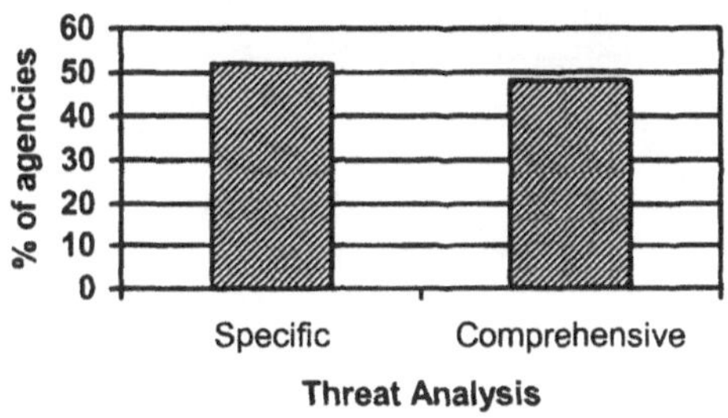

Figure 5.4 Agencies, by per cent, that examined threats in terms of specific threat analysis or comprehensive threat analysis (Where number of respondents = 25)

identified threats. The resulting buffers were of a homogenous nature with similar policies applied throughout the buffer area. A statement from one of the respondents revealed the gap between design and management:

> Usually a buffer will be established in response to a potential threat, but it is managed as far as possible to eliminate a wide range of influences.

The majority (56 per cent) of the respondents identified feral animals (e.g. foxes, cats and rabbits) as the main threat to core areas (Figure 5.5). Two agencies had introduced buffers to remove foxes preying on yellow footed rock wallabies and malleefowl populations. For example, malleefowl were raised in the protected area and released into a buffer zone where baiting was applied to eliminate foxes. Less intensive baiting by surrounding landowners provided an additional buffer zone. Other frequently mentioned threats to natural areas included fire, vegetation degradation and loss (clearing), and human pressure through development. Less frequently mentioned threats included visual pollution, drainage changes, noxious plants, micro-climatic changes, pollution (herbicides and weedicides) and damage to soil.

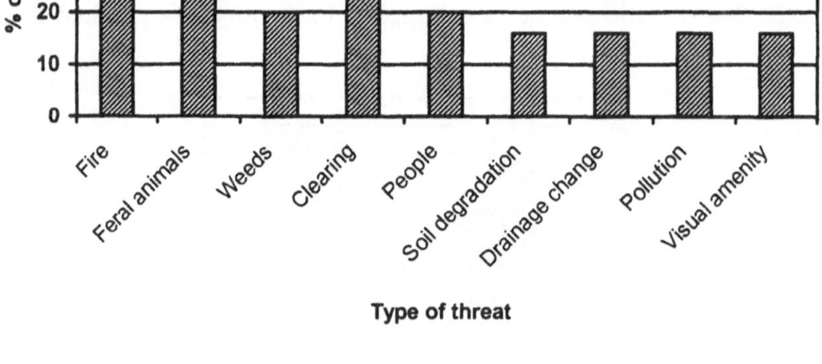

Figure 5.5 Types of threatening processes considered in buffer design in Australia
(Where number of respondents = 25)

Consideration of Threatening Impacts of the Core on its Surroundings

Agencies did not usually examine the relationship between the core area and the surrounding landscape or consider the possible negative impacts that the core could have on the surrounding community and its land uses. Only one agency indicated that the impacts of the core on the surrounding community were considered in its buffer management strategies.

Consideration of Ecological Principles

Several important principles or criteria underpinned buffer design in Australia (Figure 5.6). These included:

Prescriptive (or set) distances In 56 per cent of agencies, the buffer boundaries were derived by establishing a fixed distance, to be applied from the periphery of the core outwards into the surrounding matrix (refer to Table 5.1 for examples of wildlife buffers designed using set distances). Such distances usually did not reflect ecological needs, but rather were based on 'experience' or guess work. Buffer widths varied within agencies (e.g. one agency indicated that its buffer distances varied from 20 metres to

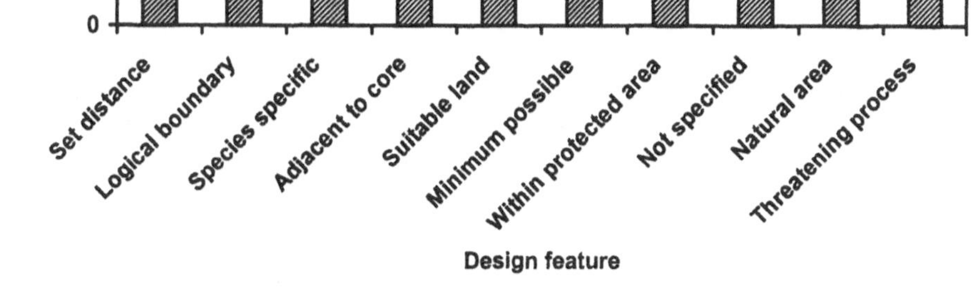

Figure 5.6 Principles used to guide the design of Australian buffers
(Where number of respondents = 25 and respondents could select more than one category)

Table 5.1 Design features of buffers developed for wildlife species in Australia

Species	Buffer design
Hastings River mouse *Pseudomys oralis*	40m buffer around creeks and soakages with sedges (NSW).
Superb parrot *Polytelis swainsonii*	Buffers or 'exclusion zones' of 100m around the nest site in which logging and silviculture occurred (Victoria).
Powerful owl *Ninox strenua*	Buffers of 100m radius around nest trees (NSW).
Yellow bellied glider *Petaurus australis*	Buffers of 100m radius around the most active 'V' notched trees (NSW).
Leadbeater's possum *Gymnobelideus leadbeateri*	50m buffer around optimal habitat sites that have a minimum area of one to three hectares (Victoria).
Tiger quoll *Dasyurus maculatus*	For sites of importance for scientific research a 200m exclusion zone is proposed around the quoll's den sites, precluding any new activities inconsistent with maximum conservation of the tiger quoll (Victoria).
Osprey *Pandion haliaetus*	Buffer of 20m radius around nest trees (NWS). Minimum radius of 150m around nest site (NSW).
Yellow footed rock wallaby *Petrogale xanthopus celeris*	10km buffer (SA). 5km buffer (SA).
Marbled frogmouth *Podargus ocellatus plumiferus*	Buffer area was included with the core special management area, where timber harvesting was restricted. The required core and its buffer were calculated to need an area 30 per cent larger than that determined sufficient to sustain a viable population of marbled frogmouth (Qld).
Koala *Phascolarctos cinereus*	Buffers of 100m around trees where a koala or recent evidence (such as dung pellets or claw scratchings) of a koala was found (NSW).

500 metres). One agency's code relating to forest practices required buffers to be a minimum of 20 metres wide on linear-shaped patches of rainforest, such as along gullies and streams, and 40 metres elsewhere. Another forest agency stated that they 'routinely established ... protection strips of 10 metre width outside national parks and designated special management areas'. A local government indicated that a 'beautification buffer zone' with a width of between 10 metres to 20 metres had been established along the frontage of all allotments adjacent to a protected area. The buffer strip was to be ceded to the local government and preserved in its natural state to help conserve 'the amenity and nature of the area'. One agency with responsibility for a coastal zone suggested a 100 metre buffer for the

coastline, major waterways and wetlands, with the size of the buffer depending on the 'type of development'.

Specific mention was made of riparian buffers, where in one region these varied from 5 metres to 20 metres depending on the 'flow regimes'. Two agencies stated that widths were variable depending on the stream order. A Forest Service indicated that in areas in which harvesting was restricted, buffer strips or 'watercourse protection zones' were to be retained on each side of a watercourse, 'generally commencing at the top of the banks confining the normal flow of the watercourse', although these distances could be varied at the discretion of the district forester.

A number of agencies were less specific about buffer sizes, with one agency stating that buffer widths were variable depending on the 'size of the feature'. This presumably meant that the bigger the core area, the wider the size of the buffer. Another respondent stated that buffers should be of 'sufficient size to protect small sub-populations of plants and their associated species'. This approach to buffering, which fails to address the nature and extent of threatening processes and merely designates a set buffer distance, is a significant deficiency in many Australian buffer designs.

Logical boundaries 'Logical boundaries' were cited by 40 per cent of agencies with buffers as being important in buffer design (Figure 5.6). These included ridge lines, creeks, catchment boundaries, transport lines as well as other administrative boundaries (e.g. fire management units). For example, in urban areas a buffer width was identified as 'whatever minimum the developer can get away with; [or] alternatively topographic or other physical constraints'. A Forest Service respondent stated, 'Logical boundaries such as ridge lines and gullies are used where possible' to delimit buffer boundaries. Other responses included the location of possible fire breaks as a boundary for buffers, which were usually established to minimise the impacts of fire on protected areas. Such fire breaks or buffer boundaries could coincide with major road routes.

Location In relation to the positioning of buffers, 28 per cent of agencies with buffers stated that they were usually located within protected areas (Figure 5.6), as such buffers were thought to more adequately protect the core. One respondent stated:

> In our Region, we have found it more practical not to develop buffer zones because of people's perception that the laws for the buffer should be different e.g. less severe than the laws for core areas. It becomes more practical to do away with buffer zones and include them in the total core/management zone....

> Our basic philosophy is that any 'buffer zone' is included in the protected area and treated as fully protected. This simplifies the legal management.

However, 44 per cent of respondents considered that it was important for the buffer to be adjacent to the core, in order to extend its area of protection. For example, one respondent stated that the design of buffers centred on including 'nearly all our neighbours that surround the park'. The remaining 32 per cent of agencies did not highlight the positioning of the buffer as an important criterion in their design.

Land suitability/tenure About one quarter of the agencies regarded the 'suitability' or tenure of the land surrounding the core as important in buffer design (Figure 5.6). Suitable land inferred a lack of private ownership, consisting, for example, of forestry land, general rural land, or government owned land. One respondent stated that suitable land for buffering included any 'reserve which has available surrounding land'. Another respondent cited the 'significance of adjoining public land; is it National or State Park', while yet another stated the requirement as being 'whatever Crown [government] land is available around the core'. Respondents believed it would be difficult to implement buffers on private land because affected landowners might need to be compensated for any loss of their existing use rights.

Natural values Forty-four per cent of agencies located buffers on areas with important natural values (Figure 5.6). One respondent defined buffer zones as 'undisturbed vegetation adjacent to special management areas'. Another considered that 'rare, significant wildlife habitats (including migratory birds), rare, endangered and relict plant species and significant, and/or vulnerable vegetation communities' should be used to define the location of buffer boundaries, and that it was important to include 'gorges, waterholes (and) moist, floristically diverse habitats'. One respondent indicated that the dimensions of buffer zones were dependent on the 'natural values' of the area surrounding the core and that buffer widths could range from a 100 metre building setback to a 500 metre setback. Another respondent cited the lack of buffers in their planning area as being due to the lack of rare or threatened species on private land adjacent to core natural areas.

Species specific Twenty-eight per cent of agencies designed their buffers to improve the conservation of specific species (Figure 5.6). Respondents identified that wildlife specific buffers were in place for the Hastings River mouse *Pseudomys oralis*, yellow bellied glider *Petaurus australis*, koala *Phascolarctos cinereus*, leadbeater's possum *Gymnobelideus leadbeateri*, yellow footed rock wallaby *Petrogale xanthopus celeris*, tiger quoll *Dasyurus maculatus*, brushtail possum *Trichosurus vulpecula*, superb parrot *Polytelis swainsonii* and osprey *Pandion haliaetus*. Although 56 per cent of agencies indicated that buffers were important to conserve animal wildlife (Figure 5.3), only 28 per cent of agencies designed species specific buffers. In general, boundaries were defined on the basis of prescriptive distance measures.

Other factors Twelve per cent of agencies considered the nature of the particular threatening processes that were acting on the core area and designed their buffer to minimise the impact of these threats. Sixteen per cent of agencies designed buffers on the basis of the minimum that could be negotiated.

Effectively Include Local Communities

Although human pressure was a commonly cited threat to core areas, local communities were generally not included in 'on-ground' planning, design and management of buffers. In only two of the 25 agencies with buffers were the local community effectively involved. For example, one respondent stated in relation to a buffer for the malleefowl, that 'nearly all neighbours that surround the park regularly take part in this program'. Another stated that buffer planning outside protected areas was '(a) very sensitive issue... We need to establish better relations, understanding and trust with these groups and individuals (landholders and local councils) before proceeding down the road of buffer zones'.

Responsive to Change and Use an Effective Methodology

Only one agency reported the need for buffers to eliminate future threats by stating, 'conservation must allow for the core area to behave dynamically'. Hence, buffers in Australia were applied mainly reactively in situations where existing threats were evident.

No respondents indicated the use of a specific methodology in the design of buffer zones. Most zones were site specific. One respondent stated that as, 'no formal program of buffer establishment is in place, they are variable in size, location and design'. One respondent identified 'a need for research to provide information on the effectiveness of a range of buffer sizes and management regimes for different management objectives'.

Buffer Policies and Legislative Foundations

Where buffers had been established the respondents were also asked to indicate the legislative or other management arrangements in place to conserve the values of the core. This issue was examined both in areas where buffers had been formally identified and in areas adjacent to protected areas and other ESAs, where buffers were not formally recognised.

Regulatory Mechanisms in Areas with Formally Identified Buffers

Respondents were asked to indicate the types of restrictions or policies applied to the land uses and activities in identified buffer zones within their jurisdiction (Figure 5.7). Clearing controls and fire policies were implemented by 32 per cent of agencies, while 24 per cent had in place

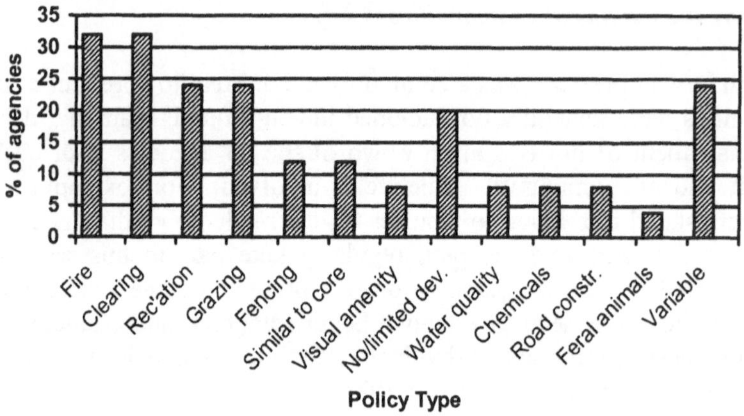

Figure 5.7 Regulatory mechanisms/policies applied within buffer zones
(Where number of respondents = 25)

policies to minimise the impact of grazing and/or recreation. Other policies related to fencing (12 per cent), visual amenity (eight per cent), water quality (eight per cent), the use of chemicals (eight per cent), road construction and maintenance activities (eight per cent) and feral animals (four per cent). Twenty per cent of agencies managed their buffers as non-development or very limited development zones, while 12 per cent applied the same policies within the buffer as in the core area. Twenty-four per cent of agencies stated that the policies in the buffer were not fixed, but rather varied with particular local circumstances.

Respondents were also asked to describe the legal status of the buffer zones and land use policies applied within them (Figure 5.8). Almost 90 per cent of agencies provided a legislative basis for their buffers, with a variety of legislation being utilised. In 36 per cent of agencies planning legislation was used to implement buffer policies. In 28 per cent of agencies, the buffer was within a legally protected area and was subject to the protected area legislation of the relevant state or territory. In such situations the buffer policies were included within the park management plan, where one existed. Forestry legislation was used by 24 per cent of agencies and 4 per cent used fire related legislation.

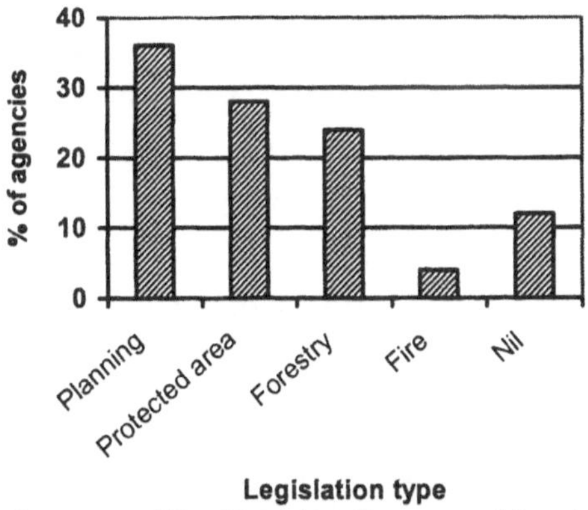

Figure 5.8 Legal status of buffers implemented by Australian agencies
(Where number of respondents = 25)

Forty per cent of agencies used a voluntary or advisory approach to buffer implementation. They indicated that where a buffer was delimited and not legally designated, local governments could consider the intents of the buffer in their planning process and by relevant statements within planning documents. One respondent indicated that where a buffer had no legal standing, management was based on 'good neighbour relations'. Another respondent indicated that these good neighbour relations extended to some financial assistance for fencing protected areas to minimise intrusion of grazing animals.

Regulatory Mechanisms in Areas where Buffers are not Formally Recognised

In situations where formal buffers were absent, fifty-seven per cent of agencies relied on complementary legislation and policies, applied to land uses and activities in areas surrounding the ESA, to help minimise threats to these areas (Figure 5.9). Thirty per cent of agencies however, responded that they did not utilise such a framework.

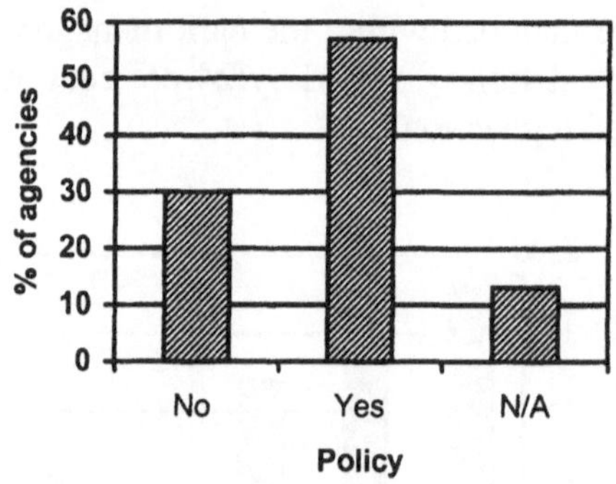

Figure 5.9 Agencies with policies applying to land surrounding core natural areas, where buffers are not formally recognised
(Where number of respondents = 54)

Respondents were also asked to indicate the status of such policies (Figure 5.10). In the 31 agencies with complementary policies in place, 58 per cent indicated that planning legislation was utilised to minimise impacts to the core area. A further 42 per cent utilised native vegetation legislation, 16 per cent forestry legislation and two per cent nature conservation legislation. In the state of Queensland, formal buffer zones were not created around protected areas. One respondent stated, 'Our intention is to promote the creation of buffer zones, however our power to enforce zones is zero'. The respondent added that the main role of staff was to better manage external impacts on protected areas through input into local government strategic plans and development control plans and that staff encouraged buffers around parks, between aesthetically or environmentally incompatible land uses and along river systems and drainage channels, but added, 'We have no way of ensuring that they are created'.

A comment from one respondent highlighted the lack of policy development in areas surrounding important core natural areas in their region. The respondent stated that no policies were in place and that, 'in fact the opposite is the case – Crown land and reserves, including national parks are continually being encroached upon by tourism developments and other pecuniary interests. Subdivisions occur on reserve boundaries at a very high rate. Crown land is sold off on a regular basis'.

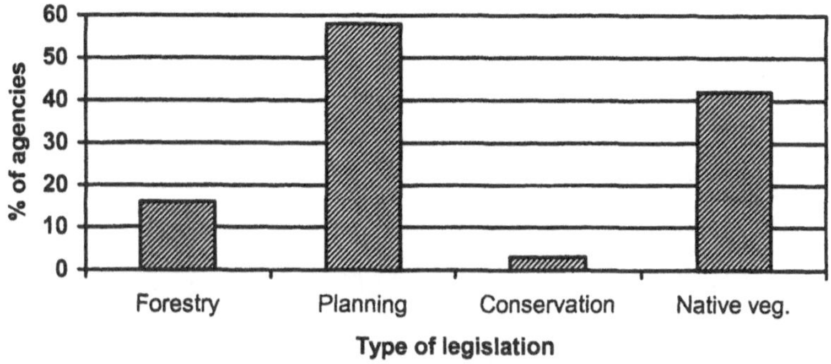

Figure 5.10 Legislation used to enhance the conservation of core natural areas in situations where there were no buffers
(Where number of respondents = 54)

Evaluation of Australian Buffers

The following evaluation of Australian buffers is based on the responses to the issues outlined in the previous sections and addresses the purpose and importance of buffers, their design features and legislative foundations.

Purpose and Importance of Buffers

Buffers in Australia were designed primarily to enhance the protection of wildlife and their habitat (56 per cent of agencies), protected areas (36 per cent) and other ESAs (40 per cent). Although only 46 per cent of respondents had introduced buffers, approximately 70 per cent viewed buffers as an important tool in natural resource management and planning. However, about one quarter of the agencies did not support the introduction of buffers. These reservations related to the following issues:

- *Buffers fail to consider a wider landscape approach*
  Several respondents preferred a wider landscape strategy, rather than relying on individual buffers to protect specific natural areas. These comments are valid and reflect the current principles that underpin

integrated landscape management approaches. Buffering individual core areas without consideration of the wider landscape context is unlikely to ensure that biodiversity values are conserved. Such individual buffers may merely increase the size of the core area and fail to integrate and connect it with its surrounding landscape and communities. In response to this criticism, it is recommended that buffers should be planned within a bioregional or landscape context, by ensuring that important natural areas are of sufficient size and connectivity and are buffered from threatening processes by including appropriate policies that are negotiated with the relevant stakeholders.

- *Buffers are only needed in some circumstances*
  Several respondents indicated that a buffering strategy was only needed in certain limited situations, such as for natural areas adjacent to urban areas or areas under development pressure. While such areas may be important sites for buffers due to the greater range and intensity of threats from urban type developments, frequently the ability to implement effective buffer policies is limited due to the often fragmented landscapes and the complexity of threatening processes. Hence, it is recommended that planning should adopt a more proactive approach and consider the implementation of buffers, in a landscape context, before areas come under development pressure.

- *Buffers attract potentially threatening developments*
  One respondent expressed concern that buffers eventually 'give way to development'. These views are pertinent, as buffers in many developing countries attract large numbers of people who wish to take advantage of the programs promoted within these zones. Increased human populations within the buffer may lead to greater environmental stress on the very environments that are to be protected. This issue indicates the importance of developing strategies within the buffer, which address either the potential carrying capacity of the area, or its limits of acceptable change.

- *'Advisory' buffers may be less effective*
  Buffers that are based on advisory policies may be unable to restrict developments occurring on the boundary of the core area. Several protected area agencies indicated that where buffers were on land outside their legislative control and were advisory in nature, that there was the real threat that development could mitigate the conservation benefits of the buffers. These agencies suggested that a legislative basis

for the buffer, for example through relevant planning and natural resource related legislation, was necessary to enhance the effectiveness of the buffer. This implies that the buffer implementation strategies should include the 'right mix' of mechanisms, such that they appropriately reflect local circumstances. This may include regulatory strategies, as well as a raft of more voluntary incentive based mechanisms.

- *Better to fence core areas*
  One respondent suggested that protected areas would be better off if they were fenced, rather than have time and money spent on developing buffer zones. This 'fortress' approach is naive given the varied nature of threats that are currently in evidence in many protected areas, in particular that of pollution (air, water, soil and visual) and deliberate human intrusion. Again it is recommended that sensitive natural areas be integrated within their surrounding landscapes and that a buffer strategy that effectively deals with existing and potential threatening processes be implemented. This process will help to expand nature conservation actions across the landscape and help to minimise the fragmentation of remnant ecosystems.

Design of Buffers

Although all responding agencies viewed buffers as important in reducing external threats to the core area, only 20 per cent of agencies with buffers examined in any detail the specific relationship between the important elements of the core area and its surroundings, as a means of determining the nature of the external threatening processes. Overwhelmingly the process used by agencies was to consider the core area as a whole, without specific analysis of its individual components. Although this latter process is less complex and perhaps produces a buffer plan more quickly, there may be reduced buffer effectiveness, especially in complex ecosystems where a thorough understanding of ecosystem elements and their interactions may be necessary in order to understand the specific nature of the threatening processes and their cumulative impacts.

Half of the respondents indicated that buffers were designed in response to one or a few specific threats and did not consider the heterogeneous nature of the environment surrounding the core and the entire range of threatening processes that could impact on the core's elements, values and features, while the other half adopted a more comprehensive approach, examining a wide range of threats. This design process may reflect a limitation in resources and time available to agencies involved in buffer design. However, an effective buffer should consider a wide range of threatening process to the elements of the core. Where only one or a few threats are considered and managed, the core's values may be diminished as a result of the cumulative effects of other less obvious threatening processes.

In general, Australian agencies did not define accurately the area of influence of the identified threats, and policies and controls tended to be applied uniformly throughout the buffer area. Hence, the buffers may not reflect the threatening processes acting on the core area and they may result in the unnecessary imposition of restrictive land use policies in areas where particular threats are not at play, and a failure to include some areas within the buffer that are the source of particular threats. It is recommended that when developing a buffer, the source area(s) of the threats should be accurately defined and relevant policies implemented to mitigate or lessen their impact. Such an approach is more likely to gain community support, as the design process is logical, ecologically based and transparent.

With only one exception, respondents indicated that they did not analyse the impacts that the core may have on its surrounding community and landscape. Buffers were seen primarily as a means of minimising the external threatening processes rather than being a mechanism to integrate the core area with its surrounds. Such an approach is unlikely to produce long-term effective results. It is recommended that a more holistic approach be adopted in designing buffers, one that examines the nature and extent of the two way interactions between core natural areas and their surroundings.

In relation to the specific design features of Australian buffers, prescriptive distance measurements were employed by 56 per cent of agencies that had instituted buffers, with the distances being determined largely on an *ad hoc* basis, although with some reference to the area or species being given additional protection. Buffer distances ranged from 10 metres outside Forestry Service lands in Queensland to 500 metres from core natural areas in one region of Victoria. Riparian buffers widths were also variable, with some being as little as five metres from the top of stream banks. However, two respondents indicated the importance of stream order in determining buffer widths for watercourses. Buffers for specific wildlife species were variable, with smaller species such as the Hastings River mouse *Pseudomys oralis* having buffers of about 40 metres from creeks and soakages with sedges, to buffers for a number of bird species such as the superb parrot *Polytelis swainsonii* and osprey *Pandion haliaetus* being 100 metres and 150 metres, respectively, from nest sites. Larger mammals such as the yellow footed rock wallaby *Petrogale xanthopus celeris*, which have wider home ranges, had buffers from 5 kilometres to 10 kilometres. Thus although the buffers were based on prescriptive distances, there was some consideration of the habitat needs, home ranges and movement patterns of particular species in the development of wildlife specific buffers. The use of prescriptive distances in many Australian buffers does not reflect sound ecological principles. This approach will either include too much or too little of the areas necessary to ensure the effective conservation of important core areas and perhaps place unnecessary restrictions on land holders.

Forty per cent of agencies used logical boundaries to define their buffers. Such a strategy is useful to aid in the clear definition of the buffer boundary on the ground and thus alert management staff and the community to the required actions to be conducted within the buffer. However, the use of such boundaries without prior consideration of the location and extent of threatening processes that affect the core may be of limited value in defining an ecologically sound, integrated buffer zone.

Seventy-two per cent of agencies considered the location of the buffer as an important criterion in design, with 59 per cent of these agencies stating that buffers should be positioned adjacent to parks and other core areas. The remainder (41 per cent) indicated that buffers should be located within park boundaries. In the latter circumstances, the buffers do not extend the area of land afforded protection. This was a standard approach used in protected area management, where a zonal strategy frequently resulted in the identification of core areas, where minimal intrusion and development were permitted, and a range of other zones that allowed increased visitor usage and infrastructure development. Such a strategy focused management within the protected area rather than extending management strategies to surrounding lands. Where external threatening processes were recognised, it was the protected or reserved area that absorbed these impacts and attempted to minimise their negative effects on the most significant features or elements of the park/reserve. However, a majority of agencies did recognise and implement strategies to extend the area of protection of the buffer external to the protected area. Such an approach is likely to result in an increased area of land and water managed to enhance the values of the protected area or core, as well as enhance regional nature conservation values.

Land suitability or tenure was considered by 24 per cent of agencies as an important criterion in buffer design. This approach, which views buffers as being possible mainly where land is in public ownership has several consequences:

- buffers may remain within protected areas (28 per cent of agencies in the current study); or
- buffers may not be applied to privately owned land, or perhaps may not be applied at all; or
- buffers may be designed to include private land tenures, with reliance being placed on advisory policies and guidelines rather than legally enforceable controls.

Where land suitability is a prime consideration, the process of buffer design may fail to take account of the heterogeneity of the environment and the possibility of individual threats operating over a wide range of tenures. It may also fail to take account of ecosystem processes and to recognise that ecosystems transcend land tenure boundaries. These sentiments relating to the suitability of land to be included within a buffer are echoed by Sayer (1991: 17) in his review of buffer zones in tropical forests:

> Where the land is state-owned, it may be relatively easy to gazette reserves or impose restrictions on use. The setting aside of communally-owned or private land for buffer zones will involve more complex negotiation and will raise the issue of compensation.

Although the suitability of land is an important consideration in the selection of the 'right mix' of buffer policies and implementation strategies, it is important to integrate conservation with development on all lands, regardless of tenure. Thus where buffers are designed to minimise a range of externally and internally occurring threats, the buffer should be delimited to encompass all necessary lands, regardless of tenure, with a variety of buffer policies applied to accommodate the different tenure arrangements. For example, private land owners could be offered incentives, both financial and non-material to enter into voluntary agreements to better manage their land; they may be adequately compensated for loss of use rights; they may trade development rights where a conservation benefit can be gained; or in the case of leasehold land, lease conditions, especially those requiring land clearance, may be altered to ensure sustainable use practices. The complexity of land tenure arrangements in many countries also necessitates that the buffer strategy considers traditional land tenure systems and ensures the continuance of these traditional practices so long as these are sustainable in the long term. In situations where settlements are encroaching onto protected areas or the land surrounding these areas, the provision of secure tenure to occupants within a buffer zone may help to ensure that conservation of resources is given consideration by the new owners. Open access to resources, without secure tenure may contribute to over-exploitation of resources and impact severely on wildlife and protected area systems. A failure to consider the implementation of buffers on privately owned land and other potentially 'unsuitable' land merely relegates the concept to the 'too hard basket'.

Land with important natural values was considered by 44 per cent of agencies as an important buffer criterion. This approach may increase the area of protected natural habitat and may help to separate the core area from incompatible surrounding land uses. However, unless the buffer area encompasses the sources of threat to the identified core, it is unlikely that the buffer will be fully effective.

Particular wildlife species were the focus of buffer strategies in 28 per cent of agencies. Where such an approach includes a comprehensive analysis of the ecosystem in which the species is found and the range of threats to this entire ecosystem, the buffer may be effective. A failure to consider broad scale ecosystem processes and functions may not ensure the conservation of the target species, or the long-term viability of its ecosystem.

Buffer Policies and Legislative Foundations

There was a preference for a legislative basis to policies implemented within buffer zones in Australia. Almost 90 per cent of agencies relied on legislation relating to protected areas, land use planning, forestry and fire. Protected area legislation was utilised by those agencies that had located buffers within the protected area's boundaries. The inclusion of buffers within parks frequently occurred because protected area managers had little or no power to control lands external to protected areas. As the IUCN's categories of protected areas have been widened to include areas where sustainable use of resources is permitted, these new classes could be used to implement buffer type zones incorporating sustainable use of resources and thus contribute to a high level of protection of the core and buffer areas. Thus protected area legislation may continue to make an important contribution to the development and implementation of buffer type zones. Planning legislation plays a significant role in buffer management with 36 per cent of agencies relying on relevant legislation as their primary means of implementing buffer objectives and management policies.

Forty per cent of agencies used a voluntary approach to policy implementation. As Sayer (1991:1) comments:

> .. legal protection is rarely sufficient to guarantee the continuing integrity of conservation areas... Laws which are resented by the majority of the population are difficult to enforce.

Reliance on voluntary mechanisms reflects the changing nature of planning, where statutory instruments have been slow to keep pace with social, economic, political and environmental changes and where, as a result the statutory planning instruments are supplemented with a range of alternative non-statutory planning tools (policies, manuals, guidelines, planning studies and local area plans). The non-statutory mechanisms vary in terms of their form and content, as well as the methods by which they are produced and the time taken in preparation. There may be no requirement for public consultation and government approval and hence these mechanisms can often be produced in a relatively short time frame.

The evolving non-statutory tools represent an important mechanism to help planners deal with emerging biodiversity issues. However, there may be several drawbacks. Firstly, many of these instruments may be of limited use in an appeal situation. However, King Cullen (1993) believes that provided such documents are soundly based and have been consistently applied over a period of time, this may not be the case. Secondly, the non-statutory documents may not be subject to public involvement in their development and may be seen to be subject to political interference. Thus, if a buffer plan is to be developed by relying on non-statutory mechanism, it is critical, in relation to the acceptance and success of the buffer, that the community is effectively consulted and involved in implementation and management.

In situations where formal buffer zones may be inappropriate due to community, political or other reasons, the ecological principles behind the concept of integrating ESAs and protected areas with their surroundings should, where possible, be the basis of regional and local planning. Hence formal buffer zone designation may not be necessary, so long as the underlying threatening processes and performance outcomes are well understood and can be integrated into alternative planning mechanisms. The Integrated Conservation and Development Program (ICDP) concept (Chapter 4), for example, is an important mechanism for the implementation of buffer zone strategies, and can be accomplished without formal buffer definition. However, it is important that the link between conservation and development is very clear and that conservation strategies are seen as a means of contributing to development and its associated improvement in the standard of living and quality of life of surrounding local populations. In addition, as statutory and non-statutory measures adopted within a particular jurisdiction can apply only within this jurisdiction, it is also important for planners to adopt a wider regional perspective and to seek cooperation with adjoining regional authorities (e.g.

local government, state or province) to ensure that the values of the important natural areas are fully protected.

Due to the importance of integrating conservation objectives into the areas surrounding ESAs, the buffer zone concept and methodology of planning may form the basis for a range of statutory and non-statutory policies to ensure long-term ecologically sustainable land uses are in place in these areas. Informal agreements and arrangements with protected area managers, local government, and communities are an important buffer implementation strategy.

Conclusion

The design of buffers within the Australian setting reflected a predominance of the following:

- fixed distances, which were usually based on guess work;
- incorporation of logical boundaries;
- positioning of buffers adjacent to important natural areas;
- reliance on suitable land tenures; and
- the use of intuition rather than a sound ecologically based methodology.

Many of these design criteria and implementation procedures were easy and convenient to use. Set distance measures were easy to map, especially where the buffer was planned adjacent to the core area. The use of logical boundaries usually ensured that the perimeter of the buffer was tangible, aiding implementation, monitoring and enforcement. Use of land with 'suitable' tenure as a buffer was administratively easier for implementation and less expensive than establishing buffers on private land.

An important purpose of the questionnaire was to better understand current Australian buffer design strategies. While the critical evaluation of these approaches has indicated several shortcomings, it has also provided an important basis for developing a 'best practice' integrated buffer methodology. In particular, this analysis and evaluation has highlighted the importance of buffers in natural resource management planning in Australia and also pointed to a number of important criteria and design principles. These include that the buffers should:

- be based on minimising or eliminating external threatening processes;

- be planned on the basis of sound ecological principles, with the final buffer boundaries being adjusted to coincide with identifiable features in the landscape (e.g. ridge crests and drainage lines);
- consider land suitability *(Note*: where land is in government ownership the implementation of buffer strategies may be simplified. However, failure to include other land tenures may limit the effectiveness of the buffer policies, especially where threatening processes remain in operation on these other tenures);
- extend the area of protection beyond the protected area or ESA;
- include, where possible, other natural areas outside of the core area, to enhance the conservation values of such areas;
- ensure that all relevant stakeholders are included, from a very early stage, in the planning, design and ongoing implementation of the buffer; and
- include a balanced mix of both regulatory and non-regulatory implementation mechanisms (*Note*: legislative mechanisms may provide greater planning certainty, while non-statutory mechanisms may have greater community support).

The results of this evaluation of Australian buffers provide an important set of criteria which will be further considered in the development of the Integrated Buffer Planning method.

References

Hruza, K.A. (1993), *Buffer Zone Planning. A Possible Management Tool for Fraser Island*, Bachelor of Regional and Town Planning thesis, Department of Geographical Sciences and Planning, The University of Queensland, Brisbane.

King Cullen, R. (1993), *A Model for Preparing Planning Guidelines*, PhD Thesis, Department of Geographical Sciences and Planning, The University of Queensland, Brisbane.

Peterson, A. (1991), *Buffer Zone Planning for Protected Areas: Cooloola National Park*, Master of Urban and Regional Planning thesis, Department of Geographical Sciences and Planning, The University of Queensland, Brisbane.

Peterson, A. (2002), *Integrated Landscape Buffer Planning Model*, PhD, Department of Geographical Sciences and Planning, The University of Queensland, Brisbane.

Roughan, J. (1986), *Planning for Buffer Zones – An Application of Protection Zone Planning to Nicoll Rainforest*, Bachelor of Regional and Town Planning Thesis, The University of Queensland, Brisbane.

Sayer, J. (1991), *Rainforest Buffer Zones: Guidelines for Protected Area Managers*, IUCN, Gland.

Chapter 6

Buffer Zone Approaches: An Evaluation

Introduction

Buffer zones have long been advocated as an important planning tool. The *Global Biodiversity Strategy* stresses the importance of reconciling human needs and activities with the maintenance of biodiversity and recommends the integration of protected areas into natural and modified surroundings through the use of buffer zones and habitat corridors (WRI et al., 1992). In the Australian context, Coveney's (1993) study of threats to the state of Victoria's protected areas recommended the creation of buffer zones around reserves as a means of reducing or eliminating many threats to the ecological and aesthetic values of parks.

When viewed from a wider perspective, interlinked systems of core areas, buffers and corridors are advocated to provide a landscape protection strategy:

> Systems of interlinked wilderness areas and other large nature reserves, surrounded by multiple-use buffer zones managed in an ecologically intelligent manner, offer the best hope for protecting sensitive species and intact ecosystems (Noss, 1992:10).

Buffers are also seen to play an important role in species conservation. Whitehouse's comments in relation to koala conservation in Australia are pertinent:

> We have to look at a continuum of conservation values across a range of tenures.... core values need to be protected in publicly owned lands.... But beyond that, we need to look at the multiple-use management of lands in private ownership for productive purposes at the same time as maintaining conservation values (Whitehouse cited in Lunney et al., 1990:195).

Grumbine's (1994) analysis of the ecosystem management concept concludes that new applied research must be initiated, a critical area being the design of a continental-scale biodiversity protection network built around a system of core reserves, buffer zones, and habitat corridors. More recently the Vth World Parks Congress (IUCN, 2003) recommended the development of an ecologically representative and coherent network of land and sea areas that includes protected areas, corridors and buffer zones, and which is integrated with the landscape and existing socio-economic structures and institutions.

Despite this wide ranging support for integrated buffers, in almost all instances, there has been no forthcoming strategy to indicate how to design natural area buffers. Further, while the term buffer has occurred frequently in planning and management literature, it has often been ill-defined and very broadly and randomly applied to many situations, ranging from zones to protect ESAs, to areas or physical features designed to minimise noise impacts from major transport arterials.

This gap, both in the theoretical and practical application of buffers prompted Peterson's (2002) research to identify a 'best practice' integrated buffer zone approach. She arrived at an innovative methodology, which drew on several of the aspects used in previous buffer zone applications, in particular that of BZP (Kozlowski and Ptaszycka-Jackowska, 1981).

The purpose of this chapter is to describe the benefits of buffer zones that are in place to conserve the values of ESAs, to outline several important buffer criteria that can be used to develop a 'best practice' integrated buffer planning method, to briefly evaluate several buffer approaches and to identify key features that can be incorporated into a recommended integrated buffer planning method.

What are the Benefits of Buffers?

While there is a range of differing views concerning the nature and design of buffers for ESAs, there is some agreement on the values or benefits that buffers can bring. Their major role in the conservation of ecologically sensitive areas (ESAs) has already been discussed in Chapter 1, but some of the more important benefits are further elaborated below.

Long-Term Conservation of Ecological and Cultural Values

The establishment of effective buffers will aid in the long-term conservation of the values of the core area that is being afforded some higher level of protection from threatening processes. In particular, buffers will enhance the conservation of species and genetic diversity and aid in the maintenance of ecosystem functioning and processes, all of which will contribute to improved regional biodiversity. Core and buffer areas that are integrated with their surrounding areas and linked across the landscape may help to facilitate the movement of wide-ranging and more generalist species. The main way that buffers achieve these outcomes is through eliminating or minimising the impacts of threats such as pest species, clearing of vegetation and water pollution on identified core areas and by integrating the core with its surrounding landscape and human communities.

Facilitate Long-Term Natural Processes

Many components of the landscape are subject to dynamic processes. For example, coasts and rivers are subject to erosional and depositional cycles that result in changing positions of the coast and inland waterways. Appropriate buffers to these landscape elements will ensure the continuance of dynamic processes and protect adjacent land uses, infrastructure and communities, as well as the ecological values of the particular resource or area that is being buffered. It will also result in a reduced need for often expensive erosion control devices such as groynes, sea walls and submerged reefs.

Adaptation to the Impacts of Climate Change

The buffering of natural habitats and areas subject to the impacts of global climate change, including predicted sea level changes, will enable plant and animal species to track environmental gradients and hence adapt to predicted spatial changes in habitat. In coastal environments, where the impacts of climate change may be severe, the provision of adequate buffers will enable the dynamic processes of coasts to continue, while minimising negative impacts such as erosion and the adverse impacts of flooding associated with storm tide events on coastal communities.

Economic Benefits

A range of economic benefits may result from the introduction of buffers. These benefits will vary depending on the type of buffer, its function, location in the landscape and the type of activities undertaken in the buffer and surrounding areas. One of the main benefits is that the quality of the habitat that is being buffered will be enhanced and may enable the sustainable use of the resources of the core and its buffer. These benefits may be in the form of improved surface and groundwater quality and quantity, and increased integrity of timber and forest products and fisheries resources. Flow on benefits may include reduced expenditure on erosion and sediment control mechanisms in degraded catchments, reduced risks to local communities from flooding and other natural disasters, stabilisation of shorelines and riverbanks, improved agricultural and fisheries production, and enhanced opportunities for recreational and tourism industries to establish in well managed natural environments. Landowners may also gain from improved land values, which are the result of more sustainable natural and cultivated environments.

In many countries landholders who implement buffer management actions may gain financial benefits from reduced taxes, rate rebates, subsidies for a range of activities including fencing, weed and fire management. Such benefits act as an incentive to encourage widespread participation in the management strategies promoted in the buffer area.

Social Benefits

Many benefits may accrue to communities living both within the core and buffer area. Where communities have been effectively involved in buffer design and implementation, there will be mechanisms established to effectively deal with any potential conflicting interests arising between those wanting further development and those wanting improved conservation outcomes for the area. Development projects directed to improving sustainability in buffer areas may produce many benefits for local communities. Such communities may be compensated for their contribution to enhanced biodiversity outcomes and may achieve more secure tenure arrangements as a result of buffer negotiations. In addition, where sustainable land use practices are in place local communities may have access to a range of traditional resources that will enable the continuation of traditional cultural practices. The support of local communities living within buffer areas will also build support for improved conservation outcomes.

The Concept of Buffers for Ecologically Sensitive Areas

Although the term buffer has been used in a variety of contexts, some of the more commonly used buffer definitions include:

> A zone, peripheral to a national park or equivalent reserve, where restrictions are placed upon resource use or special development measures are undertaken to enhance the conservation value of the area (Sayer, 1991: 2).

....vegetated filter strips or zones located between natural resources and adjacent areas subject to human alteration (Castelle et al., 1994:880).

...an area of land separating adjacent land uses, that is managed for the purpose of mitigating impacts of one use on another. A buffer area consists of a separation distance and one or more buffer elements (DNR and DLGP, 1997:4).

A fisheries management agency defined a fish habitat buffer as:

the area of land separating adjacent land uses from wetlands and fish habitats, which is managed for the purpose of mitigating impacts of one use on the other... (Bavins et al., 2000:17).

An effective buffer width for this fish habitat was defined as one 'that mitigates all (or most) of the impacts that have been identified, and provides benefits for each function the buffer is to perform' (Bavins et al., 2000: 11).

The term integrated buffer zone as used in this book is defined as:

...a zone peripheral to an environmentally sensitive area (ESA), where the local community and key stakeholders are effectively engaged in ensuring the long-term sustainability of the ESA and the conservation of its values (e.g. social, cultural and environmental), through the implementation of policies or measures, which are designed to eliminate, or minimise threatening processes (existing and potential) and which integrate the ESA with its surrounding landscape. The buffer reflects the area of operation of the identified threatening processes and may be adjacent to the ESA, or at some distance from it.

Such an integrated buffer zone, will enhance the values of the ESA and this may, in some areas, compensate the local population for any restricted access to and use of the reserve areas.

In order to devise a model buffer planning approach it is first necessary to establish the criteria that should underpin an integrated buffer, as defined above. These criteria, which are described below, incorporate the following:

- the principles of sustainable development (Table 2.1);
- the principles of good planning (Table 2.3);
- the key features of the 'model planning process' (Table 2.4); and
- broad ecological concepts (outlined in Chapter 3).

'Best Practice' Buffer Criteria

Recommended 'best practice' buffer criteria are explained in this section. The buffer is expected to fulfil the following criteria.

Address Heterogeneous External Threats

Buffers designed to protect the values of ESAs should identify and minimise the impact of all major threats that originate outside the boundaries of the ESA. These external threats frequently arise from land uses and activities that conflict with the values of importance in the ESA. This approach is frequently used in buffer planning, where the buffer is designed to extend the area of influence of the ESA. Typically however, such extension type buffers have focussed on minimising the impact of one or a few major threatening processes, and this may not ensure the long-term conservation of the ESA, which may continue to the affected by a variety of other threats.

Neumann and Machlis (1989) in their examination of external threats to parks found that a program targeted at reducing poaching would only partially solve the problem of declining species diversity, as a number of other land uses contributed to the problem. This concept is illustrated in Figure 6.1 where a core ESA is to be protected from external threatening processes. Part A illustrates a buffer that has been designed to minimise the impact of fire on a core natural area. It extends outwards from the boundary of the ESA and surrounds it. Such a buffer is however, unlikely to provide effective protection of all the values of the ESA. For example, the ecological functioning of the core may be negatively affected by water pollution, the source area of which is indicated in Figure 6.1B. Thus if the buffer boundary corresponded to the area shown in Part A, threats from water pollution may not be effectively dealt with and it is likely that the values of the core would diminish over time due to the impacts from water pollution. Therefore, an improved approach, in a situation where the only significant threats to this core are fire and water pollution, would be to include the areas where fire and water pollution threats originate. The resultant buffer (Figure 6.1C) would integrate the areas that are the source of fire and water pollution and in such a way would reflect and respond to the heterogeneous threats affecting the core. This approach was the basis of the Buffer Zone Planning (BZP) method developed in the 1970s by Kozlowski and Ptaszycka-Jackowska (1981).

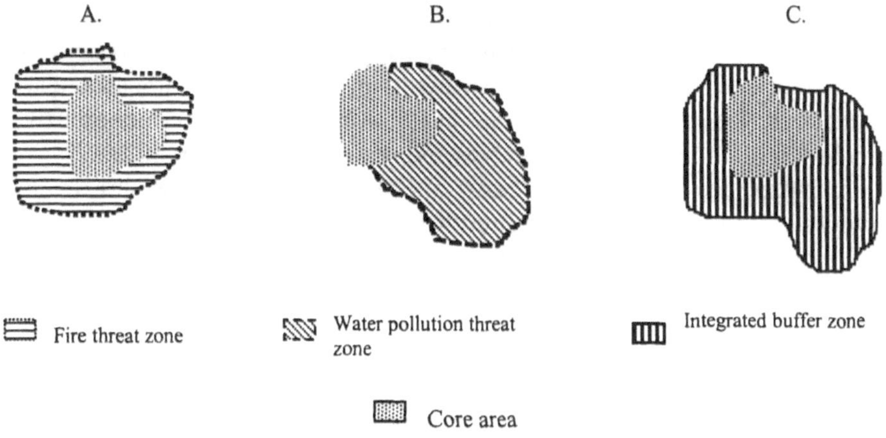

Figure 6.1 Spatial representation of threats to a core ESA: A. Fire threat zone; B. Water pollution threat zone; and C. Partial integrated buffer incorporating areas subject to threatening processes related to fire and water pollution

Buffer boundaries, in practice, are frequently based on a fixed distance (e.g. 200 metres) from an ESA, a practice that is difficult to justify on ecological grounds. In contrast, an effective buffer design should consider and respond to the areas that are the source of all major threats to the ESA. However, for threats such as water pollution, which may be the result of widespread input of nutrients and sediment, this may result in the entire catchment being included within the buffer for say a coastal wetland. Thus, it may be impractical in large catchments to formally include all this area in the buffer and emphasis may need to be placed on the major sources of water pollution threat to the particular ESA, within the catchment. The broader issues of water quality degradation may then be addressed through a catchment wide management strategy.

Good buffer design should also consider the temporal needs of species (both plant and animal) within the ESA. For instance the survival of a mobile species may be dependent on a seasonal or migratory feed source. Where threats to these areas are evident, for example as a result of clearing of important food trees, or introducing a new tourist trail, it is necessary that these threats are taken into consideration in the design of the buffer. Where transboundary sites are identified, it is imperative that these are included in the buffer and appropriate implementation strategies devised.

Once the specific threat zones are identified appropriate policies are needed to minimise the overall impact of the threatening processes. Although such policies may be difficult to implement in transboundary situations, the effectiveness of the buffer depends on the comprehensive inclusion of all threatening processes and relevant policy responses.

A buffer designed on the basis of identifying all major threatening processes and their sources will result in the final buffer being a synthesis of individual threat zones and the buffer's policies being specific to particular threats and their points of origin. As each buffer will be individually designed, its shape will be variable, dependent on the spatial variation in threatening processes and relevant control measures. In some cases, there may be no necessity for the entire buffer to surround the ESA or to be adjacent to its boundaries. The buffer, or part of it, may also be at some distance from the core habitat. In an urban context, protecting skylines is a typical example of this.

In summary, good buffer design should recognise environmental heterogeneity. Buffer boundaries should be based on identifying all major threats to a core natural area, delimiting the source of each threat and developing policies to eliminate or reduce the impact of each threat on the ESA's elements, features and values.

Address Threatening Impacts of the Core on its Surroundings

As well as identifying external processes that threaten ESAs, the buffer methodology should examine any negative impacts that the ESA may have on its surrounding environment, including its human communities. The establishment of formally protected areas frequently curtails the use rights of adjacent local communities, thus impacting on their social, economic, cultural and religious lives. Some communities living adjacent to ESAs therefore, may have a low 'wildlife acceptance capacity', especially if subject to invading native animals, feral animals and escaping fires. These events may produce negative attitudes within the community to the ESA and prevent the implementation of effective conservation strategies to enhance the values of the ESA.

A number of wildlife conservation strategies have faltered due to this failure to fully understand a community's wildlife acceptance capacity. For example, support for the conservation of the Florida panther *Felis concolor coryi* is far from universal as private landowners fear losing land use opportunities where management strategies are implemented to link and buffer protected areas (Maehr, 1990). Project Tiger in India, established parks and associated buffers to give a high level of protection to the tiger and its habitat. In many instances this involved removing local people from their traditional lands and severely limiting their access to and use of the protected areas and buffers. The original buffers were designed primarily to minimise threats originating external to the parks. However, local community antagonism to the project increased, largely due to its failure to recognise that the strategies implemented to protect the tiger were threatening the surrounding community's standard of living and survival chances. As a result, hunting of tigers increased. Gradual recognition by project leaders that the park was impacting negatively on the community and also that threats from areas outside the park were impacting on park resources, produced a change in buffer zone management to one that incorporated local villager concerns. Similar negative attitudes were also common in the region surrounding Tatry National Park (Poland), where conservation measures had negative impacts on the lucrative tourism industry, an important source of income for local communities.

Preventing and mitigating human-wildlife conflicts was a key theme of the Vth World Parks Congress (2003), which recommended that governments and conservation authorities at the local, national and international level recognise the need to alleviate human-wildlife conflicts, prioritise management decisions, planning and action to prevent and mitigate this conflict, and incorporate global, regional and local mechanism to ensure that the issues are addressed properly. Effective buffer design is an appropriate planning tool to achieve more sustainable outcomes in relation to these issues.

In summary, good buffer design should take into consideration the existing and potential impact of the ESA and its resources on adjacent local communities, land uses and activities. The buffer's policies should be designed to reduce or minimise the threat of internally originating threats on surrounding areas to ensure joint compatibility between any conflicting land uses, for the cooperation of local communities is essential to the long-term success of the buffer.

Incorporate Sound Ecological Principles

One important purpose of buffers is to ensure the perpetuation of ecological and evolutionary processes and hence such zones should be designed on the basis of generally supported principles and theories of landscape ecology, ecosystem management and conservation biology, rather than arbitrary distance measures. Although these sciences are relatively new and lacking in a comprehensive theoretical base, results from a growing number of case studies and observations are leading to general principles of resource management that may be applied to the design of buffers.

The development of an effective buffer also requires good data. However, in many countries appropriate quantitative data on regional ecosystems and their component species, on the processes operating within ecosystems and landscapes, relevant trends and limiting factors are lacking. Good quality, or even adequate knowledge or data are not always available, and insufficient time and finances frequently preclude comprehensive ecological studies being undertaken. Hence, in developing an integrated buffer, planners may have to deal with imperfect information and make recommendations before being completely satisfied with the theoretical and empirical bases of analysis, for if the problem is ignored until sufficient scientific data are gathered, future planning options may be critically reduced or eliminated, or the regional ecosystem/species may become extinct. Perhaps the most prudent course for conservation 'is to proceed on the basis of the best available information, rational inference, and consensus of scientific opinion about what it takes to protect and restore whole ecosystems' (Noss, 1992:11), and to continue to upgrade existing information and to review the buffer plan as new information becomes available.

Effectively Involves Key Stakeholders

ESAs are increasingly surrounded by diverse communities having complex ethnic, social, economic and biological influences on and interactions with the ESA. Many of the major managerial problems facing protected areas, for example, have a human component (Lusigi, 1981; Sayer, 1991a; Newmark et al., 1993; IUCN, 2003). If these complex relationships are not understood and adequately incorporated into the planning process quite disastrous consequences may result, especially in terms of their impact on ecological processes within the ESA. These social factors are as relevant to developing effective buffers as are the biological dimensions. Hence planners must begin to acknowledge that the options for conservation of important natural areas and associated buffers will be prescribed by the needs and aspirations of the surrounding human communities, and will not be based purely on ecological considerations.

An adversarial role has frequently arisen in many developing countries where local communities may be hostile to the objectives of the protected area system and in particular its wildlife (Lusigi, 1981). Local people frequently see parks as 'government-imposed restrictions on their legitimate rights' (Sayer, 1991a:1). The suggested implementation of a 200 metre buffer zone adjacent to the Wet Tropics World Heritage Area in north Queensland (Australia) resulted is extensive pre-emptive clearing by landholders in this buffer and the destruction of forests of great biological diversity. Oldfield's (1988) review of buffer zones concluded that there were few examples of stable and compatible land use systems established around ESAs in such a way that people were genuinely reconciled to the conservation function of the area. He stressed the need for greater dialogue between managers of natural areas and surrounding local people, this requiring a 'fundamental re-orientation in ... attitudes and programmes...' (Oldfield, 1988: 3). In Tanzania an important function of extension programs is to reduce conflict between people and wildlife by promoting land use practices that maintain low human density on lands adjacent to protected areas (Newmark et al., 1994). Efforts are directed at determining the needs, wants and anxieties of the communities surrounding each park, reserve or remnant habitat and creating, where possible, participatory management arrangements.

The world review of protected areas that preceded the Caracas Congress (1992) recommended that protected area management should involve partnerships with a wide range of interest groups (McNeely et al., 1994). This broad-based participation should cut across gender by ensuring that women are effectively consulted, across generations and across social strata. Hough (1988) warns of the limited success that may result if the basic requirements of the least powerful in the community are not addressed. Without this level of consultation it is likely that buffer benefits will be captured by the dominant players in the society, reducing the overall effectiveness of the buffer. However, although broadly democratic, grassroots community participation may be acceptable in some cultural traditions, it may be less successful where strong traditions of leadership exist, which emphasise age or social status over democratic participation. For example, an observed difficulty with The Wildlife Extension Project (WEP) in Amboseli (Kenya), established as a community-based participatory approach to increase Masai benefits from wildlife, was the conflict between the WEP participatory method and traditional Masai society. The project leaders frequently dealt with marginal members of Masai society and penetrated only a small section of the society and this prevented the project from addressing many fundamental wildlife and land use issues that were the prerogative of community leaders. In these situations democratic participation may be seen as culturally irrelevant or even a threat to leadership. Hence, an attempt must be made to involve relevant stakeholders.

The IUCN (1992a) recommends the establishment of 'stewardship programs' that are institutionalised in the legal framework of the appropriate jurisdictional sectors. Further, where possible, local institutions should be strengthened rather than new ones being created (McNeely, 1989). A significant difficulty with many cooperative partnership approaches is that they take a long time to develop, for attitudes and behaviours generally do not change quickly. Buffer strategies should be planned and developed within a long-term framework. Watson (1993) comments in relation to the Fitzgerald River National Park Biosphere Reserve in Australia, that local community support requires a reasonable level of community 'ownership', if not in a legal sense at least on a consultative basis or better still through self motivation. Good communication links between the managing agency and local people are essential and there needs to be opportunity for public involvement in planning and management. This includes local liaison with a variety of groups (local government, park user groups, fire brigades, neighbours and visitors), public involvement in the preparation of planning tools and in the plan approval process, establishment of an advisory committee, as well as a community information network. Watson also recognises that such strategies take time to put into place. However, this is time well spent, for community commitment and active involvement will be crucial to the success of the buffer strategy.

As people tend to support what they believe to be valuable, especially if the value they perceive accrues to themselves (Munro, 1992), it is also important that clear links are made between improved community well being and the policies in place in the buffer. These values may include the provision of employment or income, opportunity for recreation, wilderness experience and improved water quality.

In summary, effective local community and other stakeholder involvement will help to ensure that the buffer's policies will better meet the needs of these groups and will be suited to local conditions. Effective participation creates ownership of the plan and a more cooperative relationship between the managers of ESA and local people and thus enhances compliance and makes enforcement more acceptable. It is not the purpose of this book to document community participation strategies, but rather to highlight the essential role they should play in natural area buffers.

Be Dynamic and Responsive to Change

As ecosystems are continually changing, buffer planning must consider the constantly altering nature of the environment, and planners must understand not only the current characteristics of ecosystems, but also their future potentials. In addition, an important component of any buffer system must be a consideration of likely future land uses and activities within the surrounding area and anticipated social, cultural and political changes. This will help to ensure that buffer land use policies and measures respond to anticipated developments and are proactive in minimising future threats to the ESA.

The buffer process must also include continual review of plans to ensure that they are meeting their objectives. Hence, monitoring programs must also be established to analyse and evaluate changes that may be occurring within the buffer and its core. Systems to be documented over time may include soil, water, vegetation, wildlife and air as well as social systems relating to issues such as health, education and welfare. Careful long-term monitoring of external threatening processes is also critical to determine the effectiveness of buffer policies. Monitoring programs should establish base line conditions and performance indicators against which even slight changes in identified parameters can be detected. Monitoring should also ensure that any changes can be identified not only spatially, but also temporally. Monitoring may include the use of areal photographs or geographic information systems (GIS) to document changes in land use patterns and processes, particularly within the surrounding landscape. It is only through long-term monitoring that changes may be detected, thus forming the basis on which modifications can be made to the buffer plan's goals, objectives and management strategies. Where data are lacking on the status of environmental indicators, the buffer process should deal cautiously with risk and irreversibility by paying due regard to the precautionary principle.

A lack of appropriate data and the results of the monitoring program can also be considered as triggers to initiate appropriate research, which can provide new information to guide the integrated buffer planning process. The results of this research should be integrated into all stages of the planning process, including problem identification, problem solving and implementation.

Has a Methodology that Allows Effective Application

Buffers should be based on a suitable and user-friendly methodology. The use of a model approach, including a practical, working guide, involves establishing principles, basic steps and decision rules through which the planner or planning team progresses to develop the required plan or planning outcome. Some of the more important benefits of using a model process for designing buffers include:

- *Logical progression*
  By establishing an 'information processing' structure consisting of systematic phases and steps, where each phase is based on the results of preceding phases, those involved in the design and implementation of buffers are able to understand the process of design and the decision points and data input stages. A sequential progression promotes internal coherence, the ability to verify results and to introduce modifications as needed. These modifications may result from monitoring and research programs, the results of which are then included in all stages of the model process to ensure that the planning process does not end with a 'one-off' plan, but rather becomes a cyclical process of plan preparation, plan review and plan modification. It should also be noted that the plan sequence does not need to be linear, as some parts of the process may be completed concurrently, while other parts are eliminated.
- *Efficiency*
  By identifying several logical stages, complex planning problems can be placed into a more manageable framework, thus simplifying the task. This reduces the breadth of analysis and focuses attention on critical or important aspects. The integrated buffer model, supported by a guide, will help to organise the pieces into a coherent picture of the whole situation.
- *Rational data gathering*
  Use of a model helps to clarify the types of data that need to be gathered and to prevent the collection of data as an end in itself, or even the collection of the wrong data. A model/guide enables the planner to find the important pieces of information on the policy problem and what is being done about it, and this helps in identifying what pieces are important to understand and solve the problem, but are missing in the current literature (Brunner, 1995). It also helps to indicate when in the process particular data are needed, thus making in possible to rationalise the whole program of data gathering.

- *Plan consistency*
  The use of a rigorous procedural framework permits designers to achieve a consistency of plans within an organisation and, perhaps more importantly, assists in indicating the role that the buffer may have

in relation to other activities, plans and programs within an organisation.

- *Transparency*
  Use of a model allows the process of developing the buffer plan and the resultant zones, policies and land use measures to become visible and hence more accountable. The model process decreases the subjectivity involved in the development assessment process within areas identified in the final buffer plan. Use of a model process thus may lend objectivity to decision making.
- *Community and other stakeholder involvement*
  With a more transparent, open process, community and other stakeholder involvement may be facilitated. Individuals and groups may better understand the role of the buffer plan and can focus attention at various stages of the model process.
- *Effective implementation and daily management*
  Understanding the structure of the model will aid its implementation. Complex issues may also be resolved by relying on the final plan and its policies. In relation to the IBP process, it is also recommended that the final plan is formally approved and included within the statutory planning framework to aid implementation and compliance.

An Evaluation

Planning practitioners are implementing buffer zones for a range of environmentally sensitive areas (ESAs) and are utilising a variety of approaches. In this chapter several of these buffer approaches are critically evaluated using the criteria outlined above. The buffers that have been selected for review include:

- strategies advocated by international agencies and which are widely implemented e.g. biosphere reserves (refer to Chapter 4);
- strategies linked to development programs in developing countries e.g. ICDPs and the 'core-buffer-multiple use zone' approach (Dang, 1991) (refer to Chapter 4). These approaches highlight attempts to integrate conservation and development, within a buffer framework, to achieve sustainable long-term outcomes for people and nature;
- approaches that may provide a strong theoretical basis to good buffer design e.g. 'multiple use module planning' (MUM) (Noss and Harris, 1986) and ' buffer zone planning' (BZP) (Kozlowski and Ptaszycka-Jackowska, 1981);

- buffers designed specifically for animal wildlife; and
- buffers described in the questionnaire responses from Australian planning and land management agencies (refer to Chapter 5).

This evaluation of practice-based approaches, in combination with current principles of landscape management and good planning, provides a solid foundation for identifying a 'best practice' methodology for buffering ESAs.

A summary evaluation of several buffer approaches is outlined in Table 6.1. Each approach is assessed on the basis of the six suggested buffer criteria. A four-point rating scale is used ranging from 0 (the element was not present) to 3 (the element was comprehensively covered). No attempt is made to provide an additive score for each buffer, as this may be misleading. The purpose is to evaluate the effectiveness of each buffer strategy in relation to each of the six criteria, in order to identify the desired attributes of good buffer design for ESAs.

The results of the evaluation indicate that almost all buffers for ESAs attempt to assess external threats to the core area. However, only the buffer zone planning (BZP) model includes a comprehensive examination of a full range of external threatening processes to ESAs and incorporates a technique for mapping the source areas of each threat. Individual buffers developed for specific wildlife also frequently examine a range of threats to the particular species. In relation to internal threats, the biosphere reserve model and several ICDPs are based on a sound understanding how ESAs interact with their surroundings, although a model process for mapping internal threats is not available.

Ecological processes are examined in a comprehensive way in several applications of the BZP model, as well as most wildlife specific buffers. However, in the latter, processes that relate to the specific target species are examined, rather than wider regional biodiversity issues.

Many buffers incorporate consultation with key stakeholders. Biosphere reserves and ICDPs, in particular, rely on community education strategies and associated development programs to link the benefits of conservation with development outcomes. The BZP model addresses the need for the buffer planning approaches to be dynamic and responsive to changing circumstances.

Table 6.1 Summary evaluation of 'practise-based' buffer plans and strategies for Environmentally Sensitive Areas*

Buffer Approach	Buffer Criteria					
	Ext'nal Threat	Int'nal Threat	Ecological Processes	Stake-holders	Adapt-ive	Method-ology
Internal Zoning Approach (Foster, 1973)	1	0	1	1	1	1
Wildlife Conservation Unit Approach Kenya (Lusigi, 1981)	2	1	1	N/A	0	1
Buffer Zone Planning						
Poland (Kozlowski et al., 1981)	3	0	2+	3	3	3
Nicoll Rainforest (Roughan, 1986)	2	0	2	N/A	2	3
Cooloola National Park (Peterson, 1991)	3	0	3	N/A	3	3
Fraser Island (Hruza, 1993)	3	0	3	N/A	3	3
Ipswich (Vas Bowen, 1994)	3	0	1	N/A	3	3
Brisbane (Izatt, 1996)	3	0	1	N/A	3	3
Biosphere Reserve						
Global (Wells & Brandon, 1992)	1+	2	1+	1+	1	0
Australia - General (Longmore, 1993)	1+	2	1+	1+	1	0
Fitzgerald River (Watson et al., 1995)	2	2	2	3	2	0
Bookmark (Bookmark Biosphere Trust, 1995)	2	2	2	3	2	0
Kosciusko	2	2	2	3	2	0
Multiple Use Modules						
Florida - S.Georgia (Noss, 1992)	2	0	1+	N/A	2	1
Mesoamerican Biological Corridor (Marynowski, 1992)	2	0	1+	N/A	2	1
Integrated Conservation and Development Programs						
General Africa, Asia, Latin America (Wells & Brandon, 1992)	2	2	1+	1+	1+	1
Campfire Program - Omay Province (Sibanda, 1995)	2	2	2	2	2	1
Campfire Program - Geruve Province (Metcalfe et al., 1995)	2	2	2	3	2	1
Zambia Game Management Areas (Lewis et al., 1990)	2	2	2	3	2	1
Core-Buffer-Multiple Use Model - Project Tiger (Panwar, 1987)	1+	2	1	1	1	0
Dudhwa National Park (Dang, 1991)	1+	2	1	1	1	0 /contd

Buffer Approach	Buffer Criteria					
	Ext'nal Threat	Int'nal Threat	Ecological Processes	Stake-holders	Adapt-ive	Method-ology
Wildlife Specific Buffers						
Fish habitat (Bavins et al., 2000)	1+	0	2	2	1	1
Bald eagle (McGarigal, 1991)	1+	0	2+	0	1	1
Cetaceans (DEH, 1998)	2	0	2	2	2	1
Heron (Voss et al., 1995)	1	0	1	0	1	1
N. spotted owl (Bart, 1995)	1	0	2	0	1	1
N. right whale (Smullen, 1996)	2	0	2	0	1	1
Pelican (Anderson, 1988; Erwin, 1989; Rodgers et al., 1995)	1	0	2	0	1+	1+
Turtle (Burke et al., 1995)	1	0	2	0	1	1
Australian buffers						
Dorrigo National Park (NSW)	2	0	2	N/A	N/A	2
Grampians Surround Strategy (Vic)	2	1	1	1+	2	0
Environmental Protection Agency (Qld)	2	1	1	0	1	0
Cranbourne Gardens Annex Policy Area (Vic)	2	0	1	1	1	0
Campbelltown koala buffer (NSW)	2	0	1	0	1	0
Malleefowl buffer (NSW)	2	0	2	3	2	0
Yellow bellied glider buffer (NSW)	2	0	2	0	0	1
Leadbeater's possum buffer (Vic)	2	0	2	0	1	1
Tiger quoll buffer (Vic)	2	0	2	0	1	1
Superb parrot buffer (Vic)	2	0	2	0	1	1
Powerful owl (NSW)	2	0	2	0	1	1
Osprey (NSW)	2	0	2	1	1	1
Hastings River mouse (NSW)	2	0	2	0	1	1
Yellow footed rock wallaby (SA)	2	0	2	1	1	1
Marbled frogmouth (Qld)	2	0	2	0	1	1
Forest logging buffers (Qld)	1	0	1	0	1	0

(* This table does not include 'separation zones' and 'remnant habitat strips', which as argued in Chapter 4, cannot be seen as buffers in the terms of their definition in this book. Where: 0 = not present; 1 = limited coverage; 2 = medium coverage; 3 = comprehensive coverage; N/A = information not available; + has a high value within this range; NSW, New South Wales; Vic, Victoria; Qld, Queensland, SA, South Australia.)

In relation to the presence of a practical buffer methodology, most approaches are non-systematic. There are few examples of a recommended process for buffer preparation, although the BZP model offers a useful basis for design as it incorporates a step-by-step, logical planning process. Similarly, a formal review process and effective performance measures to evaluate buffer success are not in place in the majority of approaches examined. In the absence of an identified process or model, the quality and effectiveness of buffer strategies appears to be heavily dependant on the experience and professional expertise of those involved in their preparation.

A summary of the elements of best practice design for each of the selected, more promising buffer approaches (internal zoning and wildlife conservation unit are not included) and for the Australian buffers is outlined in Table 6.2. Their positive design elements will be incorporated, where possible, into the model integrated buffer planning process to be developed in Chapter 11.

The Way Forward

The term buffer has been used in a wide variety of situations. Chapter 4 (Table 4.1) presented a classification of these existing approaches, which included separation zones, remnant habitat strips and buffers for ESAs. It was suggested that the separation zones and remnant habitat strips did not fulfil many of the criteria that have been outlined in this book as being essential for buffering important natural areas. The focus of the book has thus been on evaluating buffers that have been designed for ESAs. This chapter has highlighted several elements of best practice in these approaches (Table 6.2), as well as several deficiencies. As a consequence, the way forward for an improved approach to buffering ESAs can be presented now. This proposed new approach is that of '*integrated buffer planning*' (IBP). Table 6.3 identifies the key features of the IBP approach (key features of existing buffer approaches were presented in Table 4.4).

A new category of integrated buffer zone is recommended to effectively reflect the principles of sustainable development, good planning and effective landscape management (Chapter 2). It also incorporates the elements of good buffer design, derived from an evaluation of existing buffer approaches. The term integrated buffer is suggested as a comprehensive term to describe buffers designed to conserve the values of ESAs. The approach recognises that core areas function within a wider landscape and that buffers need to minimise not only threats that originate externally from the surroundings to the core area, but also to ensure that threats which originate within the core area are also considered and their impact on surrounding communities reduced. Integrated buffers should be designed in recognition of the ecosystem processes within the regional landscape in which the important natural area is located and place emphasis on the needs of the human communities to help ensure the continuation of ecologically sustainable land use activities. Regional landscape connectivity is an important component in the design of integrated buffers, the aim being to integrate, buffer and connect the core areas within a bioregional context.

Table 6.2 Elements of 'best practice' buffer design

Buffer Approach	Important Elements of Design
Buffer zone planning	• based on a logical and transparent buffer planning process; • recognises landscape heterogeneity; • identifies all major threatening processes and their source areas; • recognises the importance of ecological, as distinct from administrative bases to planning; • encompasses proactive elements in the buffer model.
Biosphere reserve	• based on understanding the structure and functioning of ecosystems and the impact of different types of human intervention; • examines the interactions between the core and its surroundings; • emphasises continuing stakeholder participation, and in particular the use of a wide range of strategies; • recognises the need for a long-term commitment to the development of buffers and their associated development activities; • emphasises the development of sustainable land use practices within buffer and transition areas; • places a high priority on continuing research and monitoring; and • aims to establish an international network of biosphere reserves to enhance sharing of data and planning approaches.
Multiple use modules	• uses a landscape approach to natural area planning, based on the interconnection and buffering of multiple nodes across regional landscapes and even continents; • core areas and connecting corridors are selected based on an understanding of the structure and functioning of ecosystems; • considers all land tenures for inclusion within the buffer rather than relying on government-owned lands; • transboundary buffer planning process which emphasises ecological processes rather than planning within administrative boundaries; and • some strategies use a wildlife emblem (e.g. panther) as a symbol of the project and this may promote increased community acceptance of the buffer strategy.

Buffer Approach	Important Elements of Design
Integrated conservation-development program	• human communities are included in the design and management of buffers; • recognises the need for a long-term commitment to buffers and their associated development activities; • examines the interactions between the core and its surroundings; • buffer zones may include the sustainable utilisation of natural resources rather than be 'non-development' zones; • emphasises developing land uses within the buffer that are compatible with the core area, as well as with the surrounding land uses and people; • stresses the need to identify clear links between development and conservation (e.g. any benefits to the local community that are gained from their conservation efforts are recognised by this community as being dependent on the sustainable development of their natural resources); • recognises that institutional structures may need to be modified to accommodate buffer strategies, particularly at the local level; and • communities and governments are seen as partners in developing broader regional land use plans. •
Wildlife specific buffers	• monitoring is an important aspect of long-term buffer management; • knowledge and understanding of the biology of wildlife species and their functioning within particular ecosystems is essential to the design of effective wildlife buffers; and • temporal factors are frequently considered in buffer design. •
Australian buffers (questionnaire responses)	• external threats to the core area are a basis of design; • logical boundaries are used to delimit final buffer boundaries (e.g. ridges, water features and roads); • buffer design incorporates consideration of land suitability (i.e. government ownership); and • protected area and planning legislation offers a regulatory framework for implementing buffer policies.

IBP policies should encourage the development of ecologically sustainable land use activities. Such uses should be compatible with both the management plans/strategies in place for the core area and with the goals and objectives of the surrounding communities and relevant stakeholders. It is only by ensuring joint compatibility that conservation goals will be achieved. For example, where an ESA is adjacent to rural land, the IBP should encourage uses within the buffer that are compatible with the objectives of the ESA's management policies, as well as with the rural community. Thus, rather than a 'non-development' buffer, or a general open space zone, the buffer strategy should encourage the establishment and continuance of land uses and activities that result in sustainable outcomes and which have the support of the local community.

Table 6.3 Summary of desired attributes of integrated buffers for Environmentally Sensitive Areas

Attribute	Meaning
Purpose*	• to integrate an ESA into its surrounding regional landscape and to connect it with other ESAs • to minimise impact of threatening processes (external and internal threats) on ESA
Composition*	• variable, although usually contain natural environments • land uses compatible with the ESA • may contain human communities
Structure*	• both adjacent and non-contiguous lands • based on examining externally and internally originating threats • usually heterogeneous • boundaries based on sound ecological principles • linked to other ESAs by corridors
Function*	• zone of multiple sustainable land uses • land uses and activities compatible with effective ecosystem functioning
Types/ examples	IBP model (refer to Chapter 11)

(* Where: **Purpose**: the role that the buffer is intended to play within the wider landscape. **Composition**: the variety of elements in the buffer [e.g. natural and cultural features]. **Structure**: the pattern or physical organisation of the buffer. **Function**: the roles [ecological, social and economic] that the components of the buffer fulfil in driving the processes that sustain the functioning of the core.)

In summary, the IBP method is expected to effectively integrate ESAs with their surroundings. However, the IBP process should also be used as a problem analysis, information and monitoring process to help identify land use conflicts and threats associated with important natural areas and to develop appropriate policies. It also would assist in generating buffer related research. Finally it should provide a decision-making framework to assess resource development proposals in order to achieve the conservation goals related to the natural area or target wildlife that have been set by the stakeholders.

Conclusion

Buffers are a commonly used resource management tool. Almost 70 per cent of respondents to Peterson's (2002) questionnaire believed that buffers were an important means of implementing policies to minimise the impact of external threats on core areas. This responds to continuing calls made by the World Conservation Union, World Resources Institute and others to integrate protected areas and other core natural areas into their surroundings by providing buffers. Six suggested criteria, identified as being important in underpinning the development buffers for ESAs, were used to evaluate several existing approaches. Elements of 'good practice' buffer design were identified (Table 6.2), with the buffer zone planning (BZP) approach being revealed as a promising planning method. Hence, it is on this basis that the BZP method will be examined in more detail in Part 3, to better identify its specific features that may be incorporated into a best practice IBP approach.

References

Anderson, D.W. (1988), 'Dose-response relationship between human disturbance and Brown Pelican breeding success', *Wildlife Society Bulletin*, 16, pp. 339-45.

Bart, J. (1994), 'Amount of Suitable Habitat and Viability of Northern Spotted Owls', *Conservation Biology*, 9(4), pp. 943-6.

Bavins, M., Couchman, D. and Beumer, J. (2000), *Fisheries Guidelines for Fish Habitat Buffer Zones*, Queensland Fisheries Service, Brisbane.

Bookmark Biosphere Trust (1995), *Introducing Bookmark: The Bookmark Biosphere Reserve Action Plan*, The Bookmark Biosphere Trust with Andrea Lindsay, the Bookmark Biosphere Trust, Berri.

Burke, V.J. and Gibbons. J.W. (1995), 'Terrestrial Buffer Zones and Wetland Conservation: A Case Study of Freshwater Turtles in a Carolina Bay', *Conservation Biology*, 9(6), pp. 1365-9.

Dang, H. (1991), *Human Conflict in Conservation. Protected Areas: The Indian Experience*, Har-Anand Publications and Vikas Publishing House Ptv Ltd, New Delhi.

Department of Environment and Heritage (1998), *Whale and Dolphin Conservation Plan*, DEH, Brisbane.

Forster, R. (1973), 'Planning for Man and Nature' in IUCN (ed), *National Parks: Reconciling Perpetuation and Use*, IUCN Publication No. 26, IUCN, Gland.

Hruza, K.A. (1993), 'Buffer Zone Planning. A Possible Management Tool for Fraser Island', Bachelor of Regional and Town Planning thesis, Department of Geographical Sciences and Planning, The University of Queensland, Brisbane.

Izatt, C.S. (1995), 'Planning for Protection Cultural. Heritage Precincts', Bachelor of Regional and Town Planning Thesis, The University of Queensland, Brisbane.

Kozlowski, J. and Ptaszycka-Jackowska, D. (1981), 'Planning for Buffer Zones', in P. Day (ed), *Queensland Planning Papers*, The University of Queensland, Brisbane, pp. 244-38.

Lewis, D.M., Kaweche, G.B. and Mwenya, A. (1990), 'Wildlife Conservation: An Experiment in Zambia', *Conservation Biology*, 4(.2), pp. 171-80.

Longmore, R. (ed.) (1993), 'Biosphere Reserves in Australia: A Strategy for the Future', Drawn from a report prepared for the Australian Nature Conservation Agency by Dr. P. Parker, Chicago Zoological Society for the Australian National Commission for UNESCO, ANCA, Canberra.

Lusigi, W.J. (1981), 'New Approaches to Wildlife Conservation in Kenya', *Ambio*, 10(2-3), pp. 87-92.

McGarigal, K., Anthony, R.G. and Isaacs, F.B. (1991), 'Interactions of humans and bald eagles on the Columbia River (Washington and Oregon, USA) estuary', *Wildlife Monographs*, pp .5-47.

Marynowski, S. (1992), 'Paseo Pantera. The Great American biotic Interchange', *Wild Earth*, Special Issue, Cenozoice Society, Inc., Ann Arbor, pp. 71-74.

Metcalfe, S., Chitsike, L., Maveneke, T. and Madzudzo, E. (1995), 'Managing the Arid and Semi-arid Rangelands of Southern Africa: The Relevance of the CAMPFIRE Programme to Biodiversity Conservation', Paper presented at the Global Biodiversity Forum on Decentralisation of Governance and Biodiversity Conservation, Jakarta.

Newman, B., Irwin, H., Lowe, K., Mostwill, A., Smith, S. and Jones, J. (1992), 'Southern Appalachian Wildlands Proposal', *Wild Earth*, Special Issue, Cenozoice Society, Inc., Ann Arbor, pp. 46-58.

Noss, R.F. 1992), 'The Wildlands Project. A Conservation Strategy', *Wild Earth*, Special Issue, Cenozoice Society, Inc., Ann Arbor, pp. 10-25.

Noss, R.F. and Harris, L.D. (1986), 'Nodes, Networks and MUM's: Preserving Diversity at All Scales', *Environmental Management*, 10(3), pp. 299-309.

Panwar, H.S. (1987), 'Project Tiger: The Reserves, the Tigers and Their Future', in R.L.Tilson and U.S.Seal (eds.), *Tigers of the World*, Noyes Publications, Park Ridge, pp. 110-17.

Peterson, A. (2002), 'Integrated Landscape Buffer Planning Model', PhD thesis, the School of Geography, Planning and Architecture, The University of Queensland, Brisbane.

Rodgers, J.A. and Smith, J.T. (1995), 'Set-Back Distances to Protect Nesting Bird Colonies from Human Disturbance in Florida', *Conservation Biology*, 9(1), pp. 89-99.

Roughan, J. (1986), 'Planning for Buffer Zones – An Application of Protection Zone Planning to Nicoll Rainforest', Bachelor of Regional and Town Planning Thesis, The University of Queensland, Brisbane.

Sibanda, B.M.C. (1995), 'Wildlife, conservation and the Tonga in Omay', *Land Use Policy*, 12(1), pp. 69-85.

Smullen, S. (1996), 'Endangered Right Whale Protection to Increase with 500-yard Buffer Zone to Prevent Ship Strikes', Available at: http://www.noaa.gov/public-affairs/pr96/aug96/noaa96-r158.html [8 June 1998].

Vass-Bowen, N. (1994), 'A Role for Buffer Zone Planning in Urban Heritage Conservation?', Bachelor of Regional and Town Planning thesis, The University of Queensland, Brisbane.

Voss, D.K., Ryder, R.A. and Grand, W.D. (1995), 'Response of breeding Great Blue Herons to human disturbance in northcentral Colorado', *Colonial Waterbirds*, 8, pp. 13-22.

Watson, J.R., Lulfitz, W., Sanders, A. and McQuoid, N. (1995), 'Networks and the Fitzgerald River National Park Biosphere Reserve, Western Australia', in D.A. Saunders, J.L Craig and E.M. Mattiske (eds), *Nature Conservation 4: The Role of Networks*, Surrey Beatty & Sons, Sydney, pp. 482-7.

Wells, M.P. and Brandon, K.E. (1993), 'The Principles and Practice of Buffer Zones and Local Participation in Biodiversity Conservation', *Ambio*, 22(2-3), pp. 157-162.

PART THREE

BUFFER ZONE PLANNING: THE APPROACH AND ITS APPLICATIONS

Part three provides more detailed descriptions of the Buffer Zone Planning (BZP) approach. This model represents a significant advance on many of the existing buffer planning approaches and was implemented in Poland in the 1980s. Several theoretical applications using this model were developed in Australia in the 1980s and 1990s and some of these approaches are briefly described in this part.

Chapter 7

'Buffer Zone Planning': Tatry and Gorce National Parks (Poland)

Introduction

Previous chapters have provided a wide overview of buffer zone planning approaches. Chapter 4 focused on initiatives from a global perspective, while Chapter 5 evaluated recent Australian experience in the development of buffers. The Buffer Zone Planning (BZP) approach was identified as including many important features that could be incorporated into a 'best practice' integrated buffer planning model. In particular, it is a significant advance over many other buffer approaches, being based on a sequential planning framework that is relatively easy to implement. In this chapter the historical context to the development of the BZP method is described, followed by a summary of its main criteria or principles and its original applications in Poland.

The Polish Connection

'Protection zones' became a subject of interest in Poland soon after the country became independent in 1918. Sokolowski (1923) was the first to discuss how to protect the famous Tatry Mountain Range, not only from activities occurring within the park, but also from negative impacts generated outside the park. These discussions, which continued among academics and within professional circles until the outbreak of the Second World War, also attracted strong public involvement and generated several research ventures and planning studies. One of the most active in the field was Chmielewski, an architect/planner, who after the war summarised his visions, proposals and planning studies and developed the concept of the 'protective envelope' ('*otulina*') for the future Tatry National Park (Chmielewski, 1952, 1956). Around the same period, although no legal framework existed, some blocks of land around Wielkopolski and

Kampinowski National Parks, were dedicated to 'protecting' those parks from outside threats.

These early attempts were not applied further to real-life situations, and over the entire era of socialist Poland (and even beyond) the only legal way to establish a formal protection zone for national parks was through the Planning Act, which allowed the introduction of protection zones to areas formally designated as having 'protected' status. Only a very few protection zones were introduced for some national parks and a few landscape parks. As a reaction to the lack of any formal solution to this important environmental problem, academics and professional planners tried to pave the way within their own fields, by developing the concept of protection zones and other similar ideas. Ptaszycka-Jackowska (1990a) in her major research on the 'Development of Protection Zones around Natural Protected Areas' distinguished three such additional concepts:

- *Protection through the introduction of several categories of 'protected area'*, in which the guiding principle was that 'lower level' areas become protection zones for the 'higher level' protected areas. The concept was applied to Ojcowski National Park, which was to be protected by the Jurajskie National Parks (Luczynska-Bruzda, 1976);
- *'Rim character of a protection envelope'*, where the envelope for a national park was formed by successive protective rings of decreasing protection levels. For example, the landscape park was the first level; the second was an area of protected landscape; and so on. Czemerda (1983) described the application of this concept to Babiogorski National Park; and
- *'Ecological System of Protected Areas'* (ESPA), developed by Gacka-Grzesikiewicz (1978) and Rozycka (1983). The proposed system was based on the assumption that the following three categories of protected areas were to be established: areas of strict protection (nature reserves and national parks); areas of strong protection (landscape parks and areas of protected landscape); and areas of moderate protection (zones of protected landscape). It was then recognised that areas of moderate protection should be integrated into one comprehensive nation-wide, spatial system providing a protection envelope to the areas of strong protection status. This concept represented a comprehensive, wider vision, which assumed that some categories of protected areas may function as protection zones to other areas.

It is worth mentioning that the ESPA was criticised by the State Council for the Protection of Nature. It was thought that it had the potential to be an administrative nightmare, both in terms of approval processes and management. The Council also considered that protection zones should have a subordinate, service function in relation to national parks and that an independent landscape park could not possibly be reduced to such a status.

Throughout the entire socialist period the authorities of the Polish People's Republic were unable to bring about formal legislation providing for the establishment of protection zones around protected areas. It should also be noted that at the time, interest in this subject was substantial, and was shared by various scientific disciplines, professional planning and to great extent by the general public. The view that such zones, with clearly stated aims, objectives and functions, and determined by specialists, should be an integral and mandatory element of the entire system of the protection of nature, was nearly unanimous.

During the mid 1970s, while discussions about protection zones were continuing, the Research Institute on Environmental Development in Krakow was involved in the preparation of a physical plan for Tatry National Park (TNP). This was completed in 1979 and later, formally approved. It has been until very recently, the main, official tool for the management of the park (Kozlowski et al., 1979). The question as to whether a physical park plan alone, even if it was legally binding, would be sufficient to protect the park from further damage, had to be addressed in the early stages of the planning process. As the surrounding area was fully covered by statutory plans for various administrative units it was first necessary to assess those plans to determine whether their recommendations would satisfactorily contribute to the protection of TNP's ecosystems and values from externally generated human and natural threats. A special study undertaken by the Institute's team, led by Ptaszycka-Jackowska, found that the protection was only of a token character and that protection zones, or 'buffer zones' were needed for the park. The buffer was expected to introduce concrete measures mitigating or eliminating both the existing and anticipated future threats to TNP.

Due to the absence of a legal basis for establishing such zones, no reliable and tested planning method for determining buffers in practice was available in Poland and, it was quickly discovered that the knowledge gap in this field was global. Therefore, a 'do-it-yourself' approach became, by necessity, the last resort for the planning team and, within the framework of a small research program supported, among others, by the US

Environmental Protection Agency, the method of Buffer Zone Planning (BZP) was developed and successfully tested in practice for TNP (Ptaszycka-Jackowska and Kozlowski, 1978; Kozlowski and Ptaszycka-Jackowska, 1978, 1987; Ptaszycka-Jackowska, 1990a, 1990b; Kozlowski et al., 1992) and Gorce National Park (Ptaszycka-Jackowska, 1988). These two pilot studies (Figure 7.1) are presented in some detail in this chapter.

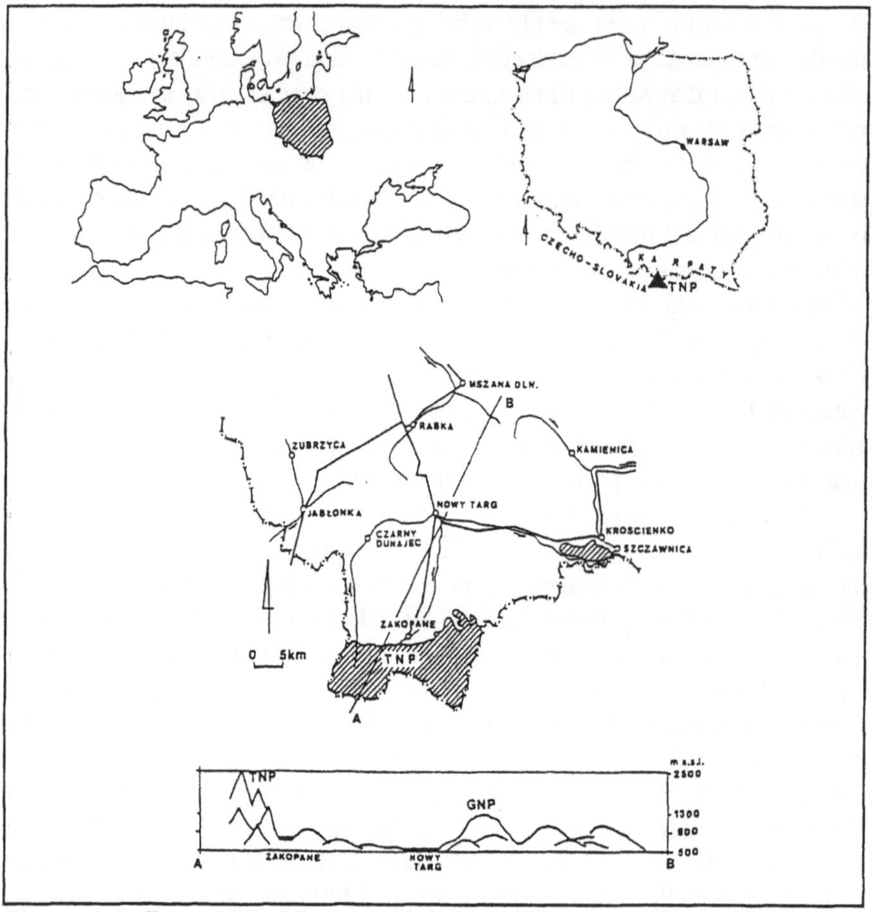

Figure 7.1 Location of Tatry and Gorce National Parks

BZP Principles and Method

The BZP method is based on the following principles:

- Legal boundaries to protected areas and reliance on internal management strategies will not provide long term protection for environmentally sensitive areas (ESAs). There is a need to eliminate or reduce externally occurring negative impacts that threaten protected areas, and to devise a set of guidelines for management within the delineated buffer.
- The nature of the environment is heterogenous both within protected areas and in their surrounds and this must be reflected in the process of defining buffer zones. This can be achieved through examining a whole range of negative external impacts on elements (such as flora or fauna) and features (such as silence) of the protected area concerned. Specific land use measures and management policies should be devised to eliminate or reduce the identified external threats.
- The needs and desires of local communities must be considered in the process of determining buffer zones and thus from an early planning stage the approach should incorporate input from the surrounding community.
- The planning procedure must take into consideration the constantly changing nature of the environment and hence existing, as well as anticipated threats to the protected area, should be incorporated to produce a more proactive form of planning.
- As the spatial distribution of both existing and potential threats to different elements of protected areas is variable, the areal extent of one negative impact (e.g. feral animals) may be quite dissimilar to that of another (e.g. fire) and hence there should be different specific zones designated to eliminate or curtail these individual impacts. Two such zones identified in the BZP method are:

  - *'Analytical' Protection Zones* (APZs), which indicate how to protect specific elements and features (e.g. hydrology, fauna or silence) of the protected area from external threats (e.g. water pollution or noise); and
  - *'Elementary' Protection Zones* (EPZs), which indicate how to protect the whole area from particular threats. EPZs can be derived through the definition of APZs, or they can be derived directly.

- Planning policies or guidelines developed for both APZs and EPZs should not be applied uniformly in the final buffer, but reflect the areas of influence of each identified threat. The final, heterogenous buffer zone should be a synthesis of the EPZs thus permitting the definition of varying protective land use measures and management policies within the buffer. These policies should, wherever possible, be incorporated into statutory development plans, which are subject to formal approval. This would establish their legal status.

Seven main steps can be distinguished in the BZP process. These are illustrated in Figure 7.2 and described in Table 7.1. (*Note*: these were briefly outlined in Chapter 4).

BZP – A Hypothetical Example

A hypothetical example of the BZP process is shown in Figure 7.3. In this example, an area of land has been provided with national park status, primarily to conserve the area's rare vegetation and fauna. The park is situated in a valley and has a relatively pristine creek flowing through it. The creek is the only source of water in the area.

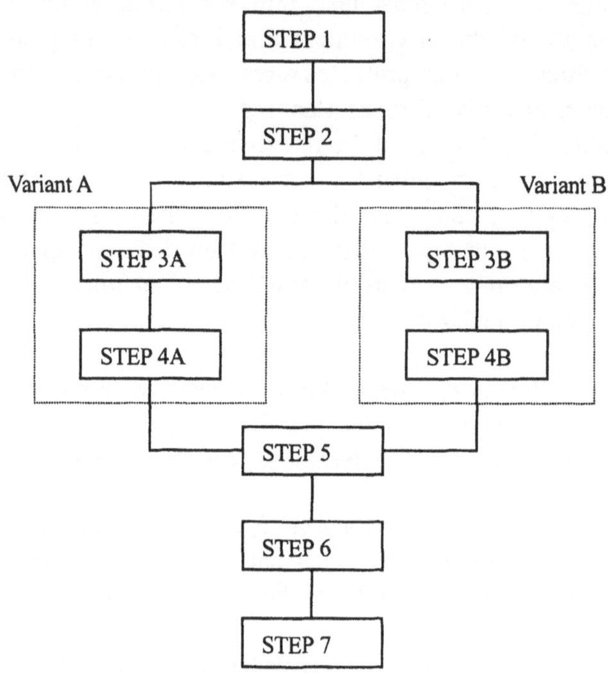

Figure 7.2 Steps in the Buffer Zone Planning (BZP) approach

Table 7.1 Explanation of steps in the BZP approach (Kozlowski et al., 1992)

Step 1: Identify the particular natural elements and features of the protected area (i.e. its essential environmental values) and the main characteristics of its surrounding territory;

Step 2: Identify the inter-relationships between these elements and features of the protected area and its surrounding territory and determine the negative impacts that this territory generates at present, or may generate in the future;

Decision Point:

Either: Define EPZs directly (variant A);

Or: Define EPZs through identifying APZs (variant B)

Variant A (e.g. Gorce National Park)	*Variant B* (e.g. Tatry case study and applications described in Chapters 8 and 9)
Step 3A: Synthesise the negative impacts.	*Step 3B*: Formulate the criteria for demarcating APZs and for defining their land use measures and management policies in relation to these negative impacts.
Step 4A: Establish criteria for the determination of EPZs in relation to these negative impacts.	*Step 4B*: Demarcate APZs and relevant measures and policies.

Step 5: Map the spatial extent of each of the EPZs for the protected area and identify the land use measures and management policies to be applied within their boundaries;

Step 6: Delineate the Buffer Protection Zone surrounding the protected area by overlapping each of the EPZs; and

Step 7: Formulate the principles or performance criteria, which are to guide the different land uses and activities within the boundaries of the Buffer Protection Zone and introduce these principles into an appropriate development plan.

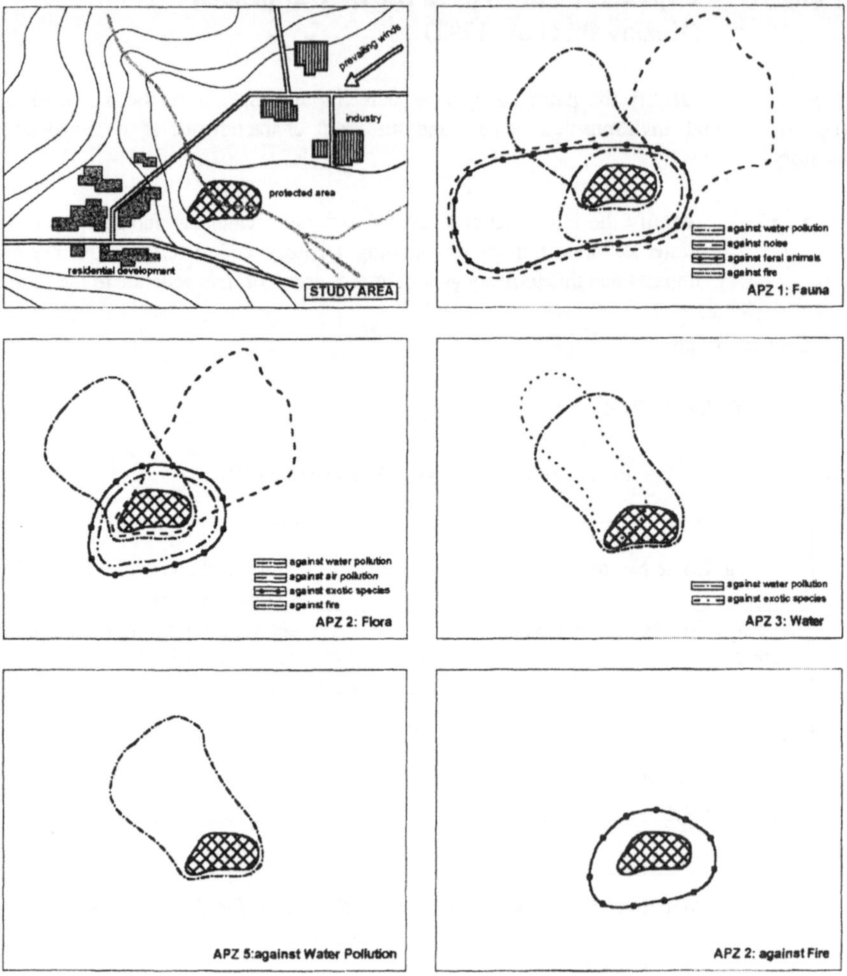

Figure 7.3 Hypothetical application of BZP

A buffer is to be designed to conserve the park's values, such as its flora, fauna and water, from potential external threats. The park is surrounded by residential development in the west and industry in the north east. The industry is a source of noise and chemical pollution, which are carried over the protected area by the prevailing north easterly winds.

As recommended by the longer version of the BZP process (variant B in Figures 7.2 and Table 7.1), APZs are first determined. Three APZs are

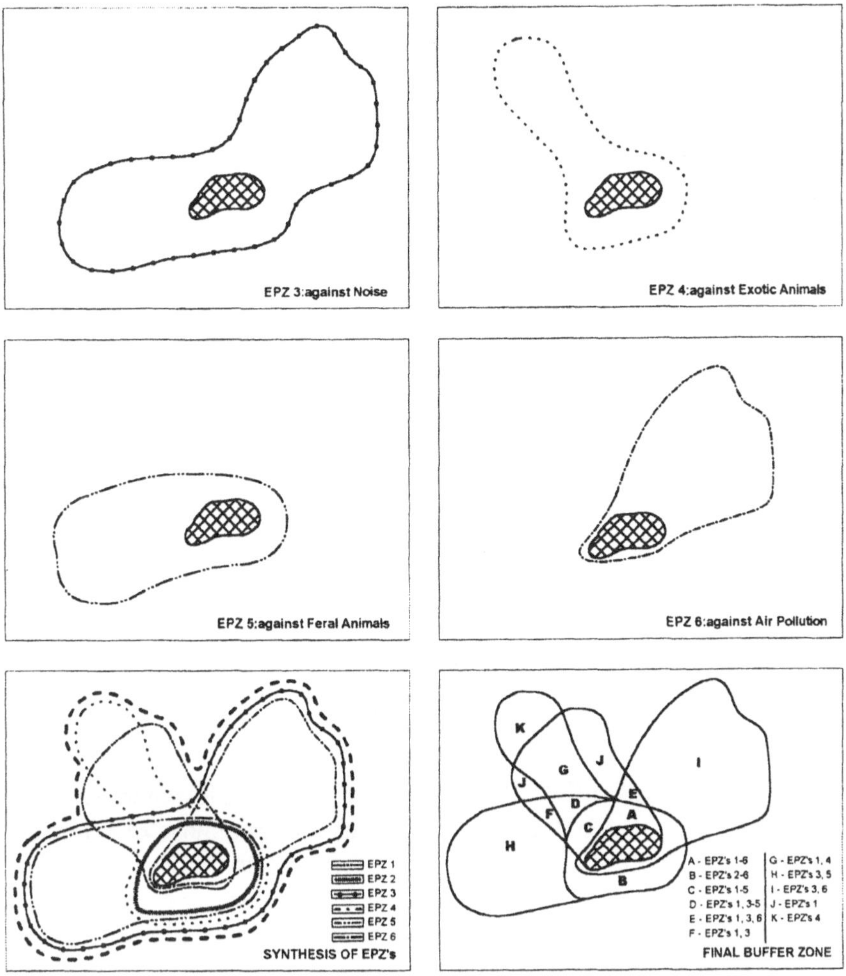

Figure 7.3 Hypothetical application of BZP (contd)

identified, for fauna (APZ 1), flora (APZ 2) and water (APZ 3). Each of the park's values is affected by a range of differing threats and the individual APZs indicate both the areas and land use policies (measures) necessary to ensure the protection of these values. As several of the threats affect each of the park's values, it is beneficial to synthesise the APZs in relation to the area of influence of a particular threat. For example, water pollution threatens the fauna, flora and water and the synthesis of these areas results in an 'EPZ against water pollution'. Using this procedure it is possible to identify six EPZs (water pollution, fire, noise, exotic plants, feral animals, and air pollution). The EPZs represent areas where a particular threat affects all the relevant identified values of the park. The synthesis of all the EPZs leads, in turn, to the definition of the final Buffer Zone, which in this theoretical example, includes eleven specific areas, or sub-zones, each with a specific set of required land use policies.

Case Study 1: Tatry National Park

The Area and its Problems

Tatry National Park (TNP) is a mountain ecosystem in the highest part of the Karpaty Range and with its alpine character and accumulation of natural values is one of the main 'natural monuments' in Poland. It is an area with a spectacular panoramic skyline, possessing outstanding rock formations, as well as sensual qualities relating to the scent of trees and plants, the hum of streams and the silence of the twilight. In addition, the area is richly endowed with an abundance of plant and animal species, many of which are rarely found elsewhere. For all these reasons, it was given national park status in an attempt to provide for the protection of its multiple values.

The southern part of TNP is bordered by the Slovakian (formerly Czechoslovakian) national park Tanap, which covers almost 75 per cent of the Tatry Mountain Range. The Polish foreland to TNP includes vast valleys (500 metres a.s.l.) and elevations (up to 900 metres a.s.l.) that are framed in the north by the forest slopes of Babia Gora (1,725 metres a.s.l.) and the Gorce Mountains (1,300 metres a.s.l.). In the north, on the border of TNP is the resort, Zakopane (30,000 population), the main centre for winter and summer tourism in Poland.

The Tatry Range, built from Palaeozoic granite and Mesozoic limestone and dolomites, was glaciated three times and has, as a consequence, unusually diverse relief and many caves. Forest, the dominant vegetation association occupies 70 per cent of the area, followed by waste land and rock (20 per cent), alpine meadows and clearings (nine per cent) and water (one per cent). Tatry's vegetation has developed in layers, including:

- lower subalpine forest (700-1,250 metres a.s.l.) of the Karpatean beechwood and fir-beech forest with spruce and sycamore;
- upper subalpine forest (up to 1,500 metres a.s.l.) of spruce wood with stone pine stands *Pinus cembra* at its upper boundary;
- dwarf mountain pine (up to 1,800 metres a.s.l.);
- mountain green and pastures; and
- peak (Rysy, the highest Polish peak is 2,499 metres a.s.l.).

In the lower and upper subalpine forests there are many clearings and several endemic mountain plant and animal species. A general layout of TNP is shown on Figure 7.1. The park covers 21,000 hectares, of which 57 per cent is fully reserved (i.e. total exclusion), and 43 per cent is partially reserved (i.e. partial exclusion). TNP is the second largest national park in Poland, although it is relatively small by world standards. It is subject to a combination of threats and pressures due to its unique scenic values, which attract exceptional tourist and recreational interest (skiing in particular). Over three million people visit Tatry each year. The structure of visitor activities during the summer season (Table 7.2) indicates an intensive general interest in this area and multiple use of the park in the form of tourism, sport and recreation. This in turn brings a number of conflicts (e.g between people and the environment, and between different user groups), which threaten the park's ecosystems. The situation is further aggravated by natural calamities such as the cyclonic type wind, called '*halny*', which is especially damaging and creates major problems. In response, during the mid 1970s the Research Institute on Environmental Development (Krakow) prepared TNP's physical plan (Kozlowski et al., 1979), which is the main, official tool for park management.

Table 7.2 Visitor activities, typical summer season[a]

Activity	People	Percentage
Hiking	700,000	31.8
Group excursions	800,000	36.4
Qualified hiking[b]	676,000	30.7
Climbing	22,000	1.0
Cave exploration	2,000	0.1
TOTAL	2,000,000	100.0

(Where: a - the volume of ski activities reaches approximately 250,000 people in the period December-April; and b - experienced hikers only) (Kozlowski and Hill, 1993)

The Approach

From the outset of the application, the main aim of the buffer zone was to protect the natural values of TNP, as these values were the main reasons why the park had been created. Although its natural environment is one ecological whole, it embodies a large range of mutually interdependent elements and features. A corner stone of the approach was the assumption that effective protection of this diverse whole required the protection of each of its basic elements and features for, due to their intricate interdependence, any significant damage to one of them might trigger a chain reaction damaging the entire natural environment as well.

Any such element or feature can be characterised not only by its different qualitative or quantitative values, but also by different ways in which the threats generated in the surrounding area can affect the element or feature. Thus the identification of the interrelations between the park's natural environment and its surrounding area was recognised as the main assumption upon which the BZP approach should be based. The planning team also agreed that the approach should proceed through two major phases. In phase one, negative external impacts on the main elements and/or features were to be identified, while in phase two, all those impacts, which often affected more than one element, would be synthesised. Relief, water, climate, soils, fauna and vegetation were the main elements and landscape, silence and tourist capacity, were the main features.

With regard to those elements and features, specialised analytical studies were undertaken to distinguish both existing and potential threats, which were generated in the park's surrounding area. In these studies it was often necessary to look separately into the various components of a given element or feature before their interrelations with the surroundings could be satisfactorily established. In the case of vegetation, for instance, it was necessary to take into account and to assess separately each of its traits. In TNP, for example, the park's five layers of vegetation were assessed.

In the park's surrounding area the natural environment, land uses and forms of human activities were surveyed and assessed not by their forms or functions (which are subject to continuous change), but by the environmental impacts they could cause. This was not limited to the existing state alone, as potential changes were also anticipated and their possible impacts on the natural environment of TNP assessed. The survey was based on the existing state of knowledge and on available research findings, as due to time and financial limitations, it was not possible to undertake new investigations. Expert advice was also extensively used.

Often it was necessary to make decisions based on analogies discovered in practice, or research conducted for other areas with similar characteristics. This was particularly useful where the impact of air pollution on various plant species was to be established. A list of negative impacts from the surroundings on the main natural elements and features of the park emerged from these specialised studies which led, to the formulation of criteria guiding the definition of protection zones (i.e. 'elementary buffers') for each element or feature. These were called 'analytical protection zones' (APZs) and included measures needed to mitigate or eliminate those impacts within the defined zones.

Examples are the APZs for soils and vegetation, which were very similar and, as a consequence, required the same land use control measures. Some APZs were, however, totally different. The APZ to protect the silence of the park was defined by delineating areas from which known, or anticipated noise could penetrate the park and be significant enough to impair this important natural feature. On the other hand, the APZ ensuring landscape protection was derived by examining the panoramic views from the surrounding areas to Tatry Mountain Range and from the range to the surrounding landscape.

The definition of APZs became an intermediate phase in the process of defining a comprehensive buffer zone, as they provided the base for delimitating the elementary protection zones (EPZs). The synthesis of present and potential negative impacts indicated, for instance, that the park's vegetation was under particular threat from the surrounding area due to air pollution and the continually growing tourist traffic, while hydrological conditions, wavering climate or the changing state of flora and fauna were subject to fewer negative impacts. Eventually, eight EPZs were determined to control the following: hydrological changes; threats to the natural silence of the park; negative impacts on corridors used by animals periodically migrating out of TNP; water management practices (responding to the demands of local population and visiting tourists); local air pollution; macro- and micro-climatic changes; the negative influence of temperature inversions; and threats to landscape values that were being strongly degraded as a result of inappropriate urban development (Figure 7.4).

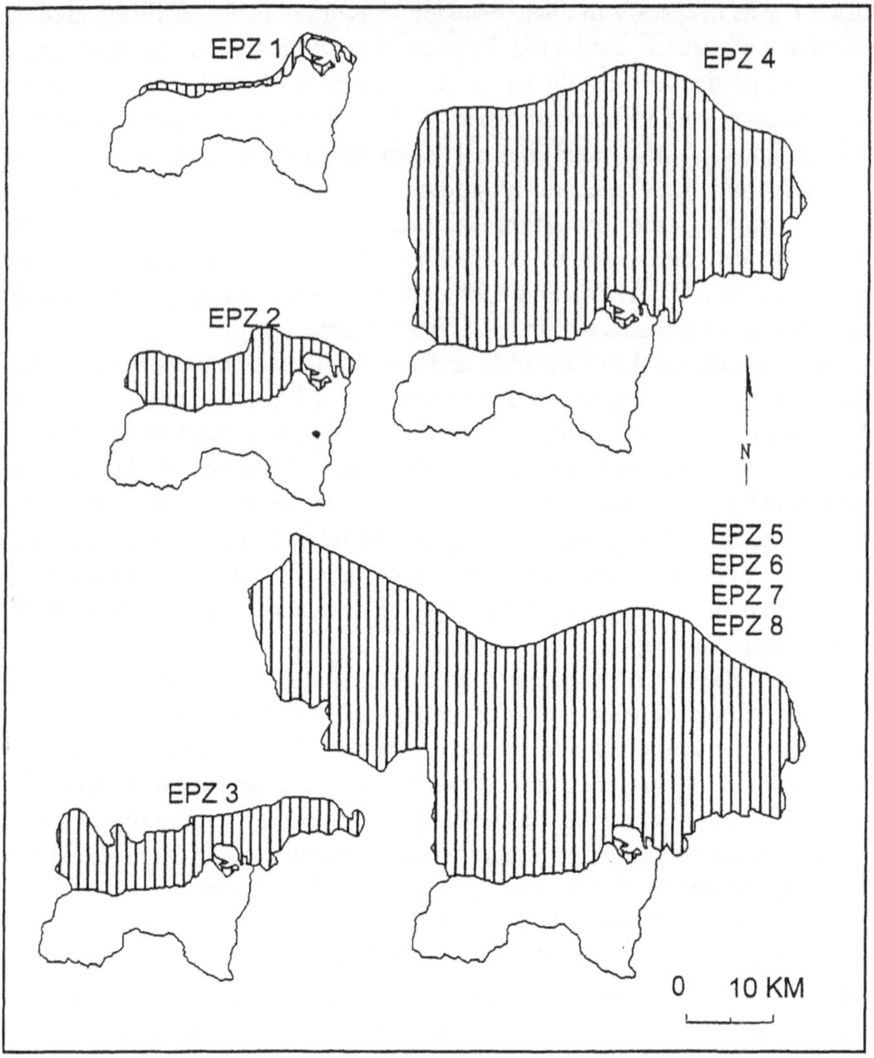

Figure 7.4 Tatry National Park's elementary protection zones
(Where EPZ 1: Safeguarding the park against the possibility of local hydrological changes; EPZ 2: Safeguarding the natural silence of the park's natural environment; EPZ 3: Safeguarding some species of the park's fauna on their external migration areas; EPZ 4: Safeguarding the park against improper water management in its surroundings; EPZ 5: Safeguarding the park against local air pollution; EPZ 6: Safeguarding the park against possible macro and micro-climatic changes on its area; EPZ 7: Safeguarding the park against the negative influence of inversions which occur in its surroundings; and EPZ 8: Safeguarding scenic views of the mountains from the surrounding area and vice versa).

Summary of the Results

The final, comprehensive buffer zone (Figure 7.5) with corresponding land use policies (measures) indicating how best to control particular negative impacts, was then derived from the synthesis of all the EPZs. It should be emphasised that the assessment - and, in consequence, development of the zones - took place only with regard to the northern, Polish part of the Tatry Mountain Range. It was not possible to gather the necessary information nor, at that time, to organise cross border cooperation with Czechoslovakian authorities responsible for the area surrounding their national park, TANAP in the south.

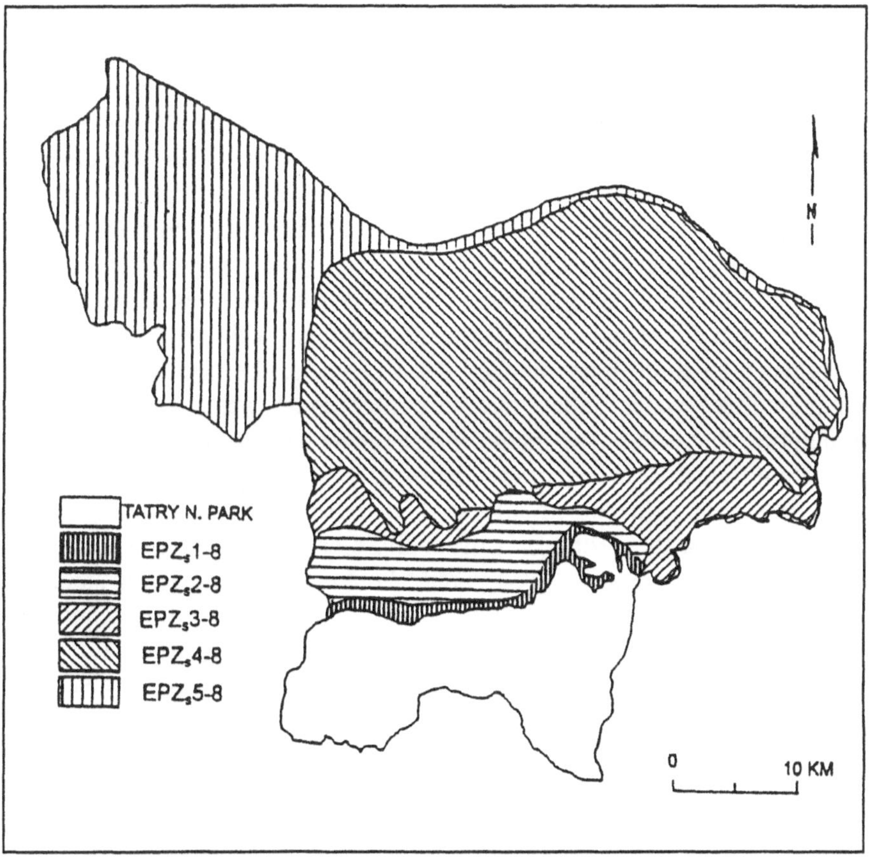

Figure 7.5 The final 'buffer zone' of Tatry National Park (Kozlowski et al., 1992)

(Where: 1. EPZs 1-8; 2. EPZs 2-8; 3. EPZs 3-8; 4. EPZs 2, 4-8; 5. EPZs 4-8; 6. EPZs 5-8)

For some threats there was insufficient data or knowledge upon which to precisely define the nature of the impact and there were insufficient resources and time to fill this gap by undertaking specific, additional studies. This made the precise definition of several EPZs difficult. Hence, it was necessary to support the process by the extensive use of environmental experts and by making hypothetical assumptions. An example is the zone defined to protect the migration corridors used by the park's fauna. Neither rangers nor forestry scientists had sufficient and reliable evidence to determine accurately the extent of those corridors and their delimitation was, at best, only 'approximate'.

A good illustration of an important contribution to be made by not limiting the planning exercise to only the existing problems was provided while determining the 'hydrological' EPZ, which was to protect TNP against a lack of water from the surrounding lands. The main threat to the hydrology was the intended development of a series of medium-sized dams on many of the creeks originating in TNP. Implementation of those projects would have produced irreversible transformations and/or serious changes in the hydrology of the entire region with unavoidable negative impacts on the park's underground and surface water systems. In addition, several relatively large areas in the park's vicinity would have become totally closed to tourists, being declared strict water reserves. This would inevitably have directed a significant part of this traffic to the park itself. To prevent the implementation of these projects, the need to develop a comprehensive water policy for the entire region was indicated on the assumption that one of the main criteria for its establishment would be to mitigate its negative influences on the park.

An important achievement of the TNP buffer zone plan was that its findings, that is, the delineation of all EPZs and their land use measures, were introduced as especially prepared '*directives*' to the existing statutory planning system. Since then the Chief Architect responsible for planning in the administrative region has been supervising the introduction and implementation of the buffer policies within the planning schemes for villages and townships situated within the buffer zone area.

Four examples of Tatry's EPZs are described in Tables 7.3 to 7.6 to better illustrate how the purpose, scope and land use policies (measures) were elaborated. Their spatial delimitation is shown on Figure 7.4.

Table 7.3 illustrates the 'silence EPZ'. 'Natural' silence means the quiet of nature, which is undisturbed by anthropogenic noise. Any natural sounds, such as those of mountain creeks, winds and storms were considered to be in harmony with nature and provided appropriate and sought after wilderness experiences. Anthropogenic noise generated by people within the park was addressed by specific measures recommended by the physical plan for the park. The 'noise EPZ' was designed to reinforce those measures by minimising the impact of noises that originated external to the park.

Table 7.3 Silence (EPZ 2), protecting Tatry National Park's silence

Purpose	To ensure that the natural silence of the park (including the sounds of creeks, wind and animals) is not disturbed by noise generated by human activities.
Scope	Due to the complexity of the relief and vegetation in the surrounding area and the lack of effective methods for measuring noise in mountain areas, the scope was established on the assumption that the impact of external, human generated noise on the park could not exceed the park's natural noise level (45dB/A during the day and 35dB/A at night).
Land use policies	• restrict motor vehicle traffic and prohibit the use of horns; • prohibit car and motorcycle rallies, choppers and planes flying lower than 4,000 metres a.s.l. (except on rescue missions); • limit the level of noise from open air concerts; and • use natural noise screens (relief and vegetation) to reduce noise.

Table 7.4 outlines issues relating to the migration routes of fauna using the park's resources. The zone included both migration corridors and temporary habitats for fauna species that did not live permanently in TNP and, thus created in these areas habitat conditions similar to those in the park itself.

Table 7.4 Migration (EPZ 3), protecting some fauna species on their migration external to Tatry National Park

Purpose	To safeguard the migration routes of animals periodically migrating out of the park and to eliminate potential negative impacts.
Scope	Areas to which TNP animals migrate periodically.
Land use policies	• restrict urbanisation and traffic and indicate places fauna crossing points; • ban harmful chemicals in farming and forestry; • safeguard clean waters, prohibit the location of refuse disposal sites and the removal (or, at least, fencing) of the existing ones; and • exclude hunting.

Table 7.5 outlines issues relating to pollution affecting the park's values. This zone was considered very important as air pollution was causing, and would continue to cause, significant negative impacts on almost the entire natural environment of the park. It directly threatened the quality of the park's waters, climate, soils and even its panoramic views, and indirectly threatened the habitat of many flora and fauna species.

Table 7.5 Pollution (EPZ 5), protecting Tatry National Park from local air pollution

Purpose	To reduce the local level of pollution (as the macro level is beyond the scope of the plan).
Scope	The territory from which the emissions of polluted air can be carried to the park area.
Land use policies	• gasification of heating in the direct vicinity of TNP (e.g. Zakopane); • exclude low quality coal for heating on the remaining territories; • restrict and control motor vehicle traffic; • provide filters to the existing industrial and service development or introduce 'clean' technologies and more restrictive standards for air pollution.

Table 7.6 addresses issues related to the landscape values of the park. This zone, due to the natural relief of the area, does not have a continuous structure as it includes only those parts from which the Tatry Mountain

Table 7.6 Landscape (EPZ 8), protecting landscape values of the surroundings (seen from Tatry National Park) and of the park (seen from its surroundings)

Purpose	To prevent further development of landscape disharmony in the park's surroundings and degradation of important panoramic views.
Scope	The visibility of the park from the surrounding land and vice versa.
Land use policies	• protect views from the surrounding lands; • protect the existing panoramic views by prohibiting development; • reduce the number of overhead power lines; • protect views from the park; • restrict development on areas seen from the park; • promote regional architecture; and • extend afforestation to areas selected by ecological, landscape and climatic criteria (air pollution plays limits visibility in a significant way).

Range could be seen and those that could be seen from the range. Non-conforming development (e.g. overhead power lines, industrial estates, and high rise residential development, in particular) were seen as the main threats that could also seriously damage the beauty and natural harmony of the existing scenery. Vernacular architectural style was considered to have considerable potential in mitigating those impacts.

Case Study 2: Gorce National Park

The Area

Gorce National Park (GNP) in the Polish Karpaty Mountains was created in 1980. It is a small park with an area of 6,000 hectares, containing the upper part of the Gorce Range (500-1,300 metres a.s.l.). It is a separate mountain range being part of West Beskidy (Figure 7.1). Ninety-five per cent of the park is covered by Karpaty beechwood, forest stands that are typical of the Karpaty Range; acid beechwood; fir-trees in the lower subalpine layer ('*regiel*'); and spruce forest in the upper subalpine layer. Forest clearings represent four per cent of the park. Numerous vegetation types are protected and endemic in the Karpaty Mountains.

The forest in Gorce was heavily exploited before the park was established. This left behind an extensive network of forest roads. Cattle and sheep grazing were also major activities in the area. At the same time, very attractive scenery led to intensive tourist activities with accompanying services, such as a well developed web of tourist trails and one major tourist hostel on the Turbacz Ridge.

GNP is surrounded by a number of villages with a strong urban character. The natural environment of the park is additionally influenced by industrial pollution from distant sources. Although the greater part of GNP is state owned, there are pockets of private or communal land and several land holders are interested in continuing their previous activities in the park.

The main forms of use in the GNP are:

- reserve management of the state forest and game;
- logging in the communal ownership areas;
- cattle and sheep grazing on the private meadows and clearings;
- tourist activities such as hiking and skiing; and
- scientific research.

The use of the park, except for reserve management and scientific research, resulted in several conflicts with the park's nature conservation management objectives. The managers of the park tried to find practical ways for solving these conflicts within the park, assisted in their endeavour by a physical plan prepared in the early 1980s by the Research Institute of Physical Planning and Municipal Economy in Krakow.

The surrounding area is characterised by mixed uses dominated by forest and agricultural uses, with two towns (30,000 and 15,000 population), one resort (20,000 population) and scattered rural settlements with numerous hamlets. Residential development in the area is fairly strong (including so-called 'second homes' built by people from large cities). It is worth noting that the entire region is of particular importance for the supply of water.

Problems and the Approach

The specialists involved in the assessment of the park's natural environment indicated, during the preparation of its physical plan, that GNP was seriously threatened by several external impacts generated in the surrounding areas. They considered these external threats as some of the main problems impacting on the park, and pointed to the need to control the negative climatic and hydrological changes, in order to prevent further atmospheric pollution and, in particular, to arrest development activities leading to the gradual disappearance of underground water in the park and its vicinity. In the latter case the key problem was to maintain the present level of water retention and to improve the water management in the region to eliminate the necessity to build new water infrastructure. Further research disclosed several negative impacts on fauna, especially on hunting game, in the areas where animals periodically migrated outside of the park's territory. Finally, due to the progressive degradation of the landscape values in the surroundings of the park, it was necessary to extend appropriate landscape management to these areas.

To address these external threats the BZP method was applied and a comprehensive buffer zone, including relevant land use policies and measures, was determined by the Institute, following a similar approach to that applied earlier for TNP (Ptaszycka-Jackowska, 1988, 1990a). The experience gained developing the TNP buffer permitted several rationalisations of the planning process. The phase in which the negative impacts were identified, for instance, was shortened considerably to focus the process on the critical interrelations between the park's natural environment and its surrounding area. Only four elements were targeted:

climatic conditions; the hydrology; fauna; and scenic values. To speed the process, 'analytical protection zones' (APZs) were not defined. The results derived from the analysis of interrelationships between the park and its surroundings lead directly to the determination of four EPZs (Figure 7.6).

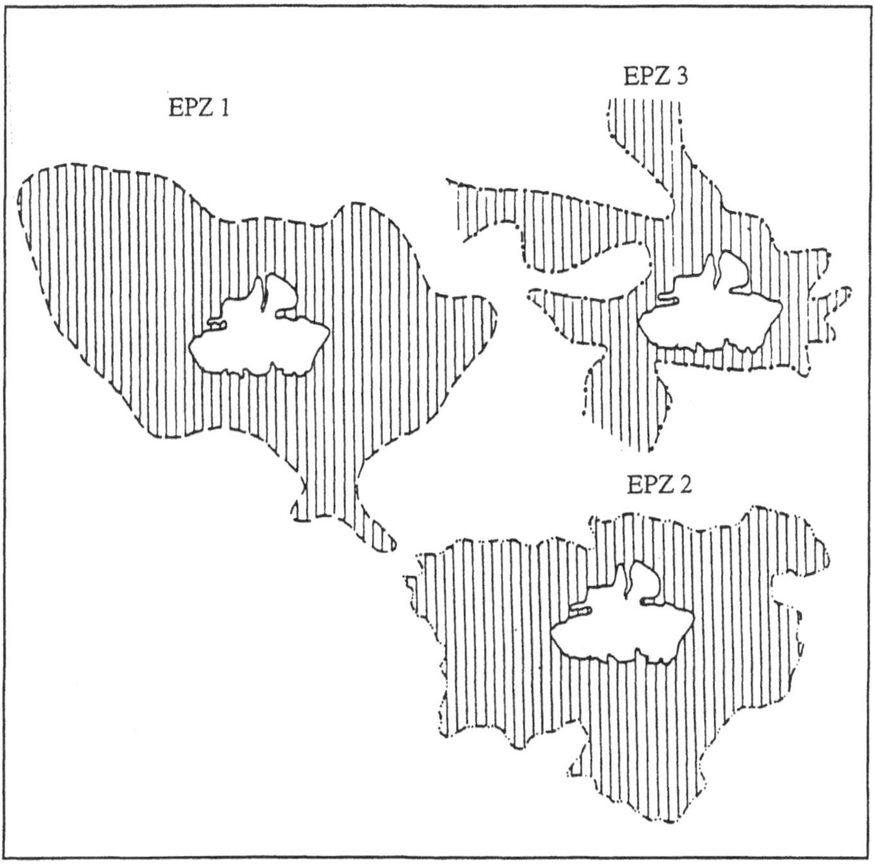

Figure 7.6 Gorce National Park's elementary protection zones

Summary of the Results

The final, comprehensive buffer zone that emerged from the process (Figure 7.7), included a synthesis of all the EPZs designed to protect the park from the main external negative environmental impacts. Two of these EPZs are presented in Tables 7.7 (climate) and 7.8 (hydrology), which detail the purpose, scope and land use policies to be applied in the relevant EPZ area.

To prevent or minimise the impact of climatic change means not only maintaining favourable conditions for many plant species, but also fighting against continually increasing air pollution, one of the main threats to the entire natural environment of the park. A large portion of this pollution is generated by sources far removed from the park, often situated in other countries and hence completely beyond the reach of any measures introduced in the area surrounding GNP, however generously it could have been defined. This is certainly a wider problem facing several, and not only Polish national parks.

Table 7.7 Climate (EPZ 1), protecting Gorce National Park from the changes in macro-climatic conditions

Purpose	To preserve undisturbed climatic conditions and the existing mezzo- and micro-climatic forms; and To safeguard the park against atmospheric pollution from local sources.
Scope	The territory that interacts climatically with the park.
Land use policies	• preserve 'a loose development' policy in the valleys to facilitate air flows; • enlarge the forest area and the 'green' uses; and • shift from coal heating to gas heating.

Table 7.8 outlines policies to mitigate the impacts of changes in the hydrology of the area surrounding the park. Most fresh water springs are primarily located in GNP. Creeks and streams flow from within the park to its surroundings. As a consequence, the park's water quality can only be marginally affected by external impacts (mainly by air pollution). Yet, inappropriate water management in the area surrounding of the park may have serious, damaging impacts on the park's flora and fauna. Lowering underground water levels and extinction of many springs may produce disastrous consequences not only for GNP itself, but also for large areas benefiting from water provided by the rivers generated in the park.

Table 7.8 Hydrology (EPZ 2), protecting Gorce National Park from changes in hydrological conditions

Purpose	To preserve the existing hydrological conditions in the park; and To integrate the protection of the park with regional water management strategies.
Scope	A part of the catchment area of rivers in the vicinity of GNP.
Land use policies	• maintain the 'class I' water quality; • ensure the natural state of water retention; • improve water management (building waterworks and treatment plants); and • eliminate local refuse dumping grounds.

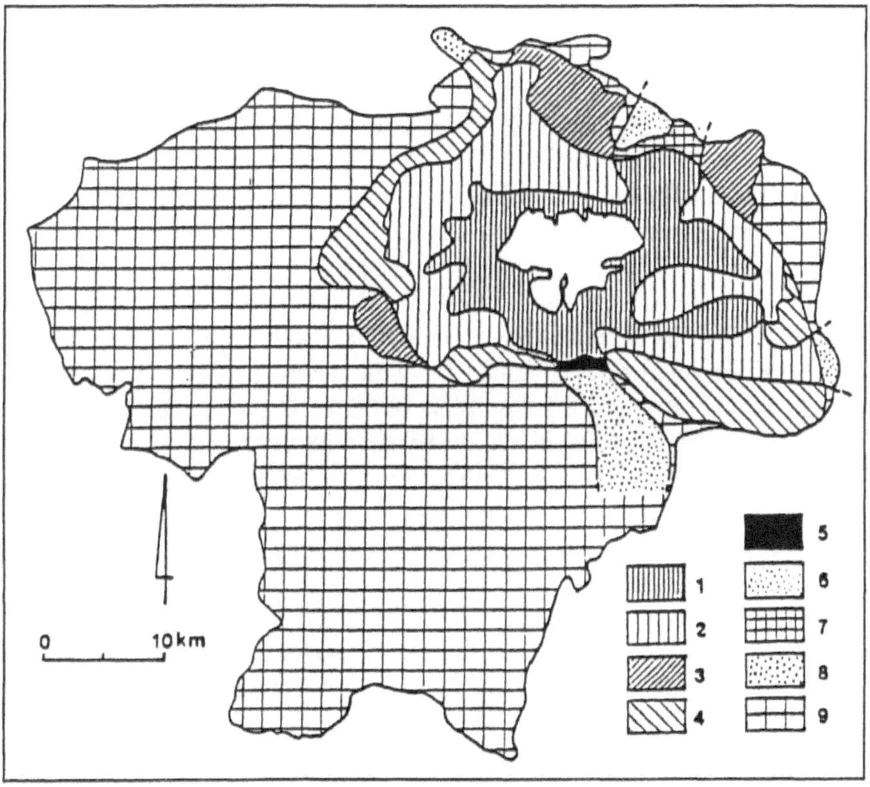

Figure 7.7 The final 'buffer zone' of Gorce National Park (Kozlowski et al., 1992)

(Where: 1. EPZs 1-4; 2. EPZs 1,2,4; 3. EPZs 2,4; 4. EPZs 1,4; 5. EPZs 1,3,4; 6. EPZs 3,4; 7. EPZs 2,3,4; 8. EPZ 1; 9. EPZ 4)

Conclusions from the Polish Applications

The two main principles guiding the definition of comprehensive buffer zones to protected areas were, according to Ptaszycka-Jackowska (1990a:55), the introduction of only those measures that 'are absolutely needed and only on these areas that really have, or may have, to play a role in the protection of a protected area...[and that]...the buffer area need not be covered by the same measures'. The validity of these principles was convincingly confirmed in both applications.

The TNP application of the BZP approach, its first real-life test, revealed also that EPZs could be defined directly from the analysis of threats to the park's natural environment without first defining APZs to its respective elements and features. As a consequence, APZs were not defined in the Gorce application and as Ptaszycka-Jackowska claimed, this simplified the procedure. Whether this conclusion could be extended to all future applications of BZP is, however, questionable. APZs may prove to be useful in some cases, not only because they facilitate the definition of specific threats, but also because they may make it easier to explain to decision makers and the communities involved, the reasons supporting the introduction of particular land use (or other) measures.

Both applications also confirmed that the approach can be successfully used in situations where there is incomplete knowledge of the precise impact of the identified existing, or potential threats. Such cases are not infrequent and are a universal problem in planning practice throughout the world. In the Polish applications, the experts involved were persuaded that it was better for them to provide their informed input, even if it was based on probabilities, hypotheses or guesswork, than to leave the decisions to planners, politicians and/or developers, who many not have any in-depth ecological knowledge.

The results can also be credited with providing a concrete and positive contribution to the protection of the two important Polish national parks. Ptaszycka-Jackowska (1990a) identified several future applications and issues including:

- the protection of important natural areas through effective buffers may require their introduction on a nation-wide or landscape scale (particularly, where air and water pollution is concerned);
- most of the principles of such buffer zones, once implemented, would benefit not only the protected areas, but also may improve the quality of life of people living in or visiting the areas situated within the buffers;

- implementation of the buffers' objectives may necessitate restrictions on land use, which may cause a negative reaction from the local population and require the introduction of moderating governmental policies; and
- to minimise potential conflicts, any restrictions must be limited only to those absolutely necessary, and should be complemented by an environmental education program targeting not only the general public but also politicians and decision makers of all kinds.

An important, expected achievement of the TNP Buffer Zone plan was that its findings, that is, the delineation of all EPZs and their land use measures, were introduced as *'directives'* to the existing statutory planning system, allowing the Chief Architect responsible for planning for the entire administrative region (voivodship) to supervise the introduction and implementation of the buffer policies within the planning schemes for villages and townships situated within the buffer zone area. Similar outcomes were achieved for GNP, where several towns and villages received such directives for implementation (Ptaszycka-Jackowska, 1990a).

Unfortunately, as it is often the case when new ideas are tested in real-life, this well structured implementation process somehow faltered. There was not sufficient positive motivation to make it work at the local level, and this was compounded by a lack of political will/interest at other governmental levels, and last but not least, subsequent planning legislation failed to incorporate a legal framework for buffer zones. This may also be a reason why recent attempts to find any follow-up publications on the post 1990 history of both TNP's and GNP's quite outstanding BZP case studies were unsuccessful.

References

Chmielewski, J. (1952), 'Skalne Podhale, elementy zagospodarowania przestrzennego' (Skalne Podhale, elements of the spatial development), *Prace Instytutu Urbanistyki i Architektury*, **1** (4), Warsaw.

Chmielewski, J. (1956), *Wytyczne do kompleksowego zagospodarowania regionu Tatr i Podtatrza. Uwagi metodologiczne* (Guidelines to a comprehensive development of the Tatry and Podtatrze regions. Methodological comments), Instytut Urbanistyki i Architektury, Individual work series, **46**, Warsaw.

Czemerda, A. (1983), 'Ochrona krajobrazu w regionie babiogorskim', in K. Zabierowski (ed.), *Park Narodowy na Babiej Gorze. Czlowiek i Przyroda*

(Babia Gora National Park. Man and the Nature), PWN, PAN, Zaklad Ochrony Przyrody i Zasobow Naturalnych, Warsaw-Krakow.

Gacka-Grzesikiewicz, E. (1978), 'Ekologiczny System Obszarow Chronionych' (Ecological System of Protected Areas), *Przyroda Polska*, No.1/2.

Kozlowski, J., Baranowska-Janota, M., and Ptaszycka-Jackowska, D. (1979), 'The Tatry National Park', *Architektura*, No.383-384, pp. 28-35.

Kozlowski, J. and Ptaszycka-Jackowska, D. (1979), 'Jak ksztaltowac otuliny obiektow chronionych: propozyoje metodyczne' (How to plan Buffer Zones: methodological proposals), *Aura*, No.2, pp. 8-11.

Kozlowski, J. and Ptaszycka-Jackowska, D. (1987), 'Planning for Buffer Zones', in P. Day (ed), *Planning and Practice*, Department of Regional and Town Planning, The University of Queensland, Brisbane, pp. 200-215.

Kozlowski, J., Ptaszycka-Jackowska, D. and Peterson, A. (1992), 'Buffer Zones for Protected Areas: A Planning Approach'. *Paper* presented at the IV World Congress on National Parks and Protected Areas, Caracas, Venezuela, p. 29.

Luczynska-Bruzda, M. (1976), 'Jurajski Park Krajobrazowy strefa ochronna Ojowskiego Parku Narodowego' (Jurajski Landscape Park as a protection zone to the Ojcowski National Park), *Chronmy Przyrode Ojczysta,* No.1.

Ptaszycka-Jackowska, D. (1986), 'Otulina Tatrzanskiego Parku Narodowego' (Buffer Zone of the Tatry National Park), *Czlowiek i Srodowisko*, Vol.10:3, pp. 305-323.

Ptaszycka-Jackowska, D. (1988), 'Otulina Gorczanskiego Parku Narodowego' (Buffer Zone of the Gorce National Park), Instytut Gospodarki Przestrzennej i Komunalnej, *Biuletyn Informacyjny*, No.2-3; pp. 15-20.

Ptaszycka-Jackowska, D. (1990a), *Ksztaltowanie stref ochronnych przyrodniczych obszarów chronionych* (The development of protection zones of protected areas), Instytut Gospodarki Przestrzennej i Komunalnej, Warszawa.

Ptaszycka-Jackowska, D. (1990b), 'Protection zones of natural protected objects', Polish Academy of Sciences, Proceedings of the Mining and Geodesy Commission, *Geodesy*, No.35, pp. 77-82.

Ptaszycka-Jackowska, D. and Kozlowski, J. (1978), 'Strefy ochronne obiektów chronionych' (Protective zones of protected areas), *Czlowiek i Srodowisko*, **8**(3-4), pp. 57-77.

Rozycka, W. (1983), 'Parki krajobrazowe a koncepcja ekologicznego systemu obszarow chronionych' Conference Paper (Landscape parks and the concept of Ecological System of Protected Areas), *Problemy gospodarki przestrzennej in landscape parks*, Conference TUP/IKS, Tuczno 20-22 October.

Sokolowski, S. (1923), 'Tatry jako park narodowy' (Tatry as a national park), *Panstwowa Komisja Ochrony Przyrody*, No.4.

Chapter 8

'Buffer Zone Planning': Great Sandy Management Area (Australia)

Introduction

Following the successful application of the buffer zone planning (BZP) model to Tatry and Gorce national parks in the 1980s, several theoretical applications were developed in Australia in the late 1980s and 1990s, the purpose being to apply the model, and evaluate and refine it, where necessary. This chapter provides a brief overview of the application of the BZP model within the Great Sandy World Heritage Area, one being to the Cooloola section of the park (Peterson 1991) and the other to Fraser Island (Hruza 1993).

Case Study 1: Cooloola Section of the Great Sandy Management Area

Background

The Cooloola sandmass system has been a focus of attention since the 1960s. Following a protracted struggle among disparate groups, who saw in Cooloola opportunities for the continued extraction of minerals, the expansion of forestry activities, and the development of land for grazing, resort and residential uses, a national park of 23,000 hectares was gazetted in 1975. Issues related to the park's management re-surfaced in the 1980s and by 1990 the park was expanded to 55,000 hectares. The application of the BZP approach provided a regional perspective for understanding how to manage this biological and culturally diverse protected area, including a mechanism for examining how the park meshed with its surrounding landscape and community, and for developing sustainable land use practices that have little or no negative impact on the park's values.

The management plan for Cooloola National Park recognised the need to 'protect the park from the adverse effects of erosion, pollution, exotic species and from any other use or misuse by man (sic) and other agencies both within and from without the park area' (QNPWS, 1979:36). Although this plan highlighted the need to identify the causes of 'unnatural change and mitigate their effects within the national park and within as much as possible of the park's environs' (QNPWS, 1979:40), the plan did not indicate the strategies to be implemented to combat the identified threats. The main approach had been one of extending the boundaries of the park, so that it encompassed additional ecologically important habitats, rather than ensuring that the park was managed as part of its wider biotic region.

Location

The park is a triangular-shaped piece of land, situated about 160 kilometres north of the State's capital city, Brisbane, and about 60 kilometres east of the smaller regional town of Gympie (Figure 8.1). It is one of the largest parks in

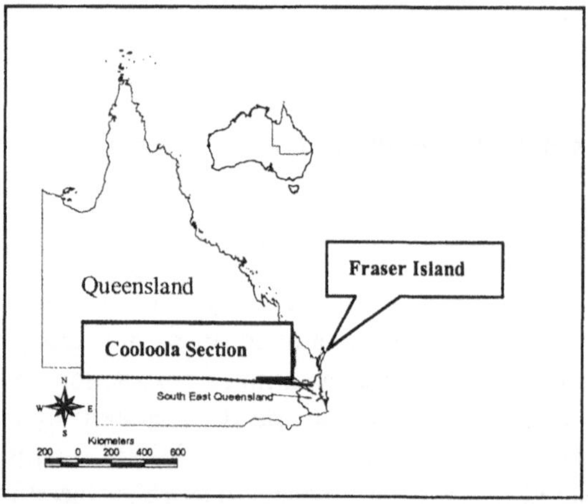

Figure 8.1 Location of Fraser Island and the Cooloola section of the Great Sandy Management Area

Queensland. It consists of extensive ocean beaches, high dunes incorporating rainbow coloured sand cliffs, the Noosa River and its tranquil lakes and tributaries, wet plains of wallum heath forming a mosaic of wildflowers in spring, woodlands of banksia and scribbly gum, blackbutt forests, vine forests and the slopes and plains of the Noosa River catchment. There is great variety in the area's landforms, vegetation, fauna, climate and hydrology, and this combination of features has produced a unique area of great scenic beauty, with recreational, educational, scientific and ecological values that are appreciated locally, regionally, nationally and globally.

Resources Within Cooloola National Park

The resources of Cooloola can be categorised into three broad classes: physical (the physical systems operating within the park); sensory (how the physical resources impinge on our senses); and cultural heritage (the important sites and structures of significance to both the Indigenous Traditional Owners and European cultures) (Table 8.1).

Table 8.1 Resources of Cooloola National Park

Physical	Sensory	Cultural heritage
Geomorphology	Silence	Indigenous Traditional Owner
Hydrology	Wilderness	European
Soil	Aroma	
Vegetation		
Fauna		
Climate		

Physical Resources

Geomorphology (and geologic structure) The outstanding geomorphic feature of the park is its high sand dunes, developed during Quaternary sea level fluctuations. During glacial periods, when the sea level fell, large areas of the sea bed became exposed. Intense onshore winds swept the accumulated sand onto the windward side of the existing dune systems, creating a new sequence of giant dunes (Sinclair, 1990). Eight periods of dune building have been recognised extending back over 100,000 years in age (Walker et al., 1981). The Cooloola region contains more episodes of transgressive dunes than have been recorded anywhere else in the world (Sinclair, 1990). The rocky headland of Double Island Point is important in trapping sand moving northward along the coastline. With time, vegetation has stabilised the dunes and soil profiles have developed.

Perhaps the most spectacular feature of the park is the coloured sands that form the core of the sandmass. They are exposed by erosion along the ocean beaches, but are also found beneath the loose oceanic sands over most of the sandmass. Geologically the coloured sands are a form of soft sandstone in which the action of peaty acids has oxidised iron materials in the sands producing a multiplicity of shades, e.g. pink, yellow, red, ochre, pale grey, brown and others. In places spectacular badland topography results where erosion has occurred on sands consolidated with silt and clay (Stanton, 1975). In terms of geomorphology, Cooloola possesses outstanding value. Together with Fraser Island, the sandmass contains the greatest representation of dune types found anywhere in the world.

Hydrology The surface and groundwater systems are the two significant water movement cycles within the park. In relation to the surface water system, the northern part of the Cooloola sandmass is drained by creeks flowing into Tin Can Inlet and on into Great Sandy Strait. To the south, the Noosa River obtains surface water from creeks flowing westwards from the sandmass and eastwards from the Como scarp. The Noosa River is a slow moving river with an average gradient of about 0.2 metres per kilometre (Toovey and Lau, 1987), and contains two large lake systems (Cootharaba and Cooroibah). Lakes Cooloola and Como are delta lakes formed from deposition by the Noosa River at its entrance into Lake Cootharaba (Reeve et al., 1985).

In the older high dunes there are a number of small freshwater lakes (Poona, Freshwater, Cooloomera, Broutha and Thannae), some of which are 'perched lakes'. For example, Lake Poona, is 100 metres above sea level and is perched in a depression located above the water table. The containing layer of perched lakes may be an impervious band of carbonaceous sands or coffee rock or the accumulation of organic matter. As the catchment is restricted to rain that falls on the lake's surface, there is little run-off into the perched lakes and hence water levels can fluctuate widely. Other lakes, such as Lake Freshwater, are 'water table window' lakes, which occur in depressions where the land surface falls below the regional water table. The level of these lakes changes as the water table rises and falls.

The particular hydrologic features of the lakes of the Great Sandy Region are not recorded in lakes outside of Australia. The waters of the Noosa River and its lakes are generally of low nutrient status and acid, a reflection of the low fertility soils of the catchment and the accumulation of plant acids in semi-stagnant marshes and swamps (Stanton, 1975). Their lack of calcium, phosphate and potash produces very soft and very pure water, being chemically amongst the freshest natural water in the world (Sinclair, 1990).

In relation to the underground water movement, the western Como scarp and the eastern high sand dunes act as aquifers that absorb rainwater and channel it into the Noosa River. On the eastern high dunes aquifer the water flows either across the beach sands in shallow streams, out of the beach through a localised bubbler development, or through the beach below sea level. To the west, most of the outflow enters the Noosa River through the water table.

The region contains two aquifers, a clearer lower sandmass aquifer and a vegetation-stained upper aquifer and these play a significant role in maintaining the complex ecosystem of the park. Stream colours range from clear through yellow to bronze, red and black. The 'black' water that gets its strong colour from organic compounds has passed through the A horizon while the 'white' water has probably passed through the B horizon into the yellow/brown sands of the C horizon, unloading its organic compounds as it passes through (Sinclair, 1990).

The western part of the park is underlain by extensive sedimentary deposits (the Myrtle Creek sandstones). The sandstones and conglomerate form a major aquifer channeling water to the Noosa River. It is these seeping water supplies, from the western Como Scarp and eastern sand mass that feed and maintain the Noosa River, a river which is unique in south-east Queensland, as it derives scarcely more than 10 per cent of its waters from the Great Dividing Range. The remaining 90 per cent enters as seepage from the sandmass. The water is held in the groundwater for great lengths of time, perhaps 100 to 200 years (Sinclair, 1990).

Soils The geologic structure of Cooloola has determined that soils everywhere are extremely low in plant nutrients. However, there is considerable variation in the 'relative fertility' (Stanton, 1975) of these soils and their ability to support various plant communities according to their size structure, depth and topographical position.

Soils of the western catchment are formed on sandstone and its derivates, being predominantly infertile sandy loams lying over mottled clay subsoils. Soils of the Noosa Plain were laid down in water and probably built of both marine and freshwater sediments (Stanton, 1975). The soils are seasonally waterlogged and have permanently high water tables. Humus podzols and peaty podzols have formed on the higher areas. When organic matter has accumulated in the depressions, acid peats up to three metres thick occur.

The natural fertility of the soils is low with respect to nitrogen, phosphorus, calcium, potassium, sulphur, copper and zinc (Coaldrake, 1975).

Throughout the sandmass, podzols have formed on the dune ridges and slopes, and humus podzols have developed in the corridors, where water tables are close to the surface. All the soils are formed on siliceous sands and the depth of soil development increases from 40 to 60 centimetres on the young dunes near the coast, to more than 20 metres on old dune systems along the western margin (Bridge et al., 1985). All soils have naturally low fertility and the humus-dominant subsoils may be quite acid (pH values around 4.0). The significance of the Cooloola podzols is that their depth far exceeds the known depths of podzol development anywhere else in the world (Sinclair, 1990).

Vegetation Cooloola's vegetation is diverse both structurally and floristically. In general the vegetation pattern is a result of a close association of soil moisture aeration and minerals, nitrogen accretion, topography (in relation to sea winds, wildfires and nutrient leaching) and history (Webb and Tracey, 1975). The evolution of the plant communities is unique. The region is believed to provide the best example in the world of retrogressive plant succession. As the soils become poorer the plant communities degenerate from rainforest to stunted heaths (Sinclair, 1990).

In Cooloola, plant succession advances towards a rainforest climax community, and also degenerates to less complex and lower height communities as nutrients are leached out of the system. In general, progressive plant succession occurs with increasing soil weathering on the younger dunes (numbers one to four). Here, the youngest dunes have only a thin vegetation cover, while on the progressively older dunes there are low coastal scrub, sclerophyll forest and dense subtropical vine forest. On the older Pleistocene dunes, as weathering extends beyond the depth of rooting for many plants, retrogressive succession occurs. As nutrient reserves are depleted, or leached beyond the reach of plants, the communities deteriorate to open forest and heath (Sinclair, 1990).

The four main types of vegetation are: wet heath; low sclerophyll forest with grassy woodland; tall eucalypt sclerophyll forest; and vine forest. The minor vegetation types include: paperbark forest; sedgelands; coastal pioneer associations; estuarine fringe associations; riverine fringe associations; and freshwater flora.

Of particular importance are the wet heaths, one of the largest of the few remaining heath plains of eastern Australia. Elsewhere, much of this ecosystem has been drained for development. The rare plant species *Boronia keysii* is found in the wallum north of Kin Kin Creek. Also of importance are the vine forests located on well-protected sites mainly in interdune corridors.

Fauna The density of fauna at Cooloola does not match that of the extensive plant communities. However, the diversity and rarity of much of the fauna is significant, a reflection of the uniqueness and variety of habitat types in the park. Being an overlap area, several species reach their most southern range here, while others reach their most northern range. Hence, the associations of fauna that occur at Cooloola are found nowhere else in Australia.

Cooloola has an exceptionally high diversity of birds, representing a stronghold at the edge of the range of several bird species such as the brush bronzewing *Phaps elgans*, southern emu-wren *Stipilturus malachurus*, and little wattlebird *Antochaera chrysoptera*. The wallum is a significant habitat for nectar feeding bids such as honeyeaters and insectivorous birds (e.g. cuckoo shrikes and cuckoos), as well as closed forest birds such as wampoo pigeon, topknot pigeon, white-headed pigeon, pale-yellow robin, large-billed scrub-wren and regent bower bird. Cooloola also represents a last remaining stronghold for the ground parrot *Pezoporus wallicus*, and has large numbers of water birds in its shallow waters of the lakes. Tin Can Bay and the Great Sandy Strait are stopovers for trans-equatorial migratory wading birds flying between southern Australia and their breeding grounds in Siberia. As large numbers of these birds rest in the area, maintenance of their habitat is critical.

Mammals are characterised by low total species richness, very low diversity and low abundance (Dwyer et al., 1979). Macropods, possums and gliders are poorly represented, the main component of mammal fauna being rodents and bats. The false water-rat *Xeromys myoides* is a rare species found within the park.

Reptiles and amphibians are also an important component of the park. The Cooloola tree frog *Litoria cooloolensis*, the three-toed, reduced limb skink *Anomalopuys reticulatus*, and the four-fingered skink *Nannoscincus graciloides* are dependent on Cooloola for a significant portion of their habitat. Four species of frog are specifically adapted to and reliant upon waters of relatively very low pH for breeding, a feature found in the wallum plains of Cooloola, where the water lies over peat soils and therefore contains considerable quantities of organic acids. The 'acid' frogs depend upon the availability of acid water to facilitate the development of their larvae (Ingram and Corben, 1975).

Due to the acidity and toxicity of the Noosa River's water, the diversity of the river's fish fauna is restricted. Similarly the dune lakes are not productive fish habitat due to their filtered purity, high acidity and low levels of chemical nutrients.

Invertebrate species are widespread. An important species endemic to the area is the Cooloola monster *Cooloola propator*, a cricket. Twenty species of earthworm have been identified. One genus *Pheretimoides* is unique to south-east Queensland. The presence of earthworms in sand podzols is in itself likely to be unique, with many of the Cooloola species being the result of localised evolution, and having limited distributions and small population size. The more than 300 ant species in the region represent perhaps the greatest diversity and number of ant species so far recorded in a given area (Sinclair, 1990). Other recently discovered invertebrates include 22 species of termites, 57 species of springtails, a giant subterranean cockroach *Geoscapheus primulatus*, as well as burrowing bees, snails and freshwater crayfish. The invertebrate populations appear to be strongly influenced by soil type and microclimate. Species diversity is lowest in the youngest dunes with low, relatively unproductive vegetation, and increases in habitats where dune age and the height and mass of the vegetation is greater (Commission of Inquiry, 1990). The invertebrate fauna plays an important role in the ecosystems of the dunes.

Micro-climate Cooloola experiences a mild subtropical coastal climate with hot, moist summers and mild, drier winters. However, great variation occurs in the micro-climate due mainly to the influence of topography and vegetation. This produces unique ecosystems containing a variety of rare species of plants and animals.

Sensory Resources

The key sensory values of the park include its silence, wilderness qualities and aroma.

Silence One of the special features for visitors to Cooloola is to experience the 'silence of nature'. The upper reaches of the Noosa River with their inky, dark stained, highly reflective waterways offer serenity, peace and calm, creating a waterway of unparalleled appeal. Similarly, the tall vine forest with their understorey of palms, epiphytes and lianas, allow the bushwalker to experience the peace and quiet of nature. This quality of silence is highly sensitive to impact from a wide range of outside activities, which generate unwanted noise.

Wilderness Cooloola's wilderness qualities relate largely to the naturalness of this environment and the sense of remoteness and isolation that is usually experienced by visitors.

Aroma The natural aromas of the park contribute to the sensual experience of visitors. The wildflower displays of late winter, the blooms of the banksias and paperbarks in autumn and the myriad of eucalypt and vine forest species, when they are in flower, produce spectacular aromas that are an important feature of Cooloola.

Cultural Heritage Resources

Indigenous Traditional Owner cultural heritage Human contact with the area is likely to have begun about 40,000 years ago (Sinclair, 1990) with the Dulingbara people occupying almost all of the Cooloola sandmass (Tindale, 1974) and either the Kabi-Kabi or Chief Uwen Mundi's tribe occupying the southern Cooloola area around Noosa (Sinclair, 19990). Little of the culture and history of Indigenous Traditional Owners of the Great Sandy Region is recorded. Within about 60 years of white contact, Aborigines were 'degraded, debased, detribalised and finally destroyed' (Sinclair, 1990:65). The main remnants of the Indigenous culture are prehistoric archaeological sites, mostly in the form of shell middens, canoe and gunyah trees and other markings such as scars where bees' nests were removed.

The culture of the Indigenous Traditional Owners of the Cooloola area was based on the river and the sea. There is evidence of over 100 shell middens (McNiven, 1984, 1985), stone artefacts (Thompson and Moore, 1984) and camping and ceremonial sites (bora rings) (Sinclair, 1990). The region is archaeologically significant, although there has been little research to document these sites.

European heritage European contact in the area began around 1866 when timber getters entered the Kin Kin Scrubs. A railway was constructed in 1873 to transport timber from the high dunes to Tin Can Inlet. The tramway cuttings and jetty remnants are a feature of this European heritage.

Land Surrounding Cooloola

Important natural areas found outside Cooloola are several national parks and fish habitat reserves, a military training reserve at Wide Bay, the ocean beaches and the river and lakes' environs. Forestry and logging activities occur along the western boundary of the park. Rural activities, consisting primarily of beef and dairy cattle farming are becoming less important as urban, rural residential and tourism development increase in importance. Today tourism is the main income earner for the Cooloola and Noosa Shires and this has been accompanied by an increase in resort development and unit accommodation, shopping centre development and an expansion of the service industry and tourist operations to cater for the visitors.

Although most visitors are accommodated outside the park in Noosa Heads, Tewantin and some of the smaller townships, an increasing number are using off-road vehicles and are camping (often illegally) within the park. A large number of commercial tour operators conduct their businesses on land and water outside of the park. A variety of powered craft ply the Noosa River and lakes, while busses, trucks, off-road vehicles, horse drawn carriages, horses and camels transport passengers along the ocean beaches to the many tourist destinations.

Infrastructure systems also have potential to impact negatively on the park. The beaches act as major highways within the region and control of vehicular access rests with the local councils. A number of roads within the park are not under the control of the park management staff and hence conflict of interest may occur. Many visitors to the park use the vehicular barge service across the Noosa River at Tewantin. Tourist operators also ferry people across the river from a number of jetties at Noosaville. As no bridge exists across the southern reaches of the Noosa River, tourist numbers are more restricted than might otherwise occur.

There are no large airports in the immediate vicinity of the park. Smaller runways exist at Inskip Peninsula, Cooloola Village, Wide Bay Military Reserve and south of Teewah. Future development trends in the area indicate the possible development of large resorts at Inskip Peninsula, Noosa Heads and perhaps North Shore and the possibility of new or expanded airports.

Water is currently extracted from Teewah Creek to supply Tin Can Bay and Cooloola Village Estate. Rainbow Beach obtains water from Cameron Creek and Seary's Creek. Water is also extracted from around Boreen Point, Freshwater Lake and the Military Training Area and Toolara Forestry Camp at Tin Can Bay. No research to determine the extent and capacity of these underground water reserves has been undertaken. Any future expansion in demand for water may have serious impacts on these supplies and their interrelated ecosystems found within Cooloola.

Solid waste disposal is by sanitary landfill. Disposal of treated waste water occurs at Tin Can Bay, Rainbow Beach and Noosaville. This issue is of significance, due to the largely unpolluted waters of the Noosa River system and their low levels of acidity.

Analytical Protection Zones

By examining the inter-relationships between each of the park's resources and its surrounding environment, 16 existing and/or potential threats to these resources were identified by Peterson (Table 8.2). Ten analytical protection zones (APZs) were identified, corresponding to the main resources of the park., i.e. geomorphology, hydrology, soil, vegetation, fauna, micro-climate, silence, wilderness, aroma and cultural heritage.

Figure 8.2 illustrates the Geomorphology APZ. The main geomorphic resource of the park is the sandmass system, forming the park's distinctive northern and eastern boundary. Internal park management extends only to the high water mark located at the base of the sand cliffs. Various administrative bodies such as two local governments, and the State government Departments of Harbours and Marine, Primary Industries (Fisheries), and Natural Resources and Mines have responsibility for managing activities conducted on Cooloola's fringing beaches between the high water mark and the low water mark. In the past, the policy objectives of these agencies have been, at times, in conflict with those of the Environmental Protection Agency, which is responsible for park management. The failure of the administrative boundaries of the park to adequately reflect the biotic boundaries required for management, have produced significant threats, particularly to the geomorphology of the park.

The external threats to the park's geomorphic resources were largely due to interface processes and interactions that occurred at the inter-tidal zone between the marine and terrestrial environment, a fragile zone, easily downgraded as a result of unwise and unsustainable land use practices. The interface is also a zone of contact between the park's natural systems and the influence of humans, providing almost unlimited access to the park from the beaches and providing opportunities for a variety of harmful land use practices.

Table 8.2 Summary matrix of threats and resources at Cooloola

Resource / Threat	Physical						Sensory			CH	Total
	Geomorphology	Hydrology	Soil	Vegetation	Fauna	(Micro) Climate	Silence	Wilderness	Aroma	Aboriginal/European	
Fire	X		X	X	X						4
Noxious/exotic plants	X	X		X							3
Water pollution											
Domestic/Recr'tional		X	X	X	X						4
Rural		X	X	X	X						4
Industrial		X		X	X						3
Feral animals				X	X						2
Removal of vegetation	X	X	X	X	X			X			6
Drainage disruption		X		X							2
Lowering water table		X		X	X						3
Road construction		X	X	X	X	X	X	X			7
Recreational activities	X	X		X	X		X	X	X		7
Airport development							X	X	X		3
Excessive visitor nos.	X						X	X		X	4
Loss of Mt Bilewilam	X							X			2
Reduced sand supply	X										1
Greenhouse effect	X	X		X	X			X			5
Solid waste disposal								X			1
Air pollution									X		1
Total	8	8	4	10	8	1	4	8	3	1	

(Where CH - cultural heritage)

The geomorphology APZ was designed to reduce the negative impacts of several threats, including: fire; noxious and exotic plants; removal of vegetation; recreational activities; excessive visitor numbers; loss of Mt. Bilewilam; reduced sand supply; and the greenhouse effect (Table 8.2). The area over which each threat to the park's geomorphology operated, was mapped then synthesised. Due to the variation in the spatial range of each negative impact on the geomorphic system, a complex APZ structure was devised consisting of five sub-zones (C_1 – C_5), each with specific policies to control or minimise the identified negative impacts. There is no uniform land use policy throughout the APZ. The fringing beach interface requires the greatest level of control (seven identified negative impacts and related policies), with the western landward boundary requiring few control measures (one identified negative impact) (refer to Peterson [1991] for information on the remaining nine APZs).

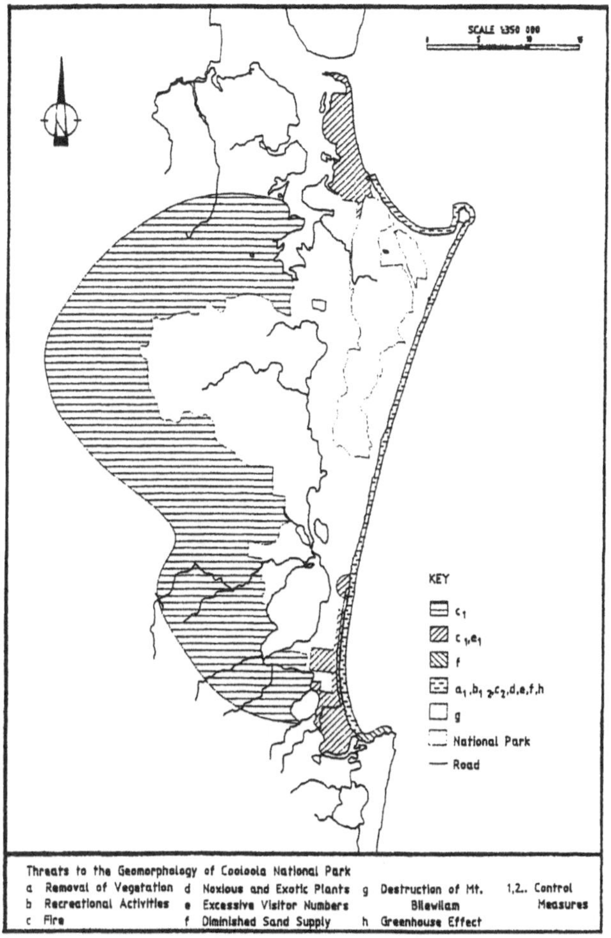

Figure 8.2 Geomorphology analytical protection zone (APZ) at Cooloola (Peterson 1991:126)

(Where threats are: a-removal of vegetation; b-recreational activities; c-fire; d-noxious and exotic plants; e-excessive visitor numbers; f-diminished sand supply; g- destruction of Mt Bilewilam; h-greenhouse effect; and 1, 2 – Control Measures)

Elementary Protection Zones

The detailed process of examining individual threats to each of the park's major elements and developing APZs ensures a comprehensive analysis of threatening processes. However, to aid 'on ground' management, the BZP approach recommends the development of elementary protection zones (EPZs), or management zones that are developed to minimise or eliminate a particular threatening process. For example, the fire threat currently affects, or has the potential to affect, the geomorphology, soil, vegetation and fauna of the park (refer to Table 8.2), while recreational activities affect seven of the park's elements/values. On the basis of the type of threatening process, 16 EPZs, each with specific land use policies were delineated for Cooloola. Examples for fire and water pollution (Figure 8.3 and 8.4) are illustrated,

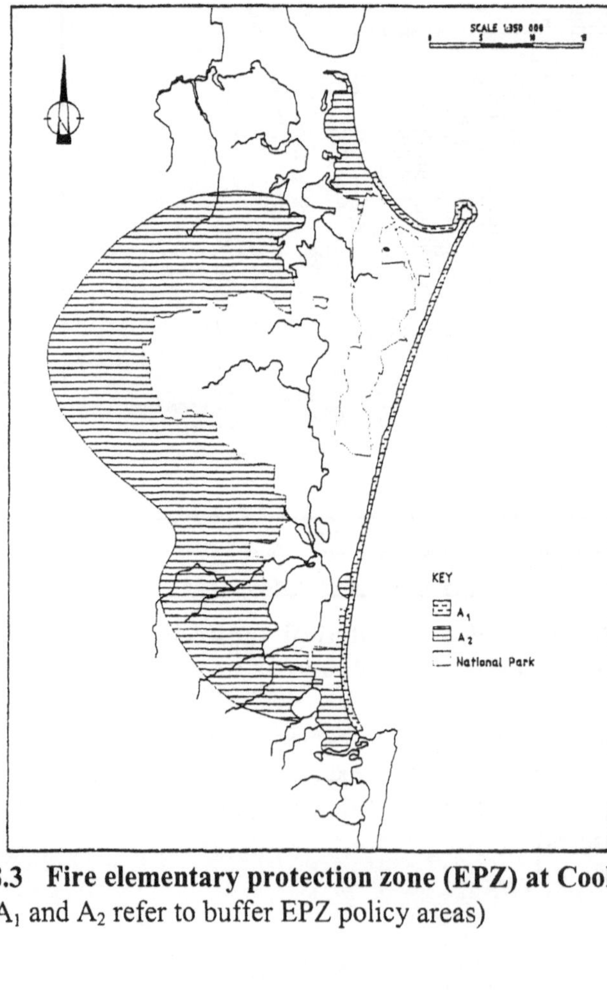

Figure 8.3 Fire elementary protection zone (EPZ) at Cooloola
(Where A_1 and A_2 refer to buffer EPZ policy areas)

Table 8.3 Fire EPZ, to protect Cooloola from fires originating external to the park

Purpose	To minimise the impact of uncontrolled fires on the park resources of geomorphology, soil, vegetation and fauna.
Scope	Land between the high and low water marks on the fringing beaches and a zone approximately 10 kilometres wide on the landward boundary of the park. The fire EPZ almost surrounds the park and consists of two sub-zones.
Land use policies	A1 • only fuel stoves permitted; • no open fires; and • foster improved fire management practices among the general public (e.g. use of signs to warn and educate).
	A2 • double fire break (50 metres) on landward boundary of the park with land kept clear of trees and shrubs to minimise fuel accumulation; • no burning without consent; • all development (especially residential and tourist/resort facilities) must incorporate adequate fire prevention measures; • limit further expansion of existing settlements along the park; and • eliminate fire escapes from solid waste disposal sites by ensuring fire prevention measures are incorporated (e.g. restricting open fires).

Table 8.4 Water pollution EPZ to minimise negative impacts on Cooloola National Park (selected examples)

Purpose	To minimise the impact of sources of water pollution on the park resources of hydrology, soil, vegetation and fauna.
Scope	Includes the catchments of rivers flowing to and through the park and contains two sub-zones.
Land use policies	C1 • routine inspection and maintenance of septic systems; and • natural vegetation is retained along stream/lake banks, with a restriction on building construction within 50 metres of the banks.
	B2 • restricted motor craft access from Kinaba Island upstream of the Noosa River and Kin Kin Creek; and • fitting of sealed sewage holding tanks on all house boats and commercial pleasure craft.

together with recommended land use policies (Table 8.3 and 8.4) to be implemented to minimise or eliminate the impacts of these threats on Cooloola National Park (refer to Peterson [1991] for information on the remaining fourteen EPZs).

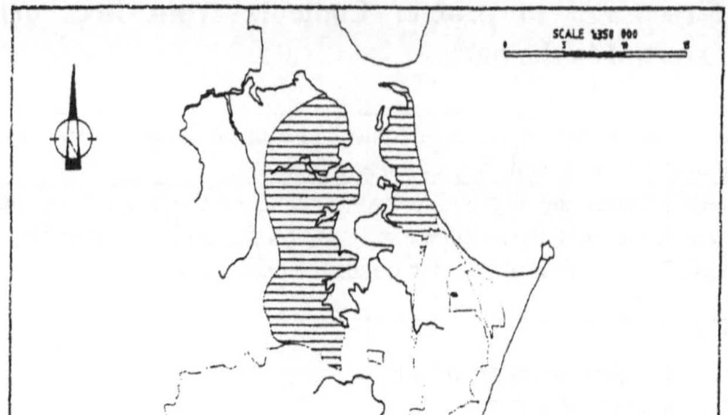

Figure 8.4 Water pollution EPZ at Cooloola National Park
(Where C_1 and C_2 refer to buffer EPZ policy areas. Policies are similar throughout each defined area)

Final Buffer Zone

Each EPZ displays a complex system of inter-relationships with its surrounds. By synthesising the EPZs and their land use policies an all encompassing buffer protection zone was produced by Peterson (Figure 8.5),

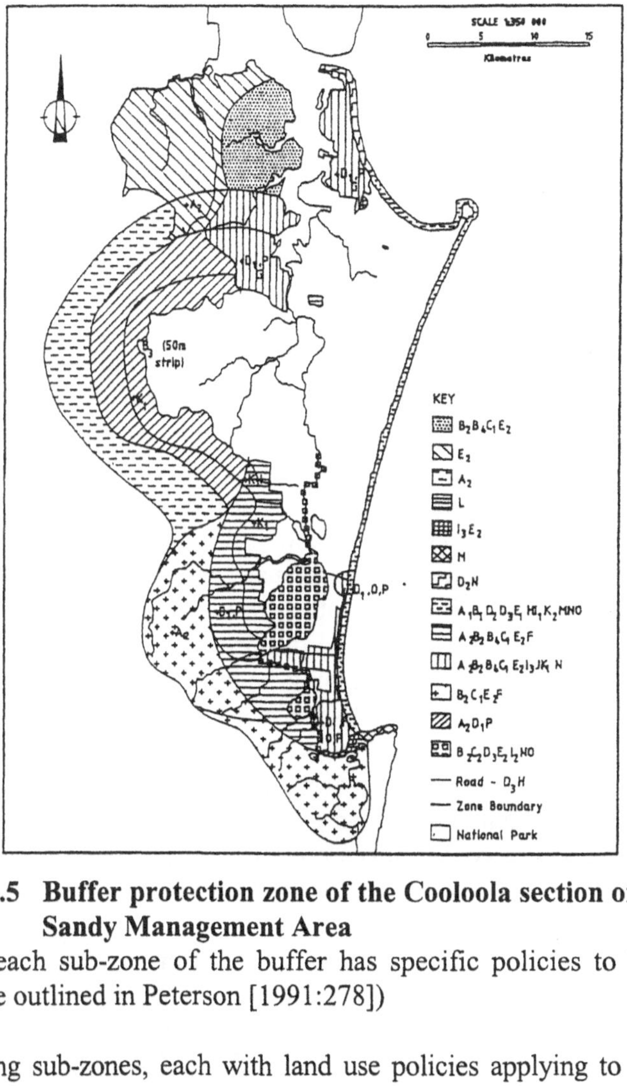

Figure 8.5 Buffer protection zone of the Cooloola section of the Great Sandy Management Area

(Where each sub-zone of the buffer has specific policies to be applied. These are outlined in Peterson [1991:278])

comprising sub-zones, each with land use policies applying to the area of operation of the particular negative impact they are designed to control.

Rather than a prescriptive distance of 100 metres being defined as the buffer for Cooloola, and in which a homogeneous set of policies are applied throughout, the BZP approach identifies for planners and resource managers the specific areas where threats are known to occur and have the potential to occur, and targets specific management strategies to these areas. In this way, a quite specific set of control measures and land use policies was devised, responding to the heterogenous environment surrounding the park, and the varied nature of negative impacts that emanate from this environment.

Cooloola's buffer clearly reveals that some areas within the buffer are 'hot spots'. These include the fringing beach zone (North Shore to Teewah), Inskip Peninsula and Tin Can Bay-Cooloola Village, all of which interface directly with the park. The fringing beach zone, although very narrow in width, is a source of several particularly threatening activities, which may negatively affect the park's values. Implementation of the suggested policies within this beach zone will help to restore Cooloola's very special sensory qualities as well as protect many of the park's physical resources.

On the landward boundary of the park, most threats are evident in the southern catchment of the Noosa River. Here much of the drainage basin is under freehold tenure. Increasing subdivision of land and changes to rural residential holdings may accelerate the pattern of negative impact already evident. It is especially important to ensure the preservation of the high quality water in the Noosa River by implementing a range of policies within this drainage catchment. To the west of the park, fewer control measures need to be implemented, for the western catchment of the Noosa River is contained within the park and water quality issues are less problematic here. However, other issues related to noxious and exotic species assume significance in this area of the buffer. Similarly to the north west in the Military Training Area, a less complex pattern of control measures is evident, due to the more compatible ways in which the land is used.

Lastly, the river system itself is a source of many threats to the park, as are a number of roads (both existing and gazetted, but as yet unformed) within and outside the park. Access routes pose special problems, for they open areas of the park to visitor pressure, and are an interface with the potential to introduce a variety of external threats.

The implementation of the buffer zone does not mean a wholesale change in the land uses and activities within the buffer, but rather that the buffer policies will guide and modify activities so that the threats to Cooloola are eliminated or at the very least minimised in their effect. The matrix (Table 8.2) plays an important role in the management of the buffer, as any proposed changes in the park's surrounds can be assessed on the basis of the relationship identified. For example, if a new road is proposed, the matrix indicates the park resources that may be affected and provides a platform for a decision on whether to allow the development and if so, where the least threatening route may be.

Summary

Buffer zones are an important means for ensuring progress towards achieving sustainable development, based on the maintenance of essential ecological processes and life support systems, and the preservation of genetic diversity. The delineation of a buffer zone for the Cooloola section of the Great Sandy National Park, which incorporates specific land use controls and policies will help to minimise the impact of surrounding land uses on the park and contribute to the goal of sustainable development within the park and wider region.

Case Study 2: Fraser Island

Background

Fraser Island, one of the largest sand islands in the world, is situated, along with the Cooloola, in the Great Sandy Region (Figure 8.1). Aborigines have had a presence on the island for about 40,000 years, with permanent Aboriginal settlements occurring over the last 6,000 years (Sinclair, 1990). To the Indigenous Traditional Owners, the Great Sandy Region was rich, diverse and productive. In 1860 the entire island was gazetted as an Aboriginal reserve. However, this was revoked following the discovery of the island's valuable timber.

European contact with the island began about 500 years ago. From the 1860s its tall timbers became much sought after by loggers and the Queensland Forest Service established a base on the island in 1913. By 1948 70 per cent of the estimated timber reserves had been removed (Sinclair, 1990). Conflict over logging the area's uniquely tall trees has been ongoing since this time.

The Great Sandy National Park was gazetted in the northern part of the island in 1971. The original park boundary was arbitrarily located on the basis that no exploitable timber was located in the north. Logging continued in the southern part of the island. Significant vegetation and hydrologic features remained outside of the national park.

Controversy developed in the 1970s concerning the continued sand mining of the island's resources. Mining leases were first granted in 1950 and 1966, with mining beginning on Fraser Island in 1971 (Sinclair, 1990). In 1984, after a protracted struggle, the sand mining leases were relinquished in the north of Fraser Island and included in the national park (Fitzgerald, 1990).

The park has been extended, on several occasions, and now comprises approximately 76,000 hectares. The whole of Fraser Island was given World Heritage Listing in 1992.

The Great Sandy Region Management Plan was later released to ensure conservation of the area's unique features and the sustainable development of land uses within the area. The plan does not allow logging to resume on any part of Fraser Island, although it does make allowances for the resumption of some mining activity.

The island's spectacular scenery, including long stretches of ocean beach, picturesque lakes and streams, tall forests, high dunes and blow outs, has made it a focus for tourism and recreation, which has been evident on the island since the 1870s. Permanent structures related to tourism activities appeared in the 1950s, with a surge in tourism to the island beginning in the 1960s and continuing unabated to the present.

Although Fraser Island is a World Heritage listed site, with the northern part of the island designated a national park, the southern part of the island, which is ecologically important is threatened from high levels of visitor use and associated development.

Hruza (1993) identified that the tall forests of the island, some of which were protected by national park status in the north, and the large area unprotected in the south, required particular management intervention to ensure their conservation and the implementation of sustainable land use practices in and around the forests. Hruza (1993) adapted the concept of BZP to the island's core of tall forests. She followed the five-step process identified by Peterson (1991). The following presents a brief summary of Hruza's buffer zone strategy for Fraser Island's tall forests.

Resources of the Core

Hruza began by identifying the most important natural areas within Fraser Island. She concluded that the central tall forests, which coincided with the ancient high transgressive dunes of the inland core of the island, and their unique lakes were worthy of further protection, especially as they were largely unprotected and subject to increasing levels of tourism and recreation. She identified this area as having regional, national and international importance (Figure 8.6).

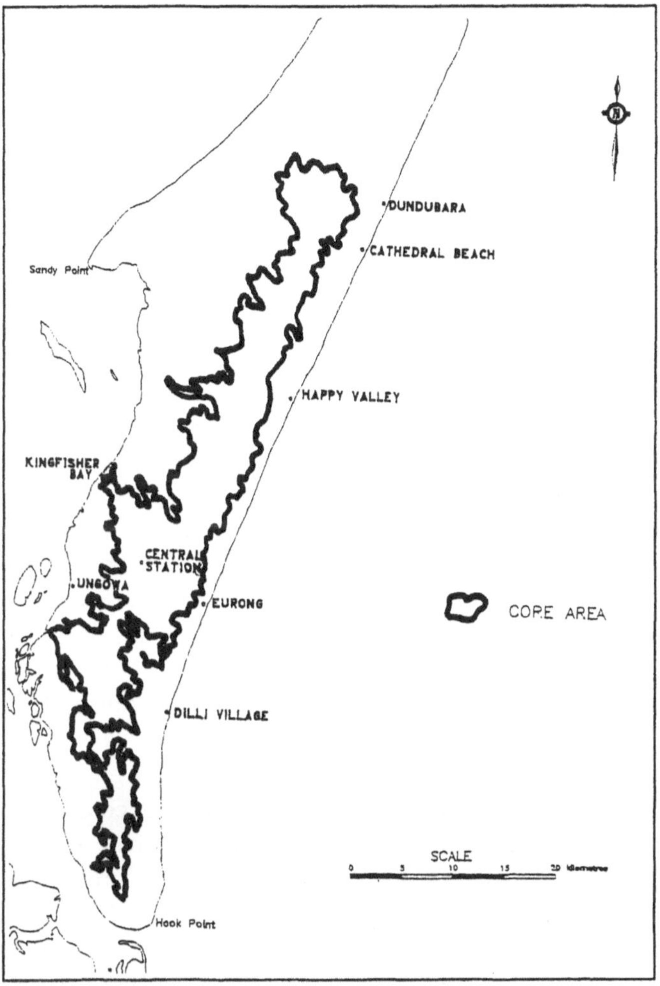

Figure 8.6 Core area, Fraser Island

As in all previous applications, Hruza identified the prime elements or values of the core (Table 8.5). These consisted primarily of the core's distinctive vegetation and hydrology. Given the similarities between Fraser Island's core and that of Cooloola, in relation to the structure and method of formation of the island's physical resources and the subsequent sensory values, only a brief outline of each resource is included.

Table 8.5 Resources of the core area, Fraser Island

Physical	Sensory
Vegetation Hydrology Geomorphology Soil Fauna	Wideness/aesthetic

Physical Resources

Vegetation Six main vegetation categories are present on the island, although the core consists of only two: closed forest (vine forest, hoop pine, brush box, satinay, and carrol scrub); and tall eucalypt forest (tall blackbutt, tall mixed forest with blackbutt, tall bloodwood and tall mixed eucalypt woodland). Most of this vegetation is in the southern part of the island, unprotected by national park status. The core's vine forests are found in well-protected inter-dune corridors. Several rare species are present in these forests, including the giant fern *Angiopteris evecta*. The tall eucalypt forests are generally over 30 metres in height and are dominated by satinay *Syncarpia hillii* and brush box *Lophostemon confertus*. Some of these forests grow on dunes which are over 200 metres in height.

Hydrology A multitude of lakes are found in the core area including window, perched, and barrage lakes (lakes formed where mobile sand has blocked water courses e.g. Lake Wabby). Numerous streams traverse the core tall forest area, flowing outwards to both the east and west coasts. Many of the core's lakes are deep (over 60 metres) and extensive. Significant ground water resources are evident, with the island forming the second largest aquifer in the world (Fitzgerald, 1990). As with Cooloola, an important aspect of the hydrology is the high quality of the water and the varied pH levels.

Geomorphology and soil The island is composed of an age sequence of sand dunes comprising nine separate dune systems (Queensland State Government, 1993), similar to those of Cooloola. The area contains the greatest representation of dune types found anywhere in the world (Sinclair, 1990). The core is located on the high dunes, which rise to 250 metres. The soils are mainly podsols, with age sequences developing from primary profiles (0.5 metres thick) to giant forms more than 25 metres thick. These latter profiles are the deepest podsol profiles recorded.

Fauna Although there is limited data on the island's fauna, there is a diversity of some species, particularly birds. The freshwater fauna, including fish, tortoises and invertebrates, are largely restricted to the dune lakes. There are also several species of acid frogs and skinks that have restricted distributions. Sinclair (1990) describes the significance of the area's flora saying that the main element is diversity and the dramatic contrasts between plant communities growing in a mosaic in such close proximity.

Sensory Resources

The sensory attributes of wilderness and scenic quality are important attributes to be conserved. Hruza (1993:85) identified the tall forests, lakes and streams as being particularly important for their sensory values.

> It is the awe of these forests which provide a unique sense of place for visitors to the island. It is an area of much beauty with its lush and diverse vegetation, and a pervading silence that is often broken only by sounds of wildlife.

The physical and sensory resources of the core have great conservation significance. To determine the strategy for protection, Hruza began by gaining a greater insight into the land uses and activities that were occurring on the land outside of the core.

Land Surrounding the Island's Core

Hruza identified that the core's surrounding lands had the potential to impact negatively on the identified physical and sensory values. She examined the land tenure situation, administrative arrangements, infrastructure and land use activities in these surrounds. To the north of the core, the land is contained within Great Sandy National Park, while the south is vacant crown land. Small pockets of freehold land also exist (Moon Point, Kingfisher Bay Resort, and Orchid Beach), as well as Crown reserves. Hruza identified tourism as a major industry, focused on activities occurring on the eastern beach and lakes and on four wheel driving along inland roads connecting scenic sites throughout the island. The island contains three major resorts (Kingfisher Bay, Orchid Beach and Happy Valley). Roads are primarily forestry tracks with a sand base. Approximately 600 kilometres of inland four wheel drive sand roads are found on the island, although the beach also acts as a trafficable zone. The island contains two aircraft landing areas and a limited services base. Electricity is supplied by local power generating plants to the three resort areas and Eurong. Solid waste is disposed of at tip sites on the island by trenching and backfilling. The main sewage disposal is septic, composting toilets, chemical toilets and pit toilets. Kingfisher Bay has an Environflow system attaining secondary level sewage treatment, the treated effluent being discharged into Great Sandy Strait.

Analytical Protection Zones

Hruza examined each resource or value (Table 8.5) and identified existing and potential threats. This resulted in six APZs designed to minimise the impact of a diverse range of threats to the values of vegetation, hydrology, geomorphology, soil, fauna and wilderness/aesthetic qualities. The threats and their interactions with the park's resources are summarised in Table 8.6.

The soil APZ is provided in Figure 8.7. Maintaining soil-forming processes is critical on Fraser Island because of its nutrient deficient soils. Processes that increase nutrient levels to aid plant growth need to be effectively managed. The important soil forming processes on the island are those that develop podsols, or humus podsols in aeolian sand (Queensland State Government, 1993). Inherent features of soil formation include the relationship between vegetation and hydrology. Changes in the soil regime may lead to alteration of some of the soil attributes on which the vegetation depends, and vice versa. The external threats to the core's soil resources were identified as removal of vegetation and water pollution. For example, vegetation removal may disrupt the nutrient processes within the tall forests. This may occur due to an absence in the accumulation of leaf litter, or a disturbance to the exchange of nutrients between plants and soil. Increased visitation and associated pollution of water also threatened the soil resources and subsequently the core's hydrology and vegetation. The area over which both the threat of removal of vegetation and water pollution operated was mapped and then synthesised to produce the Soil APZ (Figure 8.7).

Elementary Protection Zones

Ten EPZs were devised to protect the core from the identified threats. These included: removal of vegetation; introduction of plants and diseases; feral

Table 8.6 Matrix of threats to Fraser Island's core

	Physical																		Sensory			
	Vegetation						Hydrology				Geomorphology		Soil		Fauna				Wilderness aesthetics			
	Affected process						Affected process				Affected process		Affected process		Affected process				Affected process			
	Interface		Seed dispersal		Water system		Surface water		U'ground water		Interface		Interface		Interface		Habitat		Interface		Internal	
	E	*P*	*E*	*P*	*E*	*P*	*E*	*P*	*E*	*P*	*E*	*P*	*E*	*P*	*E*	*P*	*E*	*P*	*E*	*P*	*E*	*P*
Veg'n loss	Major impact		Significant impact		Major impact		Major impact			Significant impact	Major impact		Major impact	Major impact				Significant impact	Significant impact		Significant impact	Major impact
Pest plants	Low impact	Significant impact	Low impact		Low impact	Significant impact	Low impact	Significant impact														
Feral spp.																Low impact		Low impact				
Water pol'n		Major impact				Major impact	Significant impact	Major impact		Significant impact				Major impact			Significant impact	Major impact				Major impact
Litter							Low impact	Significant impact											Significant impact	Major impact	Significant impact	Major impact
Roads		Significant impact										Major impact								Significant impact		
*Recreation**	Significant impact										Low impact									Significant impact		Significant impact
Camping	Low impact					Major impact	Significant impact	Major impact												Major impact		Major impact
Visitors		Major impact				Major impact		Major impact		Significant impact		Significant impact				Significant impact		Major impact				
Mineral extraction												Low impact										

Where * motorised; E existing; P potential

Major impact   Significant impact   Low impact

animals; water pollution; litter; road construction; motor vehicle based recreation; camping; excessive visitor numbers; and mineral extraction.

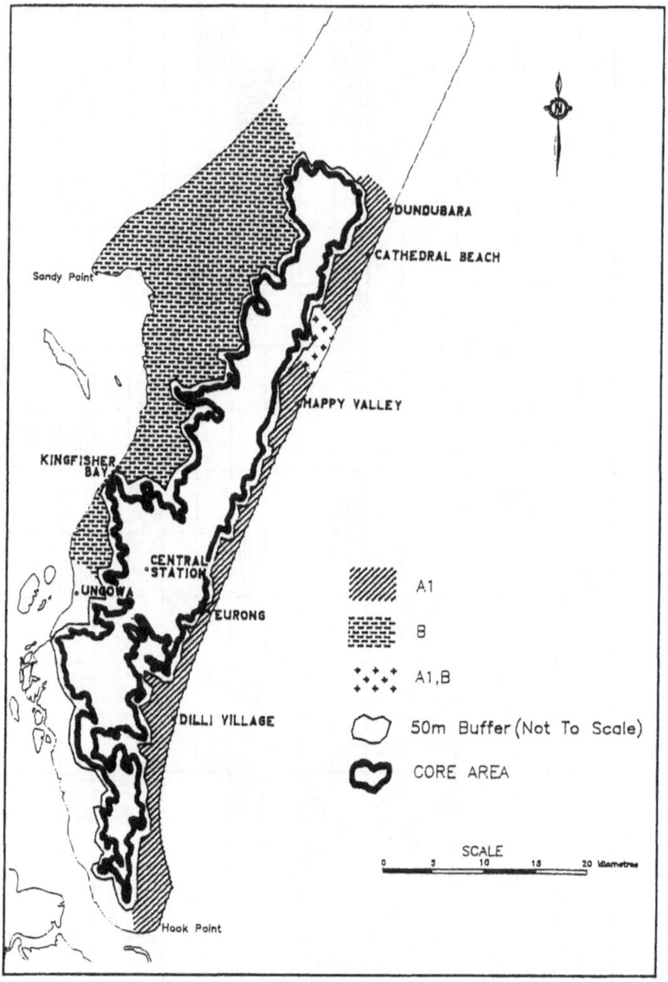

Figure 8.7 Soil analytical protection zone for Fraser Island core (Hruza, 1993:148)

Specific policies were also identified to minimise the impact of each threat. For example, to reduce the impact of introduced plants and diseases, control was required over the foredune area, catchment area of creeks, and points of access to the island and to the core area. Land use policies related to maintenance of existing eradication and surveillance programs, where they were considered to be working effectively, and the introduction of alternative programs in areas where existing practices were having little success, and the planting of vegetation native to the area.

Final Buffer Zone

The 10 EPZ boundaries were synthesised to produce a final buffer zone for the core (Figure 8.8). The whole southern area of the island adjacent to the tall forests of the core required the implementation of land use policies to ensure the long-term viability of the core. Two large areas to the west and east of the core required policies relating to almost all the threats, and thus management strategies were similar throughout. There were, however, pockets of land, namely in the south west of the island, which contained few threatening processes and which required less intensive management. The threats of vegetation removal and excessive visitor numbers were the major potential impacts on the core.

Summary

The final buffer allows for the area around the core of tall forests to be managed to reduce the level of impact on this biologically important area. However, the strategy does not address the management needs of the core area. A more comprehensive approach is needed to ensure that the core area is given a high level of protection, while the buffer area encourages a wider range of sustainable land use practices. Hruza however, identified that ecologically sensitive areas, whether protected or not, are suitable to the application of a BZP type methodology.

Conclusion

An important reason for undertaking the case studies in the Great Sandy World Heritage area was to assess the effectiveness of the BZP approach. The model displayed several strengths. The step-by-step approach provided a transparent and easy to apply methodology. It focused on understanding the values of the core area and in identifying the linkages between the core and its surrounds. Both case studies identified APZs prior to delineation of EPZs.

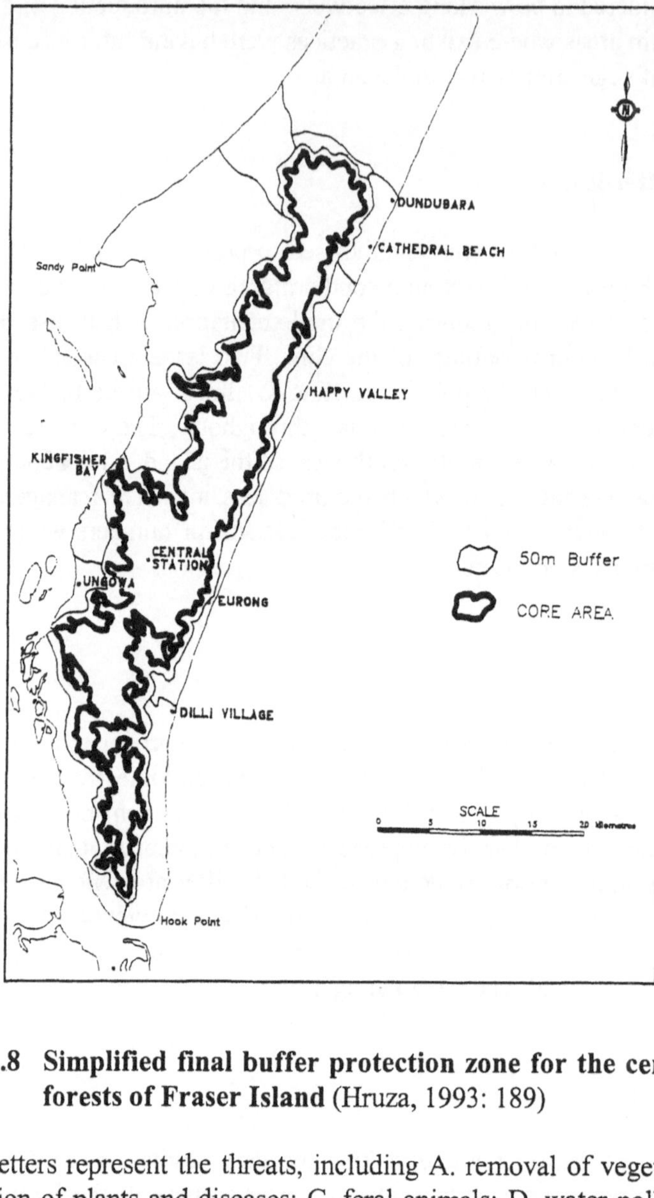

Figure 8.8 Simplified final buffer protection zone for the central tall forests of Fraser Island (Hruza, 1993: 189)

(Where letters represent the threats, including A. removal of vegetation; B. introduction of plants and diseases; C. feral animals; D. water pollution; E. litter; F. road construction; G. motor vehicle based recreation; H. camping; I. excessive visitor numbers; and J. minteral extraction; and Where numerals indicate the sub-zones of the buffer and the relevant policies that apply within the sub-zone, including 1. A, B, D-I; 2. B, F-I; 3. A-C, F, H; 4. A-C, G-I; 5. A-I; 6. A-C, H, J; 7. A-C, H; 8. A, H; 9. A, G, H, I; 10. A, B, D-I)

This process, although time consuming, provided additional information on the nature of the relationships between the elements/features of the core and the surrounding landscape matrix.

In addition to the standard approach utilised in the Polish parks, both case studies in the Great Sandy Region included interaction matrices to better understand the nature and significance of the identified threatening processes and their particular impacts on the elements and features of the core, both now and in the future. The approach allowed the identification of a range of threatening processes and the implementation of specific policies to address these threats.

The main limitation of the case studies relates to their failure to identify how the core impacted on its surrounding landscape and communities. The case studies highlight that further refinement of the BZP method is necessary to ensure comprehensive integration of the core area with its surrounding landscapes. This involves specific consideration of the core's relationship to its wider region.

References

Bridge, B.J., Ross, P.J. and Thompson, C.H. (1985), 'Studies in landscape dynamics in the Cooloola-Noosa River Area, Queensland. 3. Sand Movement on Vegetated Dunes', *CSIRO Division of Soils. Divisional Report*, No.75, pp. 1-41.

Coaldrake, J.E. (1975), 'The Natural History of Cooloola', *Proceedings of the Ecological Society of Australia*, **9**, pp. 308-313.

Commission of Inquiry (1990), *Commission of Inquiry into the Conservation, Management and Use of Fraser Island and the Great Sandy Region: Initial Discussion Paper*, Queensland Government, Brisbane.

Dwyer, P., Hockings, M. and Wilmer, J. (1979), 'Mammals of Cooloola and Beerwah', *Proceedings of the Royal Society of Queensland*, **90**, pp. 65-84.

Fitzgerald, T. (1990), *Commission of Inquiry Into the Conservation, Management and Use of Fraser Island and the Great Sandy Region*, **1-4**, Queensland Government Printer, Brisbane.

Hruza, K.A. (1993), *Buffer Zone Planning. A Possible Management Tool for Fraser Island*, Thesis submitted for the degree of Regional and Town Planning, Department of Geographical Sciences and Planning, The University of Queensland, Brisbane.

Ingram, G.J. and Corben, C.J. (1975), 'The Frog Fauna of North Stradbroke Island with Comments on the "Acid" Frogs of the Wallum', *Proceedings of the Royal Society of Queensland*, No. 86, pp. 49-54.

McNiven, I. (1984), *Initiating archaeological research in the Cooloola region, southeast Queensland*, BA (Hons) thesis, The University of Queensland, Brisbane.

McNiven, I. (1985), 'An archaeological survey of the Cooloola region, south-east Queensland', *Queensland Archaeological Research*, **2**, pp. 4-37.

Peterson, A. (1991), *Buffer Zone Planning for Protected Areas: Cooloola National Park*, Master of Urban and Regional Planning thesis, Department of Geographical Sciences and Planning, The University of Queensland, Brisbane.

Queensland National Parks and Wildlife Service (QNPWS) (1979), *Cooloola National Park Management Plan*, Brisbane.

Queensland State Government (1993), *Great Sandy Region Draft Management Plan*, Fraser Island Implementation Unit, Brisbane.

Reeve, R., Fergus, I.F. and Thompson, C.H. (1985), 'Studies in landscape dynamics in the Cooloola-Noosa River Area, Queensland. 4. Hydrology and Water Chemistry', *CSIRO Division of Soils. Divisional Report*, No.77, pp. 1-42.

Sinclair, J. (1990), *Fraser Island and Cooloola, Australia' Wilderness Heritage*, Weldon Publishing.

Stanton, J.P. (1975), *The Conservation of Cooloola* (in mimeo), Cooloola Committee, Noosa Heads.

Thompson, C.H. and Moore, A.W. (1984), 'Studies in Landscape Dynamics in the Cooloola-Noosa River Area, Queensland. 1. Introduction, General Description and Research Approach', *CSIRO Division of Soils. Divisional Report*, No.73, pp. 1-93.

Tindale, N.B. (1974), *Aboriginal Tribes of Australia*, University of California Press, Los Angeles.

Toovey, J.L. and Lau, D.H. (1987), *Report on the Noosa River Project*, Computerised Land Information Management Seminar, Noosa Heads, np.

Walker, J., Thompson, C.H., Fergus, I.F. and Tunstall, B.R. (1981), 'Plant succession and soil development in coastal sand dunes of sub-tropical eastern Australia' in H.H. Shugart and D.B. Botkin (ed), *Forest Succession: Concepts and Application*, D.C.West, Springer-Verlag, New York, pp. 107-131.

Webb, L.J. and Tracey, J.G. (1975), 'The Cooloola Rain Forests', *Proceedings of the Ecological Society of Australia*, **9**, pp. 317-321.

Chapter 9

'Buffer Zone Planning': Other Australian Applications

Introduction

In this chapter three additional case studies are presented to illustrate the application of the Buffer Zone Planning (BZP) methodology. These include a buffer for Nicoll Rainforest, where a partial application was undertaken, a buffer to better protect the habitat of *Burramys parvus*, the mountain pygmy possum in the Victorian Alps, and a buffer to conserve the Boondall wetlands in south-east Queensland.

Case Study 1: Nicoll Rainforest Buffer Zone

The buffer zone strategy for Nicoll Rainforest was developed by Roughan (1986). She undertook a partial application of the BZP method, focusing on three of the major threats to the park.

Location and Description

Nicoll Rainforest National Park (NRNP), which is located in the McPherson Range in south-east Queensland (Figure 9.1), was gazetted as a national park in 1986. Although only 24 hectares in size, the park is remarkably diverse, encompassing the whole topographic range of rainforest types found in the region. More than 200 species of trees have been identified in the park. The park has an altitudinal range extending from 40 metres to 180 metres above sea level and is one of the few remaining tracts of low altitude rainforest in the region. However, the park exists within highly modified surroundings, with clearing for agriculture (mainly banana farming), dairying, rural residential and residential development having left little of the original rainforest vegetation intact. The park was highly susceptible to a variety of externally originating influences that were degrading its natural values, and hence the introduction of an appropriate buffer zone was seen as imperative for the park's long term survival.

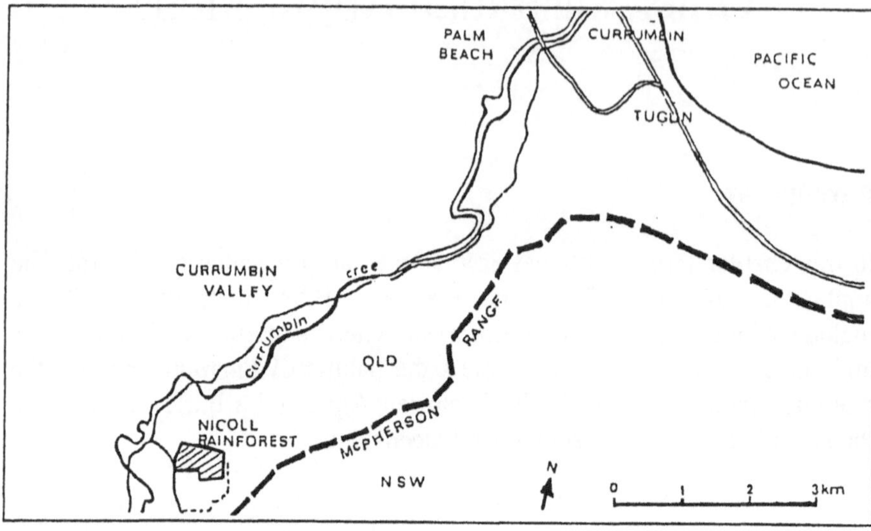

Figure 9.1 Location of Nicoll Rainforest National Park

The elements of NRNP needing protection were the soils, water, topography, fauna, dry rainforest, dry warm subtropical rainforest, transitional rainforest, warm subtropical rainforest and palm forest. Other important features included the micro-climate and landscape, and its scientific and educational values. A partial application of the BZP methodology was undertaken and a buffer zone was defined in relation to three (although perhaps the most important) of the park's elements, namely dry rainforest, palm forest and water.

The vegetation communities in the park varied in relation to slope (Figure 9.2), with dry rainforest occuring on the crest and upper slopes of the ridge, and the palm forest occurring on the poorly drained soils along the watercourse. The hydrology of the area relates to the park's position within the catchment of Currumbin Creek. Most of the headwaters of the streams that pass through the park originate outside the park's boundaries. Landuse activities in this catchment have a significant influence on the ecosystems within NRNP.

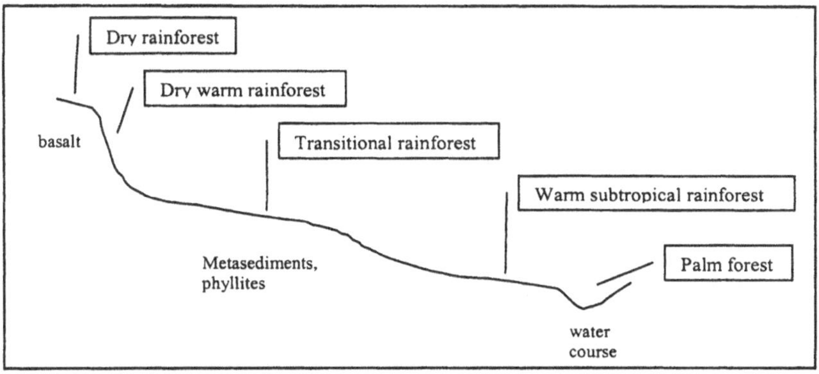

Figure 9.2 Vegetation communities in relation to land form, Nicoll Rainforest National Park (Roughan, 1986:49)
(Where Type 1: dry rainforest [Araucarian microphyll vine forest]; Type 2: dry warm rainforest [Araucarian notophyll vine forest]; Type 3: Transitional forest [Sclerophyll vine forest]; Type 4: warm subtropical rainforest [Complex notophyll vine forest]; and Type 4: palm forest [notophyll feather palm vine forest])

The land surrounding NRNP is the source of a number of threatening processes. The dominant land use on the lower slopes and valley floor of Currumbin Creek is dairy farming, with banana farming occuring on the steeper slopes of the McPherson Range. Chemical fertilizers are commonly used by land holders to retain nutrients. Rural residential development also occurs near the park. Roughan (1986) comments that local property owners do not rely on the NRNP for any specific resources, nor do they appear to be adversely affected by the park's presence. Future trends are likely to see agriculture becoming less important due to pressures for more rural residential development.

The matrix (Table 9.1) illustrates the existing and potential negative impacts derived from the interrelations between the elements of the park and its surrounding environment. The nature of the negative impacts provided a basis for defining the APZs. In the palm forest, for instance, these impacts were water pollution (agricultural and domestic); plant pests; and animal pests (Figure 9.3).

Table 9.1 Matrix of interactions between Nicoll Rainforest National Park and its surroundings

Elements / Threat	Dry Rainforest Affected Processes				Palm Forest Affected Processes						Water Affected Processes			
	Interface		Seed Dispersal		Water		Seed Dispersal		Interface		Surface Water		Ground Water	
	E	P	E	P	E	P	E	P	E	P	E	P	E	P
Plant pests														
Fire														
Animal pests														
Water pollution (agric.)														
Water pollution (domestic)														
Drainage pollution														
Erosion														
Siltation														

(Where E = Existing and P = Potential)

Major impact Significant impact Low impact

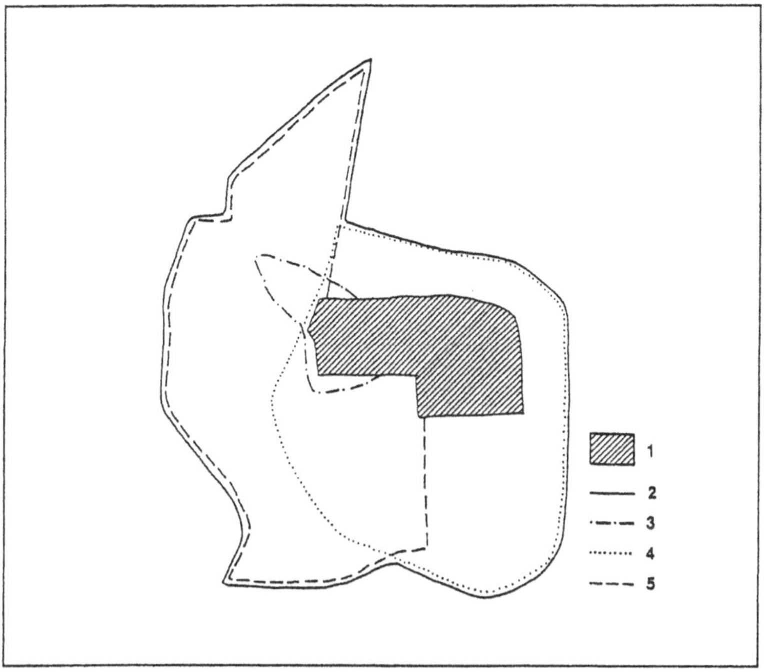

Figure 9.3 Palm forest analytical protection zone, Nicoll Rainforest National Park

(Where 1. NRNP boundary; 2. APZ boundary; 3. Noxious weeds/exotic plants; 4. Water pollution [agricultural/domestic]; and 5. Feral animals)

The definition of APZs for each of the identified elements confirmed that similar threats affected the elements of NRNP, although impacting in slightly different ways. The areas in which each negative impact needed to be controlled to protect particular elements were then defined and land use policies facilitating this control determined. For example, the policy regarding water pollution (agricultural) required control of intensive agricultural activities such as rural industry, feeding lots and animal husbandry and strategies to encourage reduced chemical input into agricultural production in this area. Each policy was accompanied by comments specifying implementation procedures.

By synthesising these threats and policies, seven EPZs were defined. Each EPZ was characterised by describing its purpose, scope and the land use policies needed to protect the different elements of the park from specific negative impacts. Two examples, that of plant pests and water pollution (domestic) are illustrated in Figure 9.4 and Tables 9.2 and 9.3.

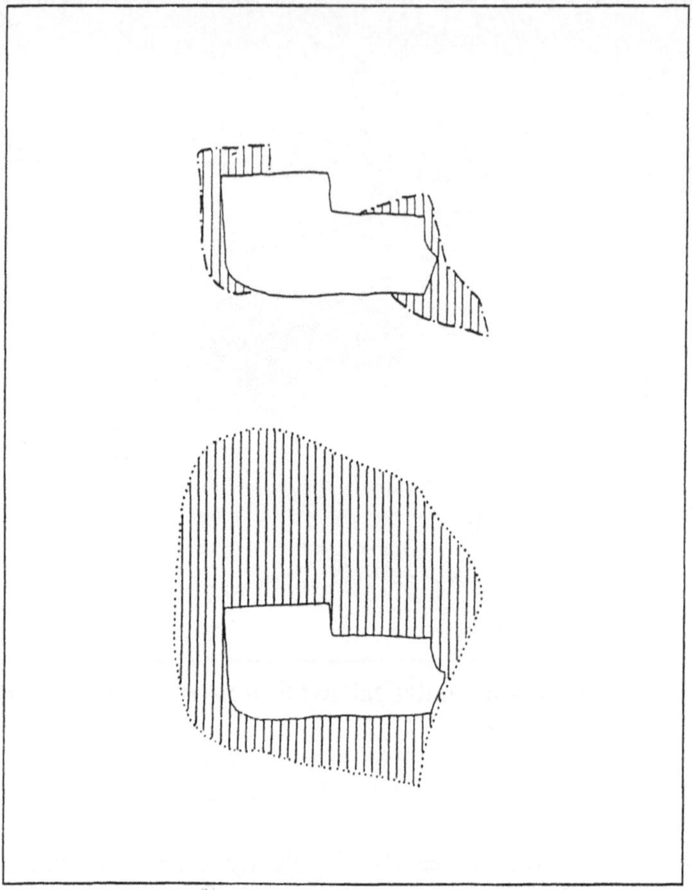

Figure 9.4 Plant pests elementary protection zone and water pollution (domestic) elementary protection zone

Table 9.2 Plant pests elementary protection zone

Purpose	To protect the park from the incursion of noxious weeds and exotic plants.
Scope	The EPZ covers land both adjacent to and at a distance from the park.
Land use policies	• to be extensively planted with native tree species common to the Currumbin and Tweed Valley region; • prohibit the planting of exotic species in this area; • clearing of native vegetation requires the consent of Council and the Queensland Department of Environment; and • regularly clear land of noxious growth and exotics.

Table 9.3 Water pollution (domestic) EPZ

Purpose	To protect the park against water pollution from domestic wastes.
Scope	The EPZ encompasses the catchment areas of Currumbin Creek tributary, which flows through the park.
Land use policies	• residential development must be connected to a town waste water reticulation system, and septic tanks are not permitted; • subdivisions smaller than the existing 5 hectare blocks are prohibited; • disposal of noxious effluent or wastes is prohibited; and • building within 40 metres of any stream is prohibited.

Figure 9.5 indicates the boundaries of all seven EPZs and illustrates how the synthesis of the EPZs results in the formulation of an overall buffer zone for NRNP.

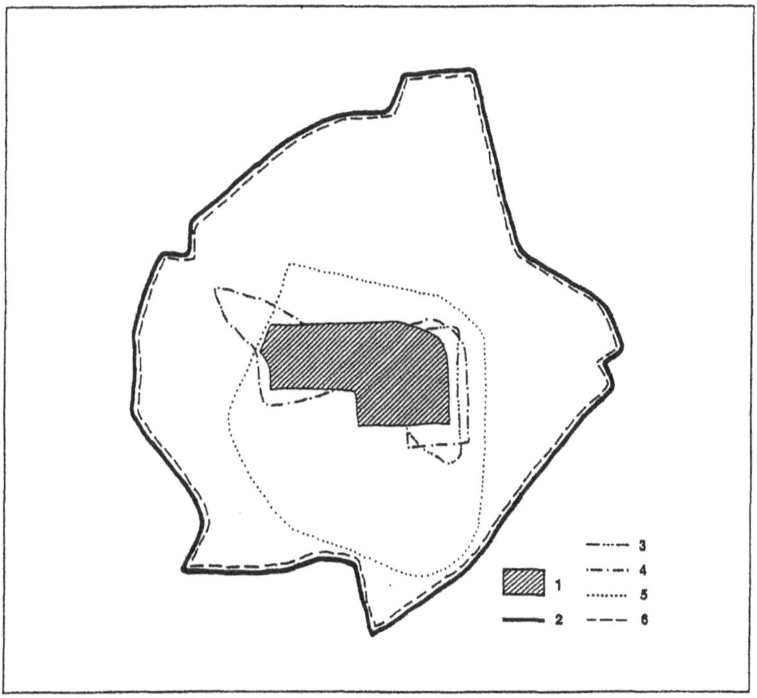

Figure 9.5 Final buffer protection zone for Nicoll Rainforest National Park

(Where 1. NRNP; 2. Buffer boundary; 3. Erosion EPZ; 4. Noxious weeds/exotic plants EPZ; 5. Water pollution, siltation, drainage pattern disruption EPZs; and 6. Feral animals EPZ)

Case Study 2: Mountain Pygmy-Possum Buffer

The impetus for designing a buffer for the Mountain Pygmy-possum *B.parvus* came from students undertaking a course in Resource Management and Planning at The University of Queensland (Australia) in the early 1990s. The students, with limited time and resources, undertook a theoretical examination of the issues, using the BZP method as their primary planning tool. More detailed work was later undertaken by Peterson and the case study is a reflection of these contributors. In addition much of the biological data on *B.parvus* is based on the research of Mansergh and Broom (1994) and Mansergh et al. (1989, 1991).

Location and Description

Burramys parvus, the mountain pygmy-possum, is a small, nocturnal marsupial whose habitat is restricted to disjunct patches throughout the Australian alpine and subalpine environment at altitudes above 1,400 metres (Figure 9.6). It was first described from fossil bones found in the state of

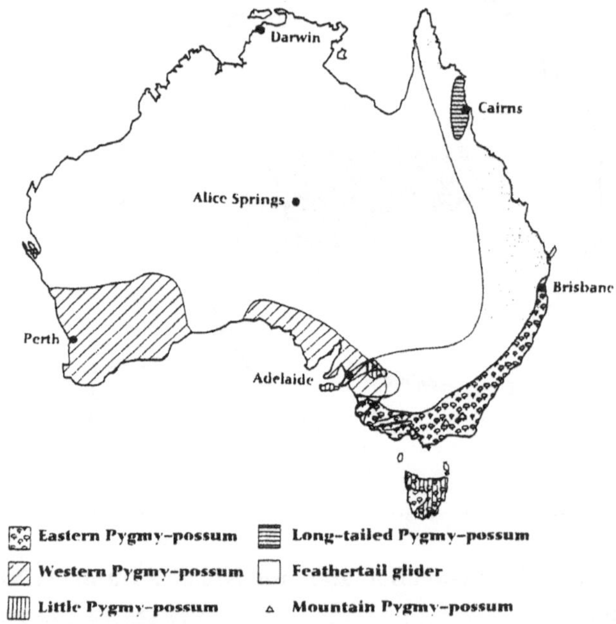

Figure 9.6 Distribution of Mountain pygmy-possum and other possums and gliders (Mansergh and Broom, 1994:63)

New South Wales (NSW) in 1895 (Broom, 1895), and in 1966 the first live animal was discovered at Mt. Higginbotham (Victoria). Burramys is classified as a 'vulnerable' species by the Australian and New Zealand Environment and Conservation Council (ANZECC, 1991), and as a 'threatened' species in Victoria's *Flora and Fauna Guarantee Act 1988.* It numbers approximately 2,600 adults and has a total known habitat of less than 1,000 hectares. Two genetically distinct populations are distinguishable: one in Kosciusko National Park (NSW); and the other in the Victorian alps in three sub-populations located at Mt. Bogong, the Bogong High Plains and Mt. Higginbotham.

Climate warming has caused the contraction of Burramys' habitat over the last 1,500 years and some populations are isolated and contain habitat too small (<1 hectare) to support viable populations. Burramys is vulnerable to a range of natural and human-induced threatening processes that have negatively influenced its survival in the past, and which continue to exert pressures that threaten its long-term viability.

In response to the threatened status of Burramys, the Victorian Department of Conservation and Natural Resources (DCNR), in 1989, developed a management strategy for core Burramys' habitat and stated, 'buffer zones.... are important to the integrity of habitat' (Mansergh et al., 1989:21). A management area comprising core habitat and a surrounding buffer zone was developed, its purpose being to:

- maintain, conserve and enhance existing Burramys populations and habitat;
- enhance the present Burramys distribution; and
- increase public awareness of the conservation status and habitat requirements of Burramys.

A range of management prescriptions and guidelines were applied throughout the management area. The homogeneous structure of the buffer area however, fails to identify for resource managers or planners, the precise areas where particular threats are present, and the specific management actions that are required to target these threatening processes.

The BZP methodology was applied to develop a heterogeneous buffer zone for core Burramys' habitat at Mt. Higginbotham. By utilizing the BZP approach management can be directed to specific areas where threatening processes operate and hence more efficiently and effectively minimise or eliminate these negative impacts. The long-term viability of Burramys depends on providing a high level of protection to the core habitat while minimising external threatening processes. The buffer strategy is based on an analysis of existing data on the species' biology and habitat requirements, and major threatening processes.

Biology

B.parvus is the largest of five species of pygmy-possums (*Corcartetus* spp.). Its adult weight varies seasonally, but averages 35-40 grams. Females with pouch young may weigh more than 70 grams. Its total length is about 28 centimetres, including a prehensile tail of about 13-16 centimetres that can be curled to grasp branches or carry nesting material. Burramys has fine dense fur which is longer (12 milimetres) on its back than its underneath (7 milimetres).

Burramys habitat on Mt. Higginbotham covers about 40 hectares and contains approximately 900 adults, with a density of up to 116 individuals per hectare on the western slopes during the breeding season. This is the highest concentration of Burramys habitat in the Australian alps.

The mountain pygmy-possum is omnivorous and feeds on Bogong moths *Agrotis infusa*, other invertebrates, seeds and fruit. The species enters a short-term hibernation (torpor) generally from March to September and caches food for use over the winter. It is the only marsupial known to store food and hibernate. Burramys makes a grass nest that it often shares communally with others.

Burramys are segregated according to sex, with adult females occupying better quality habitat (Figure 9.7). At Mt. Higginbotham the adult females occupy boulderfields above 1,700 metres. These are overlain by *Podocarpus* heathland that provides a dense shrub layer, resulting in a high quality habitat. Males move to this area from less sheltered habitat, generally at lower altitudes (1,400-1,600 metres), to breed during October to December. Males move out of the favoured areas after the young are born. By January most young are independent, and within one month all juvenile males leave their natal area. Some juvenile females also disperse. Segregation according to sex is not altitude dependent however. It occurs where good and poor quality habitat are at the same altitude. In these situations males tend to occupy the poorer quality, shallower boulderfields.

Usually only one litter is produced each year, with a gestation period from 13 to 16 days. Most males and females breed in their first year. Females usually carry from two to four pouch-young for around 30 days. They are fully weaned and become independent around nine weeks.

Survival is sex dependant, with males having a 13 per cent chance of survival to adulthood and females about a 50 per cent chance. Burramys may live to around 12 years, with females in general having greater longevity. This may be due to males seasonally inhabiting more marginal habitat, the result of forced dispersal from core habitat areas.

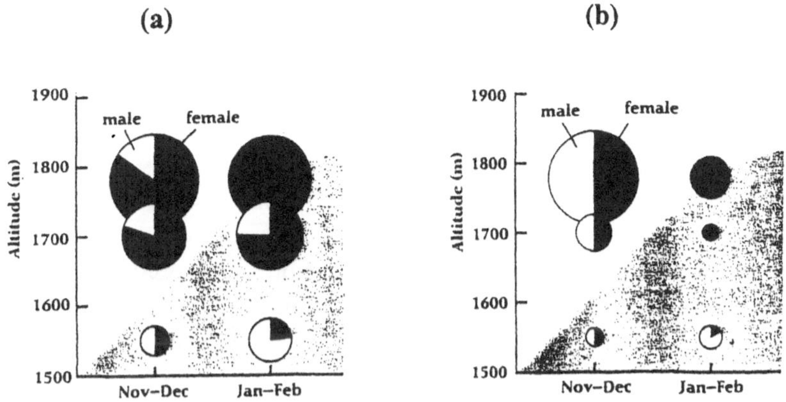

Figure 9.7 Density and population structure of (a) adult and (b) juvenile Burramys at different elevations on Mt Higginbotham (Adapted from Mansergh and Broom, 1994:70)

Females are highly sedentary with small home ranges of $589m^2 \pm 258m^2$ on the western slopes of Mt. Higginbotham. The average maximum home range diameter was 33m ± 8m. Female home ranges may overlap, comprising the habitat of between four and 10 younger females, many of whom are related. The home range may expand during the non-breeding season. Males range more widely, over distances of up to around two kilometres, with home ranges being less clearly defined.

Movement is critical for the survival of the species in terms of mating and later dispersal, and obtaining food. Mansergh and Broome (1994) observed Burramys travelling up to six kilometres through the boulderfields in search of Bogong moths. When the moths are not available (from April), Burramys movement becomes more restricted, to within 100 to 200 metres of their nest sites. Movements in winter are very restricted, although they do travel under the snow. Burramys prefer to move through areas with a dense shrub layer, rather than grassland and cleared areas.

B.parvus habitat occurs in areas that have an average annual precipitation of over 1,500 milimetres. Winter temperatures range from 10°C to below freezing and the area experiences snowfalls and gale force winds. To survive in this climate Burramys occupies a narrow niche. Its habitat is located in boulderfields (Figure 9.8) and associated corridors of rock and shrubs at altitudes from about 1,400 metres to 2,000 metres. The granite and basalt boulders provide a suitable microclimate which is snow free, affords protection from predators, winds and summer sun, and where temperatures seldom drop below freezing. The boulderfields usually occur in depressions or on the tops of gullies and they provide an adequate water supply, while the tree and shrub cover, consisting primarily of mountain plum-pine *Podocarpus lawrencei*, as well as alpine pepper *Tasmannia xerophila*, alpine mint-bush *Prastanthera cuneata*, and dusty daisy-bush *Olearia phyloggopappa*, provide food, protection from predators and a movement corridor. The boulderfield snow gums *Eucalyptus pauciflora*, while not essential Burramys habitat, provide seeds and nest sites.

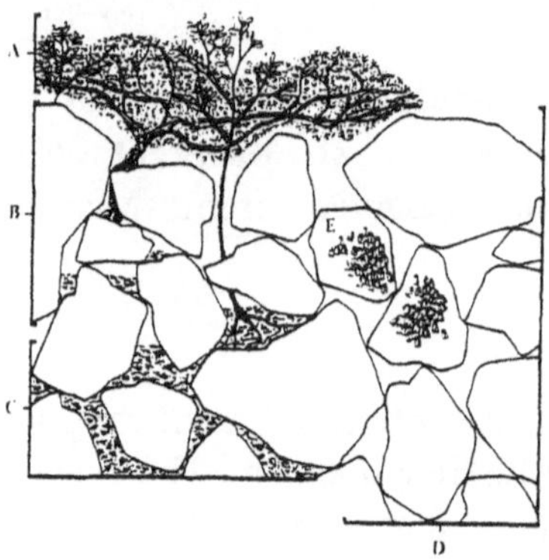

Figure 9.8 Cross-section of a boulderfield showing the important features of Burramys habitat (Mansergh and Broom, 1994:34)

(Where: A: shrub layer; B: piles of boulders; C: leaf litter and Bogong moth remains over soil and partially buried rocks; D: where the boulders are more than two metres deep, light levels are insufficient at the soil to allow plant growth; E: congregations of Bogong moths)

The core habitat at Mt. Higginbotham represents the species' essential breeding habitat and movement corridors. It is in a variety of tenures. Most of Mt. Higginbotham's western slopes are within Bogong National Park. The majority of the remaining core habitat is within Mt. Hotham Alpine

Resort, managed by the Alpine Resorts Commission. There is also a small area (0.8 hectare) of private land. The core habitat is bisected by a road, the Alpine Way, and contains a number of buildings and associated skiing infrastructure, in particular, the 'Blue Ribbon' ski run.

Threatening Processes

The very specific habitat requirements of Burramys, the fragility of its ecosystem, and conflicting land uses and activities have contributed to the species' decline in numbers. The Mt. Hotham/Higginbotham region is important for year round recreational activities, including down-hill and cross-country skiing, and summer hiking, mountain climbing and fishing. There is conflict between the recreational uses of the park and the preservation of the area's biological diversity, and in particular the survival of local populations of Burramys. There are several threatening processes affecting Burramys on Mt. Higginbotham. These are described below.

Vegetation Removal and Degradation

Vegetation, particularly *Podocarpus* heathland, is important in stabilizing the temperature within the boulderfields and providing food, safe shelter and corridors for movement. The vegetation that forms Burramys' habitat may take hundreds of years to develop (Gullan and Norris, 1981) and is sensitive to disturbance. Its removal and degradation are due to: building construction (e.g. resort development and housing); infrastructure development (primarily roads such as the Alpine Way); ski-run construction; slope grooming; oversnow transport; recreational trail construction and bushwalking; fire; generation of rubbish and litter (e.g. plastic sheeting), which may cover and kill important plants; and competitors (e.g. house mice *Mus musculus*, rats *Rattus fuscipes, R.rattus*, rabbits, hares, goats, sheep and cattle). Competitors may selectively graze *P.lawrencei* habitat, resulting in its elimination. Cattle grazing occurs in and near Burramys habitat within licensed grazing blocks.

The primary impacts of vegetation removal and destruction include:

- *habitat fragmentation and isolation*, which may impede the movement of Buramys, a species that does not readily cross cleared areas such as roads, ski runs and grassed sites. The high altitude and good quality breeding areas may be separated from the male and juvenile dispersal areas, preventing male movement to female breeding areas and resulting in overcrowding of natal areas by juveniles and males, who are unable to disperse. Habitat fragmentation may interrupt the breeding cycle and social organisation of the species, reduce fecundity, prevent recolonisation of isolated areas due to excessive distance, and lower the probability of survival of the breeding stock. Roads on the eastern slopes of Mt. Higginbotham have resulted in reduced population size, and composition of the Burramys' population. The Alpine Way is thought to have caused the local elimination of small numbers of animals from a patch of habitat above the road;

- *loss of food sources* may impact on the species' ability to cache food and survive the winter hibernation;
- *loss of shelter and exposure to predators* may affect the survival chances of individuals as the vegetation is important in protecting Burramys from predators;
- *micro-climatic changes* may produce greater temperature extremes within the boulderfields. Vegetation is important in regulating temperature and loss of the *Podocarpus* heathland may impact on the species;
- *interference with nesting* due to the loss of ground cover Burramys usually nest close to the ground and utilise nesting material found in their vicinity; and
- *other factors* which include increased runoff, soil erosion and deposition and soil compaction (see below).

Weeds

Development near preferred Burramys' habitat may result in the introduction and spread of plant species not naturally occurring in the core habitat (e.g. dumped as garden cuttings or propagules spread by birds, small ammmals and humans). Weeds may become more abundant in disturbed areas (e.g. roads, buildings, recreational areas, and along ski-runs, where exotic species have been planted) and provide less food and shelter for Burramys.

Erosion, Deposition and Compaction

The boulderfields provide habitat for the Mountain Pygym-possum. Both basalt and granite derived alpine soils are shallow and highly susceptible to erosion and slumping. As the species needs sufficient depth below the boulders to provide for its habitat requirements, soil erosion and subsequent deposition, as well as compaction threaten its habitat. These threats to the boulderfields are caused by: vegetation removal; ski-run development and slope grooming (e.g. heavy machinery and explosives are used to remove boulders and graders are used to maintain the ski-runs); skiing and over snow transport; bushwalking; building and road construction; and lack of storm water drainage.

The impacts of these activities include:

- *boulder compaction* caused by heavy machinery and constant above ground traffic (human and machine);
- *increased erosion and deposition* in disturbed areas (e.g. roads and ski-runs), which may be prone to increased amounts of water runoff, resultant erosion of soil and its deposition amongst the boulders making the habitat shallower, or even removing it completely;
- *covering of low-lying vegetation* by eroded soil, thus reducing food availability and shelter, and altering the micro-climate within the boulders; and
- *soil compaction* (especially from heavy machinery), which could affect hibernation due to the alteration in micro-climate under the snow.

Hydrologic Changes

As boulderfields generally form in depressions and gullies Burramys' habitat frequently occurs along drainage lines. Consequently, any water pollution or disruption to the drainage pattern may affect Burramys. The main causes of hydrologic changes are development for infrastructure and resort related structures and activities associated with increased water pollution, such as accidental sewerage discharges, run-off of oil from roads, litter, and the addition of nutrients (primarily from fertilizers). Impacts include diversion of drainage flows, altered water flows, increased deposition of eroded soils in waterways and boulderfields, and potential impacts on the vegetation structure and composition, resulting in altered biological processes.

Noise

Many species are susceptible to noise, which may affect nesting and movement and result in a change away from preferred habitat areas. Noise may be a potential threat to *B.parvus*. The primary sources of noise include: explosives used in road construction, slope grooming and other development; snow grooming machinery and graders; road traffic; and human activities (bushwalking and resort development).

The outcome of these activities is increased noise and vibration, which may impact on the species, especially during hibernation. Excessive arousal may deplete fat reserves resulting in death (Mansergh and Broome, 1994). Evidence is not available to confirm the possible negative impacts of noise however.

Fire

The vegetation within Burramys' habitat is sensitive to fire. *Podocarpus*, a native conifer, is flammable and slow growing. The sensitivity of the habitat to fire, the time required to develop suitable habitat, and the relatively restricted nature of the habitat make fire potentially extremely dangerous (Mansergh et al., 1989). Fires have penetrated into Burramys' habitat in the past. The fire risk increases during the summer season with an increase in the number of outdoor campers and visitors. The main causes of fire are fire escapes from campsites and resort areas, and natural factors.

Burramys may survive fires restricted to the surface vegetation, although they are susceptible to heat stress and may die if they can't escape from a fire. The impacts of smoke inhalation on Burramys are unknown. Fire may result in a loss of vegetation, alter vegetation structure and composition, reduce food availability and cause local populations to be eliminated or significantly reduced in size. Many may also die indirectly from the slow regeneration of fire-affected areas. Burned areas may also have a lower chance of being successfully recolonised and hence fire may result in local extinctions of Burramys' populations, unless reintroductions take place.

Competitor Species

Animals such as rabbits, hares, mice, goats, pigs and other livestock compete with Burramys for food. Some rodents (e.g. *Rattus fuscipes* and *R.rattus*) may colonise Burramy's habitat. Primary sources of house mice are human structures and hay brought in to mulch disturbed areas. Litter and rubbish can function as habitat and a food source for introduced animals. Hard-hoofed livestock trample the soil and vegetation, and rabbits and hares disrupt the soil structure by digging burrows. Non-target poisoning and trapping may also occur as a result of programs aimed at controlling introduced rodents (mice). Many introduced species may also carry disease, dangerous to Burramys.

Predator Species

Red foxes *Vulpes vulpes* and dogs *Canis familiaris* are known to prey upon Burramys (Green and Osborne, 1981). Cats *Felis catus* are also thought to be a predator. Predators are abundant due to the high density of small mammals and are related to the surrounding human settlements and resorts. It is likely that predators may transfer to native prey species after plagues of introduced rodents (Mansergh et al., 1986).

Climate Change

Although beyond the scope of the current buffer strategy it is important to note other important threats to *B.parvus*. Rising temperatures will have a significant impact on the species due to its geographically isolated populations and restriction to the alpine-subalpine region. Burramys is adapted physiologically and behaviourally to colder rather than warmer climates. Possible effects of climate change could include: vegetation change, including weed invasion; reduced rock scree formation (due to repeated freezing-thawing); increased rainfall that may increase erosion within or into habitat; and changes to occurrences of Bogong moths. As the species is intolerant of temperatures above 30°C, present marginal habitat (shallow scree) may become uninhabitable.

Changes in Food Supply

The Bogong moth is important in the diet of Burramys. Breeding success of Bogong months may be influenced by rainfall and pastoral activities in their breeding grounds. Any changes to the abundance of Bogong moths may impact severely on Burramys' populations (Mansergh and Broom, 1994).

Buffer Strategy

The main elements within Burramys core habitat (Figure 9.9) at Mt. Higginbotham include the abiotic elements of the boulderfields, soil, micro-climate, and hydrology; and the biotic elements consisting of *Podocarpus* heathland and Burramys' diet, hibernation, movement, nesting, shelter and population structure. An interaction matrix was developed to highlight the relationship between the threatening process an elements of the core habitat. Five EPZs were delineated reflecting the area of influence of each identified threatening process. The final buffer zone was produced by synthesising the EPZ boundaries (Figure 9.10). A brief summary of buffer policies relating to vegetation management is presented in Table 9.4.

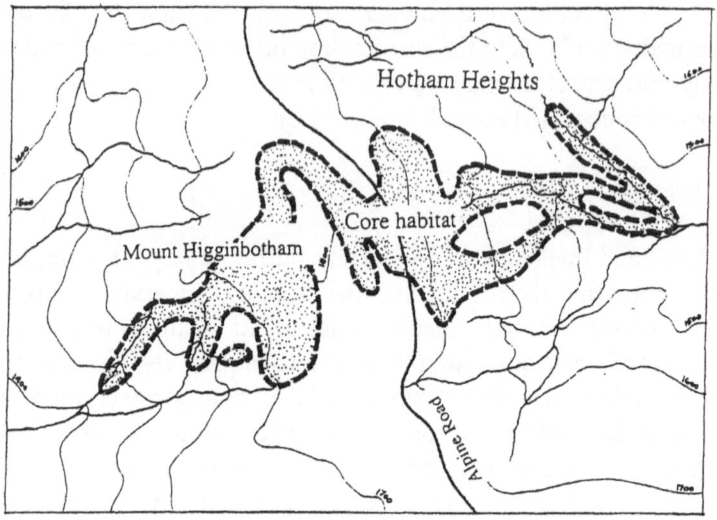

Figure 9.9 Core habitat for *B. parvus* at Mt. Higginbotham

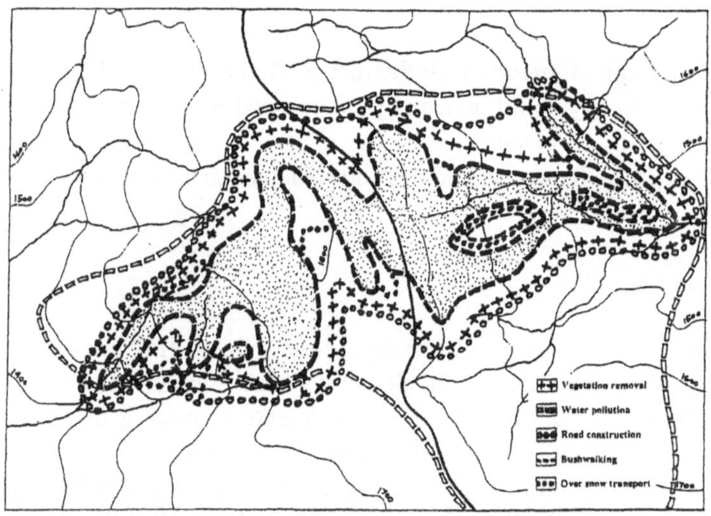

Figure 9.10 Final buffer for *B. parvus* habitat at Mt. Higginbotham

Table 9.4 Vegetation removal and degradation policies for implementation in the Mountain pygmy-possum buffer, Mt. Higginbotham (sample only)

Purpose	To minimise the removal and degradation of vegetation, especially *Podocarpus* heathland and to restore and rehabilitate degraded sites.
Scope	Land within a minimum of 30 metres from core Burramys' habitat and important movement corridors.
Land use policies	• retain the condition and extent of native vegetation within 30 metres from core Burramys habitat; • avoid, where possible, the use of over-snow transport on land within 10 metres of core habitat (except in emergency rescues, unless on designated roads or tracks); • revegetate and reestablish, using plants endemic to the Mt. Higginbotham area, multiple connections among formerly connected habitat patches. This will include establishing linking corridors, especially along waterways, across ski runs, and under roads, where necessary; • revegetate slopes modified for downhill skiing (e.g. native dwarf heathland species, such as *Hovea longifolia* or slightly taller heaths where corridors are required); • design and locate roads to minimise their impact on Burramys' habitat; • minimise the number of access tracks into the core habitat, so that human traffic is kept to a minimum; • avoid extensive cut and fill by constructing roads that follow natural ground contours; • incorporate movement corridors, where necessary, under roads; • minimise road pavement and shoulder width to reduce vegetation clearing;
Implemen -tation	Dept. of Conservation and Natural Resources and Alpine Resorts Commission

Conclusion

This case study highlights the effectiveness of the BZP model in enabling a logical and structured approach to the implemention of approariate planning and management strategies to better conserve wildlife habitat. Although the final buffer boundaries for *B. parvus* habitat may need further refinement, the associated policies provide clear direction to the management agencies on ways to achieve more sustainable outcomes within the core and surrounding habitat.

Case Study 3: Boondall Wetlands

The original impetus for the Boondall Wetlands buffer came from Richard Lloyd and Colin Wade who undertook the research project as part of their postgraduate course requirements in Resource Management and Planning, offered through The University of Queensland (Australia). Additional research and mapping was completed by Peterson.

Location and Background

Wetlands are ecosystems that support great diversity of life forms and often have a complex hydrology. Although Australia is one of the driest continents, since European settlement, approximately half of its wetlands have been 'exploited, manipulated and degraded as wasteland to be developed, a water supply in drought, or a dumping ground for rubbish' (Department of Environment, 1996a: 1). Since the 1960s greater interest has been shown in wetlands and they are now recognised as important ecosystems that need to be conserved.

In simple terms, wetlands are lands with soils that are periodically flooded (Williams, 1990). They may be defined as areas of permanent or periodic/intermittent inundation, whether natural or artificial, static or flowing, fresh, brackish or saline, including areas of marine water, the depth of which at low tide, does not exceed six metres (Department of Environment, 1996a:4). They are an integral part of the hydrology of an area, absorbing and slowly releasing floodwaters, thus conserving water and protecting against flooding. Coastal wetlands absorb wave energy and reduce erosion on estuarine shorelines. They filter excess nutrients and sediments from run-off that would otherwise be discharged into rivers and creeks. Wetlands provide a nursery for fish and crustaceans, and are an integral part of the life cycle of water-dependent wildlife, providing shelter for local and migratory species, particularly birds. Wetlands are a valuable recreational resource, providing opportunity for outdoor education and scientific study; are a significant cultural resource for indigenous people; provide for the well-being of people through landscape diversity and aesthetic enjoyment; and provide archaeological and historical benefits.

Traditional management of wetlands has focused on establishing protected areas. However, many such areas are facing problems including air and water pollution, loss of habitat and altered hydrologic processes. As open systems, wetlands are influenced by activities well beyond their boundaries. Local planning controls have had limited application to conservation and management of wetlands, mainly because many potential threats are due to gradual alteration resulting from a continuation of existing uses or from up-stream changes caused by alteration to the hydrological regime or water quality (Department of Water Resources, 1990).

This case study focuses on examining the important values and threats to Boondall Wetlands and utilises the BZP methodology to develop a preliminary, theoretical buffer for the wetland. The Boondall Wetlands Reserve is located in South East Queensland, about 15 kilometres north east of the state's capital city, Brisbane (Figure 9.11). It has an area of approximately 1,200 hectares bounded to the north east by Moreton Bay, a Ramsar site or an internationally recognised wetland site. To the east the boundary follows Kedron Brook estuary, and to the north west, Cabbage Tree Creek. Much of the western and southern boundary follows the Gateway Arterial Road. Nundah Creek and Nudgee Creek also flow through the reserve.

Figure 9.11 Location of the Boondall Wetlands Reserve

Boondall wetlands are in close proximity to residential areas, an airport and industrial areas. The surrounding region of South East Queensland is one of the fastest growing regions in Australia, having more than doubled its population in the last 20 years (Department of Environment, 1996). This rapid growth presents many problems for the effective management of this wetland ecosystem.

The Brisbane City Council and the Boondall Wetlands Management Committee produced a management plan for the reserve. It is oriented towards the control of areas contained within the legal boundaries of the reserve. Little emphasis has been placed on the management of external threats, apart from those areas adjacent to the reserve's boundaries.

A variety of management arrangements are being developed to ensure the conservation of Moreton Bay's wetlands. Under the 1993 Moreton Bay Strategic Plan, Queensland and local government agencies involved in Bay use and management cooperate to protect the resource. Conflicting demands are weighed against wise, ecologically sustainable use. Moreton Bay Marine Park was declared in 1993. The Bay's marine and intertidal areas are managed through a marine park zoning plan. A fisheries management plan is being developed to protect the fishery and a coastal management plan is being prepared in response to conflicting uses in the region and the need to effectively manage a number of significant threatening processes. Queensland's environmental protection policy for water will control waste discharges. The Brisbane River Management Group coordinates the activities of Queensland and local government agencies that operate in the catchment. An environmental protection policy and management plan for the Brisbane River should result in improved catchment management and reduced impacts on wetlands. However, as stated by Queensland's Department of Environment (1996b:4), perhaps one of the greatest threats to Queensland's wetlands is 'inadequate buffers between wetlands and surrounding lands'.

Resources within Boondall Wetlands

Boondall Wetlands comprises physical, sensory and cultural resources. The components of each category are briefly described in the this section.

Physical Resources

Geology (including geomorphology and soil) Boondall Wetlands is an estuarine system subject to the combined influence of tidal marine water and freshwater flow from rivers and creeks. It is a low-lying coastal wetland undergoing long-term sedimentation processes. In the last 200,000 years Moreton Bay has fluctuated in depth by up to 45 metres from its present level. Sea level changes have been a powerful influence on the geomorphic elements of the area. The surface layer of recent soil deposition is held together by ground cover, beneath which lies an active layer of peat sediments. Gradual inundation is not a threat to the structure of soils, although higher concentrations of run-off will weaken this structure.

Hydrology Boondall Wetlands is a transitional environment between land based ecosystems (e.g. forests and grasslands) and water based ecosystems (e.g. river, creeks, and bays), and are highly productive and diverse. It is a low-lying area subject to regular inundation and periodic drying out according to conditions and seasons. The most important influences on the wetland are tides, waves, currents, river discharge and ground-water seepage. The water table is close to the surface and its quality is influenced by changing surface conditions. The wetland is part of a larger hydrologic system and its conservation depends on the control of successional development.

Micro-climate With its low lying, flat topography and high moisture content the reserve has a unique micro-climate, which is relatively stable. However, the dynamic nature of the wetlands makes them susceptible to human induced changes that may threaten the micro-climate.

Flora The wetlands support a diverse mix of plants and animals that are intricately balanced and depend on one another for food and shelter (Department of Environment, 1996b). The reserve is the largest intact area of coastal vegetation in the Greater Brisbane Region, containing 120 different species of native flora, including rare and endangered species. The forest types represented include: eucalyptus woodland; melaleuca woodland; remnant rainforest; mangroves; ironbark forest; freshwater habitat; casuarina forest; and grassland. These species are adapted to living in wet conditions. The mangroves, in particular, are a nursery for Moreton Bay's fish, prawns and crabs.

Fauna A wide variety of butterflies (27 types), frogs (21), reptiles (37) and bird species (230) have been identified in the Boondall Wetlands. Thirty of the 43 shorebird species that visit Moreton Bay's intertidal flats are migratory species listed under the Japan Australia Migratory Bird Agreement (JAMBA) or the China Australia Migratory Bird Agreement

(CAMBA). Most migrate from Arctic or sub-Arctic regions at the end of the breeding season moving to the southern hemisphere and stopping to rest in the Bay. Waders feed here and store energy for their return trip north, to breed again (Department of Environment, 1996b). The inshore marine environment also provides a valuable breeding ground for fish and other marine animals.

Sensory and Cultural Resources

Boondall Wetlands is an important area where residents can appreciate the diversity of nature, in particular its variety of bird and frog fauna. The Boondall Management Plan places high priority on screening adjacent incompatible land uses in order to retain a sense of isolation and the feeling of being surrounded by nature.

The reserve contains evidence of past Aboriginal inhabitants and activities in the form of campsites and shell middens in the Nudgee Beach area and a burial ground on the promontory between the mouths of Nudgee and Nundah Creeks.

Land Surrounding Boondall Wetlands

Land surrounding the reserve is designated for residential, non-urban, future urban and special uses. The wetland is bordered by the Boondall Entertainment Centre, a sewage treatment works, solid waste disposal site and an industrial area. Significant existing and potential threats result from these adjacent land uses. As wetlands are frequently fragile and dynamic ecosystems, these external threats have the potential to impact negatively on all of the wetland's resources and to deflect or halt primary successional processes.

Five APZs were identified for flora, fauna, hydrology, geology and sensory resources and an interaction matrix was devised to better understand the significance of the relationhips between the external threatening processes and the resources of the core.

Elementary Protection Zones

Eight elementary protection zones were identified for the wetlands, with specific policies related to fire, noxious and exotic plant species, water pollution, altered hydrology, feral animals, vegetation removal and degradation, noise and recreational activities. Management of the threatening processes was directed at ensuring the sustainable use and development of the wetland and its adjacent land and integrated catchment management. The EPZs for fire, noxious and exotic plant species, altered hydrology, vegetation removal and degradation and noise are examined in more detail below and illustrated in Figure 9.12.

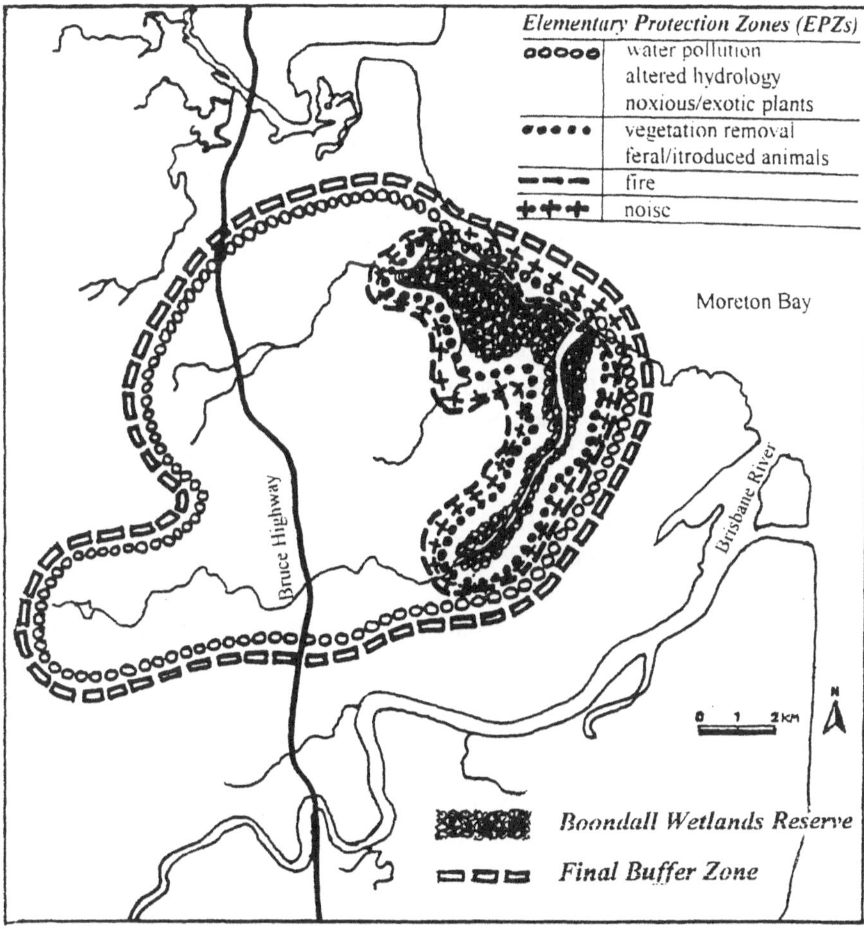

Figure 9.12 Elementary protection zones and final buffer for Boondall Wetlands

(Where A. Fire EPZ; B. Noxious and exotic plant species EPZ; C. Altered hydrology EPZ; D. Vegetation removal and degradation EPZ; E. Noise EPZ; F. Final buffer zone)

Fire EPZ

Fire is a serious threat during the wetland's seasonal drying out period (June to August). Fire may disrupt nutrient cycling, causing a loss of nutrients from the ecosystem in the form of smoke and releases into the soil, where nutrients may be leached from the system or become attached to exchange sites on the peat (Williams, 1990). Fire may destroy peat beds; change the vegetation composition; simplify habitat; reduce the organic layer; expose roots and rhizomes; increase sedimentation and water temperature; and destroy ground cover and nest sites (e.g. hollow trees). Bushfires in the upper catchment may indirectly affect the wetland by increasing the probability of erosion in the burnt out areas, and hence sediment and nutrient levels in the wetland (Department of Water Resources, 1990). The relative isolation of Boondall Wetlands from similar ecosystems may limit re-invasion of plant and animal species following a severe fire.

The sources of fire are many, including fire escapes from neighbouring residential areas, arson and natural causes. Increasing urban development and recreational use of the area may result in higher fire risk.

The EPZ fire policies relate to minimising the impact of uncontrolled fires on land within about one kilometre of the landward boundary of the park and incorporate education, as well as control strategies.

Noxious/Exotic Plant Species EPZ

Introduced plants compete with and alter the composition of the native vegetation; alter the nutrient cycle; interrupt the food chain and habitat requirements of animal species; and clog natural water courses. The main threat for Boondall Wetlands is from lantana encroaching into the reserve from waterways or by seed dispersal.

Horses, cattle and goats were periodically grazed in the reserve, and currently have access to the reserve. These species pose a threat through introduction of weeds, destruction of habitat and competition for food.

Foxes, cats and dogs threaten native wildlife by predating on them. The exotic Mallard ducks found in artificial ponds around Nudgee Golf Course have interbred with the native Pacific black duck, compromising the genetic integrity of the native duck. Cane toads eat young frogs and other small wildlife.

The recommended EPZ policies aim to minimise the impact of invasive plants on the fauna, flora, hydrology, geology and sensory resources and focus on education and awareness raising, establising a register of species that pose a threat to the reserve, and an expansion of the existing weed control programs.

Altered Hydrology (Including Water Quality) EPZ

Changes to the hydrological regime are perhaps the most fundamental and potentially detrimental to Boondall Wetlands. This may include changes in the frequency, timing and depth of inundation. An increase in available water may result from increased run-off in the catchment, brought about by watershed clearing and an increase in non-permeable surfaces associated with urban development. Inundation may favour species which are water dependent, at the expense of some invertebrate species which thrive only when the wetland has been inundated or is drying up. This may cause changes in biomass resulting in imbalances in trophic relationships (Department of Water Resources, 1990).

A reduction in available water may be brought about by upstream withdrawal due to draining land; diverting water by dam construction in the upper reaches of tributaries; flood management which confines rivers to their channels so that they seldom flood the fringing marshes. Wetlands thus experience a sediment deficit and do not accrete quickly enough to keep up with submergence.

Canals may contain stagnant water trapped by soil dumped along the canal causing localised increases in species toleration of stagnation (e.g. mosquitos). The quality of water flowing into and within the wetlands is also crucial. As the wetlands trap water-born sediments, salts, nutrients and chemical pollution from anywhere in the catchment and may receive water from ground-water, the land management over the ground-water basin's catchment will affect the wetland. For tidal wetlands such as Boondall, the estuary is also included as part of the catchment.

Water pollution may be diffuse. Farming and forestry in the catchment can lead to eutrophication, particularly an input of leached fertilisers, especially nitrates. This may result in increased productivity in the wetlands, dominance of robust species and a decrease in species diversity. Excess nutrients may encourage algal growth and alter the vegetative (and wildlife) composition of the wetland. Microbial decomposition of large amounts of organic matter (including sewage and dead algae) can result in prolonged low oxygen levels causing the death of fish and other animals (Department of Water Resources, 1990). Pesticides from agricultural runoff may also accumulate and damage invertebrate communities and sometimes the vertebrates that feed upon them (Williams, 1990). Other possible causes are urban stormwater and deposition from the air. Single point sources include sewerage treatment works, industrial discharge, and oil and other toxin spills. Excessive sedimentation due to increased soil erosion in the catchment may block natural channels, divert flows and raise the soil level so that the wetland loses deep habitats or is flooded less often, causing changes in species composition.

The recommended EPZ policies are to ensure that the wetland receives sufficient water in terms of its quality, quantity, timing and duration, to enable the maintenance of essential ecological processes and to minimise negative impacts on the wetland's fauna, flora, hydrology, geology. The policies apply to the drainage catchment of all streams and creeks flowing into the wetland.

Vegetation Removal and Degradation EPZ

Removal of vegetation in the wetland's catchment may increase runoff, erosion, turbidity and sedimentation of the wetland. Control of vegetation clearing is however, difficult to achieve. Herbaceous plants play an important part in the vegetation and are specifically adapted to the waterlogged conditions, having air channels by which oxygen can be made available to roots, thus enabling them to respire aerobically. Physical damage to these plants may be caused by trampling.

The recommended EPZ policies aim to minimise the impact of vegetation removal on the wetland's fauna, flora, geology, hydrology and sensory resources and apply within 100 metres of the reserve boundary and riparian corridors.

Noise Pollution EPZ

The area adjacent to the wetlands contains a busy international airport, major arterial road, the Boondall Entertainment Centre and industrial areas. Various recreational activities occur on land and water (e.g. water skiing, jet skiing and trail bike riding). Detrimental effects may include reduced numbers of some fauna, and altered behavioural responses.

The recommended policies aim to minimise the impact of noise on the wetland's fauna and sensory resources and apply within 500 metres of the wetland boundary and waterways within the wetland. The policies relate to restrictions on noise recreational activities near wetland wildlife refuges or bird nesting areas, reduced boat speeds and monitoring of noise levels.

Final Buffer

The final buffer is a synthesis of all eight EPZs. It is areally extensive, extending several kilometres to the west of the wetlands and encompassing much of the immediate drainage catchment. It is also complex due to the relatively large number of threats to the wetlands.

Conclusion

Formulation of a comprehensive buffer using the BZP method requires detailed data on a wide range of ecosystem processes. Much of this data were unavailable for the Boondall Wetlands and hence an integrated approach to management of the land and water surrounding the wetland is necessary. Community understanding and appreciation of wetlands may still be at a low level, but awareness is increasing. Long-term conservation of wetlands will require involvement of landholders, leaseholders, government agencies and other users of the area. Total catchment management requires a whole of government approach to ensure integrated management. Continued reliance on sectoral management will result in the wetlands being viewed by each user as a single-product system, precluding other values, and resulting in less diverse wetland ecosystems. The effectiveness of the suggested wetland buffer would be enhanced by the establishment and effective operation of a cross-sectoral management structure, involving a wide range of institutions concerned with wetlands, and resource specialists with a background in environmentally sound and sustainable management of wetlands. Queensland's proposed strategy for wetlands identifies that decisions made in isolation are perceived as the largest threat to the conservation of wetlands (Department of Environment, 1996).

To aid management, the resource and information based must be expanded to better understand wetland processes and therefore, indicators of wetland values. Education is important to build awareness of the importance of the wetland, the problems it faces and the opportunities for conservation. Other crucial issues for the Boondall Wetlands will be global atmospheric warming as a result of the build-up of carbon dioxide and other gases in the atmosphere.

Chapter Conclusion

The three case studies examined in this chapter highlight the diversity of possible applications of the BZP model to conserving the values of natural areas, the first being a partial application to a small rainforest threatened by residential expansion, the second, a theoretical application to better conserve Burramys' habitat in the alpine and sub-alpine areas of Victoria, and the third, a buffer for a wetland located in a very rapidly developing area of Queensland. The model's focus on the values of each core and the ways in which the core interacts with its surroundings provides a sound basis for indntifying the main threats and their mode of interaction with the elements and values of the core and the area over which land use policies are required.

The three case studies highlight the ease of application of the method, its transparency and focus on understanding ecosystem processes. The following chapter extends the scope of BZP by examining its application to the area of cultural heritage conservation.

References

ANZECC (1991), 'List of endangered vertebrate fauna', Ausralian and New Zealand Environment and Conservation Council.

Broom, R. (1895), 'On a new fossil mammal allied to Hypsiprymmus but resembling in some points the Plagiaulacidae', Abstract Proceedings of Lim. Soc. NSW, 1 June 26, p. 11.

Claridge, G. (1994), 'Caring for Wetlands. A Practical Guide to Urban Landcare Projects', National Landcare Program, Queensland Department of Primary Industries, Brisbane.

Department of Environment (Qld) (1996a), 'Wetlands – More than just wet land', Department of Environment, Brisbane.

Department of Environment (1996b), 'Proposed Strategy for the Conservation and Management of Queensland's Wetlands', Department of Environment, Brisbane.

Department of Water Resources (NSW) (1990), 'Issues in Wetland Management', Water Board, Sydney.

Dugan, P.J. (1990), *Wetland Conservation: A Review of Current Issues and Required Action*, IUCN, Gland.

Gullan, P.K. and Norris, K.C. (1981), 'An investigation of environmentally significant features (Botanical and Zoology) of Mt Hotham, Victoria', Min. for Conserv., Vic. Envir. Stud. Ser. Report, 315.

Machliss, G.E. and Tichnell, D.L. (1985), 'The State of the World's Parks. An International Assessment for Resource Management', *Policy and Research*, Westview Press, Boulder.

Mansergh, I. and Broom, L. (1994), *The Mountain Pygmy-Possum of the Australian Alps*, New South Wales University Press, Kensington.

Mansergh, I., Kelly, P. and Scotts, D.J. (1989), 'Winter occurrence of the Mountain Pygmy-possum Burramys parvus (Broom) (Marsupialia: Burramyidae) on Mount Higginbotham, Victoria', *Australian Mammal*, 9, pp. 35-42.

Mansergh, I. and Scotts, D. (1986), 'Habitat continuity and social organisation of the Mountain Pygmy-possum restored by tunnel', Journal of wildlife Management, pp. 701-7.

Mansergh, I. and Scotts, D. (1989), 'Habitat continuity and social organisation of the Mountain Pygmy-possum restored by tunnel', *Journal of Wildlife Management*, pp. 701-7.

Mansergh, I. and Scotts, D. (1990), 'Aspects of the life history and breeding biiology of the Mountain Pygym-possum (Burramys parvus) (Marsupialia: burramyidae) in "Alpine Victoria" ', *Australian Mammal*, 13, pp. 179-91.

Mansergh, I., Kelly, P. and Johnson, G. (1991), 'Mountain Pygmy possum, Burramys parvus', *Flora and Fauna Guarantee Action Statement*, No.2, Department of Conservation and Environment, Melbourne.

Roughan, J. (1986), 'Planning for Buffer Zones – An Application of Protection Zone Planning to Nicoll Rainforest', Bachelor of Regional and Town Planning Thesis, The University of Queensland, Brisbane.

Williams, M. (ed.) (1990), *Wetlands. A Threatened Landscape*, Basil Blackwell, Oxford.

Chapter 10

Buffer Zones in Heritage Conservation Planning

Introduction

While countries are often identified by their natural environment, nations are identified by their cultural inheritance, or 'heritage'. As a consequence, while conservation of the natural environment is a major condition for the physical 'survival' of any country and its inhabitants, conservation of heritage is a precondition for the 'survival' of national identity, which is no less important, although certainly, less tangible. The pages of history have ample evidence of people who are ready to die to protect their national identity.

The word 'cultural heritage' relates to those natural or artificial features of the environment that have resulted from the actions of humans. It was first officially recognised in the 1970s when UNESCO's Committee for the Protection of World Cultural and Natural Heritage defined heritage as the 'built and natural remnants of the past' (Davison, 1991:4), an idea later adopted by other conservation organisations. Davison draws attention to the issue of 'value' when he further defines heritage as 'what we value in the past' (Davidson, 1991: 4). He correctly makes it implicit that heritage is something of worth, in that it reflects some important aspects of the past. In conservation this value is often referred to as 'significance' and it is these elements of significance that should be conserved from the ravages of development and decay. For planners, however, it is more important to answer the question 'in what way is it valued?' as this implies understanding of the element being conserved, why it is important and, above all, how it can be protected.

At this point attention can be drawn to the similarities and differences between conservation of cultural heritage and conservation of nature. Firstly, both are associated with the protection of identified, specific values. However, cultural heritage conservation is always directed at elements of value that reflect the culture of a certain group of people, and is used in reference to environments where evidence of the cultural past may exist. Nature conservation is primarily concerned with protecting and enhancing the diversity of the Earth's biota, which may have important ecological, aesthetic, intrinsic, educational and scientific values.

Secondly, in both cases, conservation is commonly undertaken by assigning legal, 'protected' status to elements or areas of significant and recognised values, which may be threatened with damage or destruction due to various human activities. Legal protection may be followed by protective management practices designed specifically for and applied directly to those elements and areas (typically national parks, nature reserves, heritage precincts and so on).

Thirdly, a conservation strategy is unlikely to succeed if policies are limited only to the element or area given legal protection, without giving adequate consideration of the surrounding environment. This issue has been addressed in the earlier chapters of this book. The idea of surrounds impacting detrimentally upon a protected element or area is recognised in the field of cultural heritage conservation, although considerable differences of opinion exist on how damaging the impacts may be (Carlhian, 1980; Hedman and Jaszewski, 1984; Brolin, 1980). The problem of external impacts threatening protected heritage areas or elements has not been adequately addressed by either planners or architects and hence heritage areas and elements are extremely vulnerable to those external impacts.

In Australia, designation of heritage precincts and places ensures that heritage issues are included in the development decision making processes. Potential threats are, however, limited to those that directly affect a heritage place and outside threats are usually not considered. A good illustration can be found in Part 1 of the *Queensland Heritage Act No. 9* (QSG, 1992) where development which poses an existing or potential threat to the cultural heritage significance of a heritage place is linked with:

- subdivision;
- a change in use;
- demolition of a building;
- erection, construction or relocation of a building;
- work (including painting and plastering) that substantially alters the appearance of a building;
- renovation, alteration or addition to a building; or
- excavation, disturbance or change to landscape or natural features of land that substantially alters the appearance of a place.

External threats, generated in the surrounds are not mentioned in the entire document. Control of the areas outside heritage precincts or around specific elements are not subject to controls that respond to the values of the conserved heritage area. There is, therefore, potential for degradation of those areas by externally impacts. Even when any threatening development occurs in close proximity, but on a site that is not identified as 'significant' in the heritage study, its impact on the heritage site may not be readily identifiable by a planning officer who, in any case, may have no legal authority to address the issue.

However, in Australia, there are some instances where current approaches to conservation planning have, to some degree, noticed and addressed the problem of externally originating environmental threats. These include the following (Marquis-Kyle and Walker, 1992):

- Cooma Cottage (Yass), where the rural landscape setting was considered to be of heritage significance due to its association with the 1824 overland journey of Hume. Proposals to re-route the Barton Highway through part of the area were prevented on the basis that the visual intrusion would result in adverse affects on the appreciation and understanding of the historical significance of the place.
- Lanyon Homestead (Canberra), where its cultural heritage significance was also enhanced by the rural landscape setting and where, in an attempt to retain associations of the homestead with pastoral activities, an urban expansion into the setting of the heritage place was formally controlled.
- Royal Botanic Gardens (Melbourne), where to preserve its landscape qualities and amenity, the Melbourne Metropolitan Planning Scheme requires consultation with the Director of the Gardens before a permit for development is issued, thus ensuring the development will not be visible from within the Gardens, or have the potential to cause air turbulence and/or overshadowing.
- Berrima Township (New South Wales), where a by-pass was developed to eliminate its intrusion on the rural landscape setting of Berrima, thus protecting its historic values.
- The Old South Brisbane Town Hall, which was built in 1982 and symbolised the civic aspirations of the newly created municipality and where views to the site were retained, as significant heritage values, in devising the Expo-88 site plan.
- Ayres Rock (Uluru), where unplanned visitor facilities, motels and an airstrip at Yulara, were criticised for visually impacting on the landscape setting of Ayres Rock and then removed to maintain a setting that permitted appropriate appreciation and enjoyment of the heritage place.

- Shearer's Strikeout Campsite (Barcaldine), where a protection zone surrounding the shearers' strike campsite was introduced to minimise the impact of tourists on this archaeological site.
- Grace Brothers (Sydney), a site with archaeological remnants providing evidence of the history of the campsite associated with retailing in Sydney for more than 150 years. The conservation policy recognised that retail trading should continue on the site, at least in part, thereby maintaining its historic association and significance, that retention of the external facades and awnings was essential, and that any new structure must not interfere with the appreciation of significant buildings (for example, by blocking views of building facades or shop windows).
- United Church Cemetery (Chatswood South) where installation of night lighting, site fencing and on-site interpretation of the place, formed part of the conservation strategy to reduce vandalism of culturally significant gravestones.

In their summary Marquis-Kyle and Walker (1992) recognise the failure of current approaches to urban heritage conservation planning to respond to the problem of external environmental threats, by stating that,

> ... at times there is not a clear distinction between the place and its setting – only rarely is a culturally significant place self-contained within definite boundaries, without some visible link to the world around it. If the cultural significance of a place relates to its visual attributes – such as its form, scale, colour, texture and materials – its setting is of special importance... (Marquis-Kyle and Walker, 1992: 38).

They strongly advocate, therefore, that care of the heritage place sometimes must extend outside its immediate, legal boundaries.

Izatt (1995) reinforced criticisms of current approaches to cultural heritage conservation planning, which maintain development and conservation as mutually exclusive objectives. Following extensive research, Izatt confirmed that limited recognition is given to the problem of externally originating (existing and potential) environmental threats, the heterogeneous nature of the environment and the varying requirements of protection that should stretch, as a rule, to areas that envelop a heritage place or cultural heritage precinct.

Clearly, the relationships between development the heritage values of the protected area and any development in the surroundings that may threaten the heritage area should be known both to the developer and to the decision maker. This implies a need to develop planning tools to address this issue. Tools are needed which allow:

- the translation of the information contained within a heritage study (usually undertaken for an area or element that has been given a protected status) and from other sources, to a form which spatially defines the value of a heritage place and details the relationship between the value of what is conserved and its surrounds;
- the definition of areas in the surrounds, in which certain types of activities or development, both existing and/or expected in the future, may reduce or destroy the value of the conserved area; and
- the application of controls or guides, in the areas surrounding heritage areas, to mitigate the undesired impacts of such development activities on the heritage area. The areas in which controls are to be implemented need to be spatially defined and should not represent areas where all types of development are precluded.

As heritage conservation can range in scale of application, the planning tool should be equally applicable to the conservation of a single building, a group of buildings, or an area.

To substantiate these arguments, a brief review of selected plans relating to heritage conservation areas in Australia, undertaken by Vass-Bowen (1994), is summarised on Table 10.1. The results indicate that planning techniques are primarily based on protection of the character of a defined precinct with controls applied, as a rule, solely within that defined area. No technique was found that attempted to apply wider control over threatening impacts originating outside the place or precinct. The only exception is Melbourne's Carlton study, which calls for the allocation of unspecified building envelope controls outside the protected area through 'other' unspecified planning processes. There is absolutely no reference made anywhere to the possible physical degradation that can be caused by inappropriate land uses or activities in close proximity to that which is being conserved.

This evidence reinforces the previous claim that a planning tool is needed for assessing the existing and future impacts of development outside the bounds of a protected heritage area or building and that mitigating or eliminating those impacts may prove very useful in the formulation and implementation of local governments' heritage conservation strategies.

Table 10.1 Selected approaches to heritage conservation in Australia

Plan	Approach
Adelaide City Plan (1991)	Identified precincts with comprehensive policies covering a broad range of issues. Heritage places (buildings) within are identified and guidelines provided for maintenance of the significant character of precincts.
Adelaide Plan 1986-1991	Five major precincts divided into a further 87. Each precinct provided with a policy statement to maintain 'desired future character'. Accompanying set of non-statutory guidelines (Hammett, 1987).
Draft Development Control Plan Central Sydney (1991)	Controls within identified precincts to ensure that the precinct's special character is maintained.
Latrobe and Given Terrace, Brisbane (Draft 1992)	Development Control Plan concerned with land use allocations within a defined area and maintenance of that area's character. Outside links limited to maintenance of views from within the area, through controls placed in the defined area.
Carlton, North Carlton and Princess Hill, Melbourne (1984)	Designates Urban Conservation Areas with controls within. Recommends Building Envelope Controls for areas outside proposed conservation areas be implemented through other planning processes. Provides a building envelope map but only covers the identified study area.
Melbourne Planning Scheme: Central City Planning and Design Guidelines (1991)	Identified core area with specific guidelines introduced which are cross-referenced with other scheme guidelines. The guidelines are aimed at maintenance of the character of areas.

Source: Adapted from Vass-Bowen (1994)

The Proposal

The best way to obtain a new tool may be to develop it afresh and ensure that it complies with predetermined criteria. This, however, is time consuming and may require substantial resources. Another, perhaps less ambitious way, is to 'borrow' something that already exists and adapt it to the problem in question. This was suggested by Vass-Bowen (1994), who argued that the fields of natural and cultural heritage conservation are closely related, with common problems associated with activities external to protected areas that threaten the very values for which these areas were granted their protected status. As a consequence, after reviewing a series of techniques Vass-Bowen (1994) concluded that the Buffer Zone Planning (BZP) approach could be adapted to fill the gap in heritage conservation planning.

The approach, discussed at length in the previous chapters of this book, is particularly promising for heritage conservation purposes, because it recognises the inherent heterogeneity of the environment and continuing interrelationships that exist between any protected area and its surrounds. In the particular case of heritage conservation, it is irrelevant whether an entire area or a specific building is to be protected, for both areas and individual buildings contain a range of diverse values for which simplistic control practices are totally inadequate to ensure heritage conservation. Different impacts affect heritage values in different ways and therefore, different protective 'measures' (policies, restrictions, sanctions or incentives) are needed to ensure that important values are effectively safeguarded.

In the process of adaptation of BZP to the field of conservation of historic heritage, the first task was to find how heritage values, which may be different from those of the natural environment, should be determined. From literature associated with the conservation of the built and modified environment five such values are identified:

- *Visual Character Values*, which are created by the fabric and physical elements of a heritage place and which contribute visually towards its character. They can occur at two distinct levels: 'macro', where they are significant at an *urban* scale; and 'micro', where they are only important or distinguishable at *streetscape* or *built* levels. As a consequence, there are also two levels relating to the observation of the character of a place. 'Macro', that is, at a distance (including, for instance, 'skyline' values), and 'micro' relating to elements such as buildings, where detailing is more meaningful. Certain values such as the scale of development may be important at both levels, although in different ways. For example, a large scale building at the micro level of observation may dominate the smaller scale of a heritage place, whilst at the macro level of observation the large scale building may impact upon the landmark value of a heritage place through creation of a competing focal point.
- *Visual Association Values* created by visual linkage between the protected place and associated places in the setting and surrounds, such as the linkage between a historic church and the area with which it was associated. These are, primarily, specific elements or features which are visually significant because of their prominence within the landscape. Such values can be damaged by breaking the viewing channels linking the protected place and surrounds or degrading their landmark status. This, again, may take place at two levels 'macro' and 'micro'.
- *Structure Values* derived from the materials and/or building forms of the protected place. These might be threatened or lost by chemical pollutants, vibration, use of contemporary (instead of traditional)

building methods or other activities originating in the surrounds causing physical deterioration, for instance, chemical weathering of the stonework.

- *Functional Values* derived from the original role of the protected, cultural heritage place, that is, those associated with its activities and functions both within its setting and surrounds. The understanding of the place might be obscured where activities in the surrounds mask its original role, for instance, where an original and functioning fishing village is transformed by the location of a tourist facility in close proximity.
- *Sensory Values* derived from olfactory or auditory qualities of a place such as silence within a place of worship which may be degraded by noise producing activities in the surrounds.

The literature available in relation to the identified values varies greatly, with emphasis placed more on character (visual) values than on the other four values. Many of the values identified, and the potential threats to them, are a matter of some debate, particularly in relation to impacts on the physical integrity of buildings (Feilden, 1982). Such emphasis reflects the perceived importance of these values in contributing to the significance of a heritage place. It may not always be necessary or justifiable to protect all of these values from degradation through external development or activities. However, recognition and analysis of the main threats to these values is essential if comprehensive protection of any place is to be undertaken.

Each of the identified values may vary greatly in significance and other values may be also found in specific cases. Therefore, individual assessment and identification of values for protection from outside threats is a priority task. Once these values are identified any protection measures should be flexible enough to address potential impacts upon those values and to allocate the control measures at various scales, such as local or regional, dependant upon the source of the existing or potential threat.

Conservation of values should primarily include the following:

- *maintenance*, which involves the continuous protective care of the physical fabric, contents and setting of a heritage place or object;
- *preservation*, which involves maintaining the existing state and retarding deterioration;
- *restoration*, which means that the existing state is brought back to a known earlier state by removing accretions or by reassembling existing components without the introduction of new material;

- *reconstruction*, which means returning a place (or an object), as closely as possible, to a known earlier state, through the use of such materials as necessary (that is both old and new); and
- *adoption*, which is usually seen as the modification of a heritage place (or object) in a way that would suit proposed compatible uses (e.g. those that do not involve change to the culturally significant fabric, content and setting and that are substantially reversible) (Marquis-Kyle and Walker, 1992; QSG, 1992).

This is certainly valid, not only within a given heritage area, but anywhere in its surrounds where threats to conservation of its identified values are, or may be generated. Two case studies are presented in this chapter, which trial these new ideas in relation to the application of buffer zone planning for areas of cultural heritage significance. The first involves a case study of the Ipswich area and the second, the Red Hill – Paddington area of the city of Brisbane, both located in Queensland, Australia.

Case Study 1: Ipswich

Vass-Bowen (1992) undertook a limited application of the BZP methodology to a group of buildings within the central business district of the City of Ipswich, situated 40 kilometres south-east of Queensland's State capital, Brisbane (Figure 10.1). The buildings were located within a central, larger heritage precinct identified in the Ipswich Heritage Study (Satterthwaite, 1992). The precinct faced high development pressure due to the region's rapidly expanding population and the city's designation as a key growth centre under a regional planning framework (RPAG, 1993). This first attempt at regional planning in South East Queensland included the State Government taking a lead in promoting the growth of 'Key Strategic Centres' through the relocation of government offices to these locations to help stimulate development. Despite provisions being incorporated into Ipswich City's Town Plan to protect heritage places, it became evident that the onus was on planning officers to recognise the potential impacts of development particularly in the surrounds of heritage places.

Vass-Bowen's (1994) application of the BZP approach to the protection of the cultural heritage values of a sub-group of buildings in the heritage precinct of Ipswich, adopted the seven step methodology of the original BZP approach. They are summarised below.

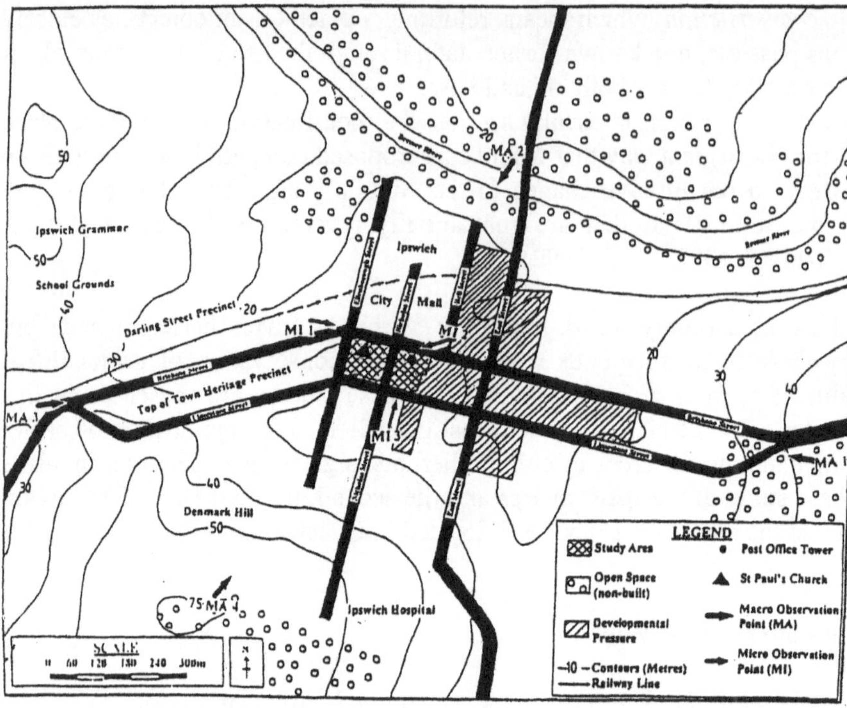

Figure 10.1 The key elements contained within the Ipswich heritage area (Kozlowski and Vass-Bowen, 1977)

Step 1. Identification of elements and characteristic features of the protected place and its surrounds

Step 1 focussed on identifying the heritage values of individual buildings, as the elements requiring protection. The values were derived from the Ipswich Heritage Study (Satterthwaite, 1992) and modified on the basis of a visual analysis (Figure 10.2). The results are summarised in Table 10.2.

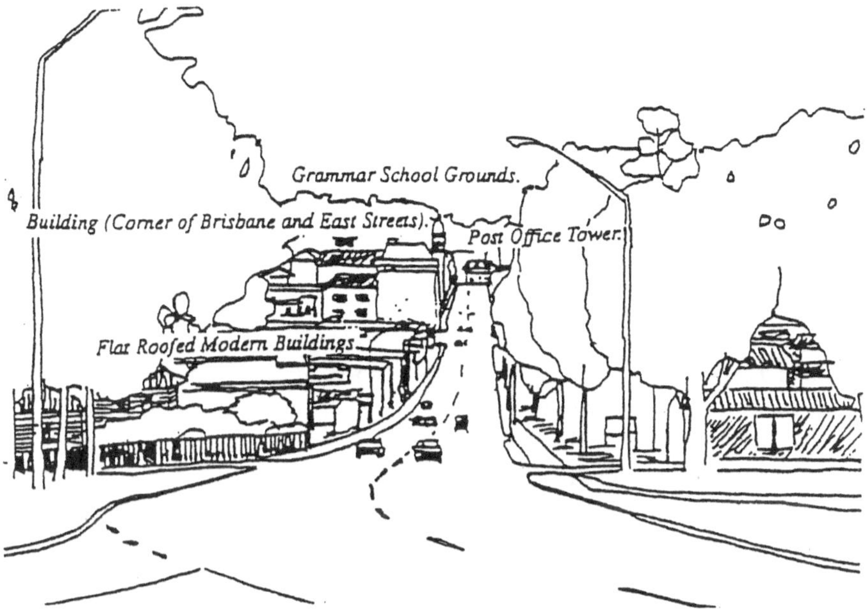

Figure 10.2 View analysis from Macro Observation Point 1 (see Figure 10.1) (Kozlowski and Vass-Bowen, 1977)

Table 10.2 Heritage values of selected buildings in Ipswich

Basic Values	Values in Ipswich area
Visual Character (Macro observation level)	Skyline Colour
Visual Character (Micro observation level)	Scale Colour Architectural detail Street alignment
Visual Association (Macro observation level)	Views towards protected area (landmark)
Visual Association (Micro observation level)	Views towards protected area (landmark) Views away from protected area to associated site
Structure	Building integrity
Functional	None identified
Sensory	Quite

Step 2. Identification of relationships between the protected place and its surroundings and determination of the negative impacts the surrounds generate on the protected place, present and future

Two basic forms of relationship between the values of the building sub-group (identified in Step 1) and their surrounds were evident:

- *The visual relationships among elements within close visual proximity to the protected place.*
  Development in visually juxtaposed areas has the potential, when not undertaken in a sensitive manner to: degrade the protected place, by obscuring landmark values such as those of the Post Office Tower; change the place's original role as the historic commercial core of the city; and visually conflict with the place's aesthetic values.
- *The structural integrity of the protected place.*
  The structure of the protected place may be physically linked to activity in the surrounds. This may include activities that create vibration or pollution and which threaten the materials of the buildings and structures themselves.

From a simplified assessment of each of these relationships the existing and potential impacts were identified to indicate the degree to which they threatened the heritage place's values. As specific data on the physical threats to the buildings by activities in the surrounds was not available, this was substituted by using commonly known threats, related to traffic and vibration in close proximity to buildings (Feilden, 1982; Smith, 1988; Watkins, 1981).

Each value was then cross-tabulated with identified threats to create a matrix in which the significance of the threats was indicated (Table 10.3). This matrix explicitly recognises the way in which, and the degree to which, each impact/threat affects the values. Impacts that were likely to occur as a result of a certain type of development in the future, were also foreshadowed.

The Ipswich Heritage Study (Satterthwaite, 1992) highlighted some potential threats. Other threats were derived from a visual analysis, such as that from the Macro Observation Point 1 (M1) (Figure 10.2). M1 is located on Brisbane Road one of the entry points to Ipswich City. Clearly visible is the Post Office Tower, an important landmark. Behind the Tower, the vegetated grounds of Ipswich Grammar School are an important contrast to the landmark values of the study area. The distant skyline also reflects the

Table 10.3 Matrix of existing and potential treats to heritage values

	Macro Visual Character				Micro Visual Character				Macro Visual Association				Micro Visual Association				Structure			
Values	A		B		C		D		E		F		G		H		I		J	
Impact	E	P	E	P	E	P	E	P	E	P	E	P	E	P	E	P	E	P	E	P
Height/ mass	Major	Major	Significant	Significant	Significant	Significant	Significant						Major	Major	Major	Major	Major			
Roof form	Low	Significant																		
Landscaping			Significant	Significant			Significant	Significant												
Signage													Significant	Significant		Significant				
Traffic vibrat'n																				Significant
Pile driving																				Significant

(Where: Values include A. Skyline; B. Colour; C. Scale; D. Colour; E. Architectural detail; F. Alignment; G. View towards protected place [landmark]; H. View towards protected place [landmark]; I. View away from protected place; and J. Structure integrity; and threats include E. Existing threat; and P. Potential threat).

Major Impact	Significant Impact	Low Impact

historic nature of the study area in contrast to the flat rooved modern development in the foreground.

The landmark value of the Post Office Tower is reduced by the larger mass and height of the building in the foreground (located on the corner of Brisbane and East Streets). Vass-Bowen (1994) considered this a major impact, in terms of building height and mass, one which degraded the landmark value of the Post Office Tower. Similarly, the skyline of the city centre is also dominated by the same building from viewing point M1, although its horizontal dimensions are not sufficiently extensive to completely obscure the skyline. The Grammar School grounds behind are visible creating a contrasting backdrop to the skyline. The impact of height and mass on the skyline value was therefore considered significant.

Due to time and resource limitations, the next five steps (3 to 7) were undertaken for only the most significant values. They were the macro character values of skyline and colour, the macro association value of views towards the protected place (landmark), and the structure value of structural integrity. In a full application of BZP in an urban context, the extent of threats would be highly variable in form and the relationship between the place and its surrounds highly complex. Given this constraint Vass-Bowen decided to use the variant B of the BZP methodology, in which the derivation of APZs precedes that of the EPZs.

Step 3. Preliminary formulation of the criteria for demarcating APZs and for defining their land use measures and policies

The previous step provided the basis for the preliminary formulation of criteria for demarcating the APZs by indicating which values are (or may be) degraded by the various impacts. In previous Australian applications of BZP (Roughan, 1986; Peterson, 1991), the demarcation of APZs was derived from spatially definable areas such as a watercourse catchment, which in turn defined the area where water polluting activities could impact on the hydrologic features of the protected area. In demarcating the APZs in the urban environment, although structural integrity can be viewed as being linked to physically definable elements such as an area where geotechnical structure accentuates the impacts of activities creating vibration, visual relationships can not be defined in such a manner. The visual catchment of the place therefore determines the extent within which visually degrading impacts may occur. However, impacts will vary inside this catchment, being dependant upon other characteristics such as topography and viewing channels created by the built form.

Within the APZs, measures to negate potential impacts need to be determined and applied. In the Ipswich application the preliminary threats to be addressed by land use measures in relation to the four values being investigated were defined as:

- *Mass and height of buildings* This requires not only direct restriction of the dimensions of buildings through tools such as building envelopes, but also control of factors that contribute to larger buildings such as subdivision of land (and lot amalgamation) and plot ratio, which in combination with topography, will vary the impact of a building's height and mass in relation to the various values.
- *Clearance of vegetation* In the case study vegetation clearance would impact on the protected place by reducing the visual contrast between the landmark values of the protected place and the dominance of colour created by the vegetation. Vegetation protection measures in these areas would help ensure that a contrasting background, to observe the place's landmark values (particularly at the 'macro' level of observation), would be maintained.
- *Signage* Signs may obstruct the viewing channel created by streets within the urban environment and may require control where potential impact on important viewing channels exists.

- *Roof form* This includes impacts through incompatible design that is in conflict with the existing building form and requires land use measures to address building design.
- *Vibration* This originates from traffic and construction activities in the surrounds and threatens the integrity of buildings and structures in the protected place. This requires control by land use measures.

It should be noted that in a more complete application of the technique, the criteria would be more complex, particularly if physically degrading impacts on the physical structure through air pollution and vibration, with corresponding assessment of such issues as hydrology, geology and atmospheric conditions, were included.

Step 4. Demarcation of APZs (and land use measures)

For each of the four values identified, APZs were demarcated using the basic criteria determined in Step 3 (refer to Figures 10.3 and 10.4). The final APZ demarcation is a synthesis of the varying spatial extents of the threats to the specific value which is being afforded protection. The land use measures are then developed specifically for these sub-areas in direct response to the specific threat that development in that area may have on the protected place.

Step 5. Demarcation of EPZs for the protected place and definition of the land use measures within their boundaries

The four APZs defined in Step 4B were synthesised to produce five EPZs (Figures 10.5 and 10.6). Each EPZ reflects the aerial extent of the influence of the particular threat and the land use measures to be implemented to minimise the impact of the threat. When synthesising the extent of the threats to different values (e.g. visual character values and visual association values), distinct sub-zones developed, where the potential threat, such as height and mass of buildings, differed in its potential to impact on the protected value. Each sub-zone contained land use measures specific to the particular threats identified, resulting in a mosaic of sub-zones and related policies.

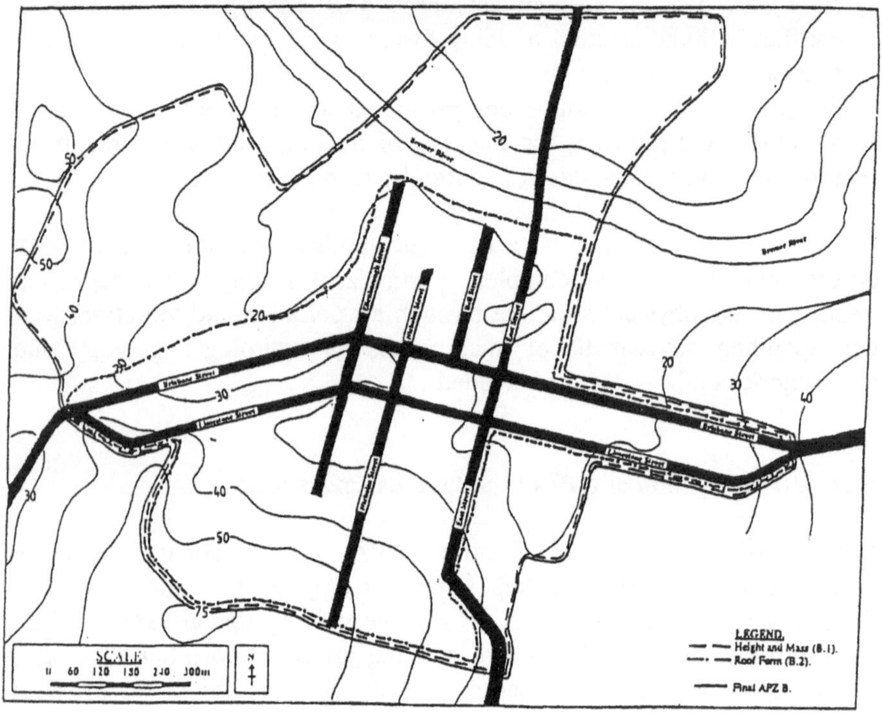

Figure 10.3 APZ B: Areal extent of threats to the Macro Character Value of Skyline (Kozlowski and Vass-Bowen, 1977)

Step 6. Delineation of the Buffer Zone surrounding the protected area by overlapping the EPZs

The final delineation of the buffer zone resulted in two maps. The first shown on Figure 10.7 was a synthesis of the five individual EPZ boundaries. The second, presented on Figure 10.8 indicated the specific land uses and activities to be applied throughout the buffer. This provides the development planner with a map to aid in the assessment of development applications within the places, setting and surrounds of the heritage area. From this 'reference' map the planner can identify the individual EPZ maps that comprise the final buffer and in turn the land use measures that need to be incorporated within the approval process. For instance, Figure 10.7 indicates that sub-area 4 contains threats in relation to

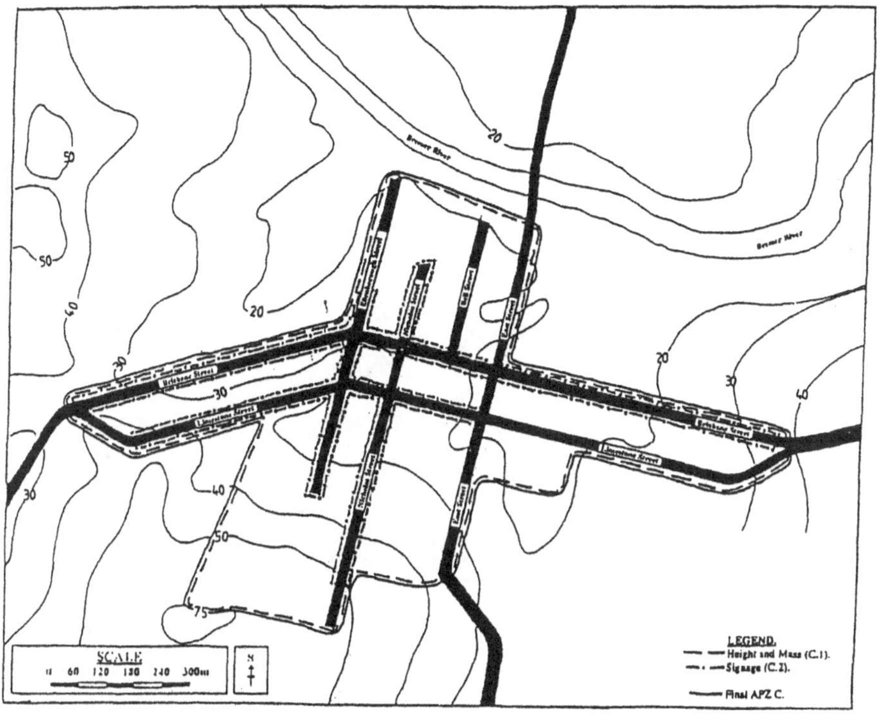

Figure 10.4 APZ C: Areal extent of threats to the Macro Association Value of View Towards the Protected Place (Landmark)
(Kozlowski and Vass-Bowen, 1977)

EPZ's 1, 2 and 3 relating to potential impacts that development within this sub-area may have on the protected place through height and mass of buildings, vegetation clearance and roof form. By reference to the relevant EPZ maps, specific land use measures relating to the control of any proposed development within that area can be applied to any assessment and approval process.

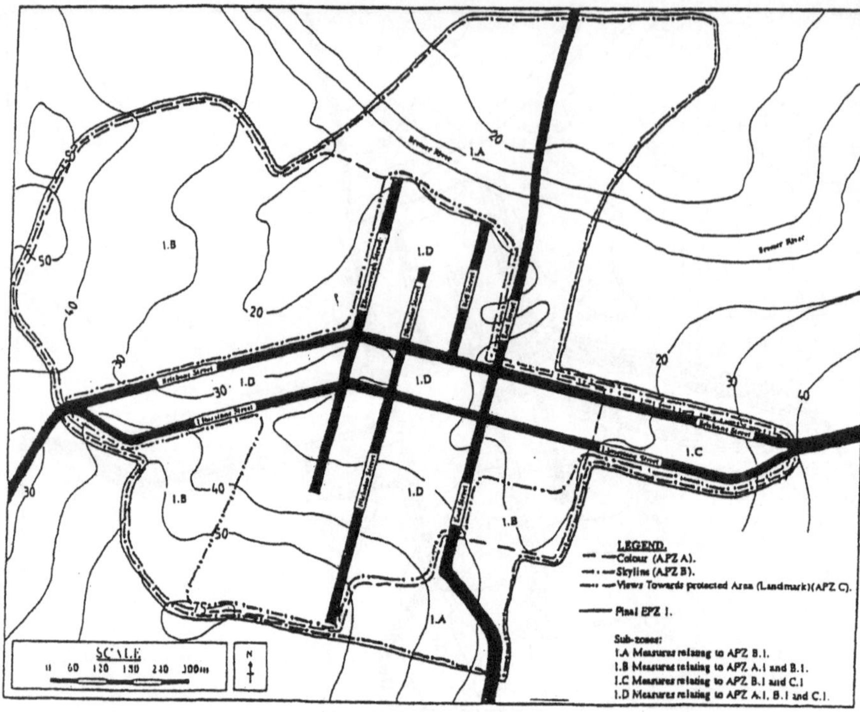

Figure 10.5 EPZ 1: Areal extent of the threat of Height and Mass of Buildings (Kozlowski and Vass-Bowen, 1977)

Step 7. Formulation of the measures guiding different land uses and activities within the boundaries of the Buffer Protection Zone and introduction of these measures into an appropriate development plan

Due to the emphasis on visual relationships in this pilot application of BZP to heritage conservation, land use measures guiding development primarily relate to built form. However, in a more comprehensive application where structural, functional and/or sensory values are addressed, control of the use of the land itself may become more important. The ways that land use control is achieved will vary depending upon the provisions of the controlling legislation. However, three common forms of measures can be identified, namely indicative guidelines, prescriptive controls, and policies

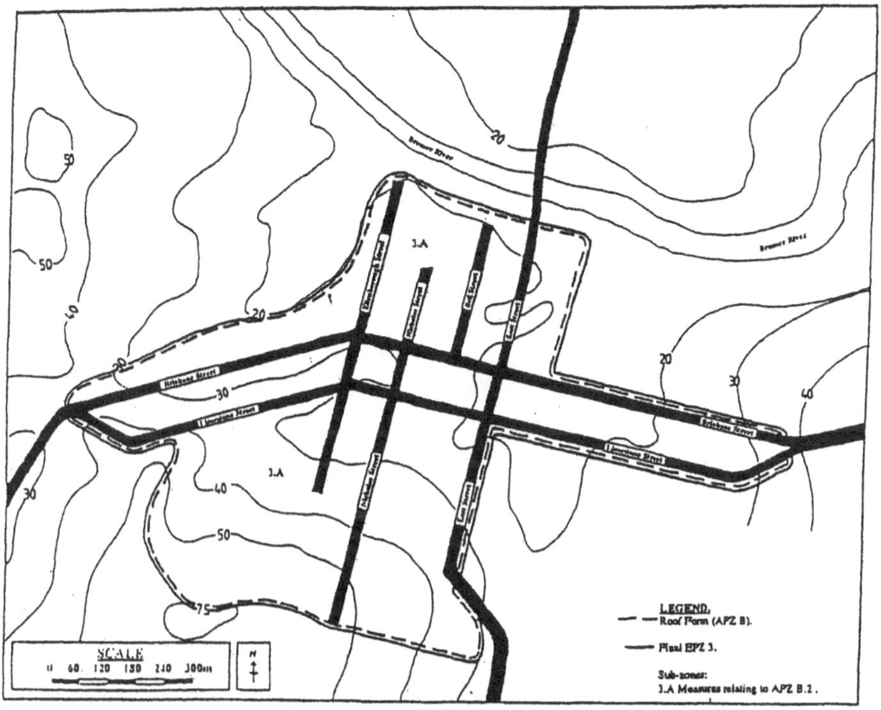

Figure 10.6 EPZ 3: Areal extent of the threat of Signage (Kozlowski and Vass-Bowen, 1977)

which are incorporated within a Town Planning Scheme. As all of these measures need some form of spatial delineation to determine the extent to which the measures apply, in Australia they are best implemented through a statutory plan. Whichever measures are adopted, the plan must be legislatively enforceable such as through incorporation within the planning scheme of a local government. For example, a regulatory map may be a useful tool to define areas where the controls apply. The map, supported by the BZP application, could be included as part of the local government's planning scheme with the area the map covers determined by the extent of the threats to the protected place identified in the BZP process. However, if impacts occur from a wider area, as in the case, for instance, of a point source of pollution some distance from the protected place, then implementation of the technique will require an inter-governmental approach with controls applied on a regional or even national scale.

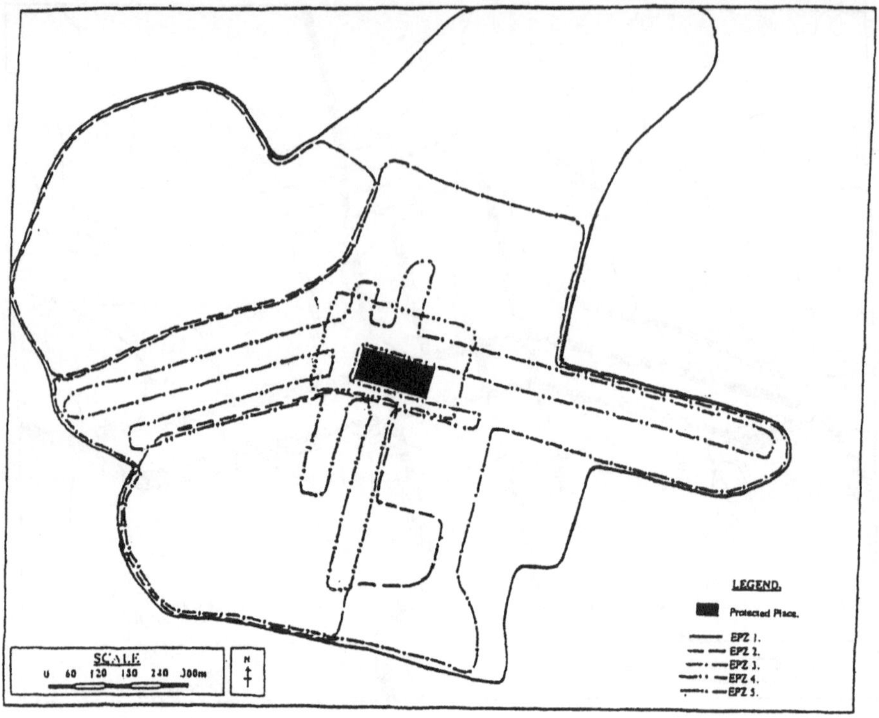

Figure 10.7 Final delineation of EPZ boundaries (Kozlowski and Vass-Bowen, 1977)

Case Study 2: Red Hill-Paddington (Izatt,1995)

The Red Hill-Paddington Special Character Area has been selected to illustrate another adaptation of the BZP method to the field of heritage conservation. The Special Character Area (Figure 10.9) is located in Brisbane's central business district and has 'strong topographic and landscape characteristics combined with a high concentration of late nineteenth and early twentieth century buildings ... known and appreciated by a wide cross-section of the metropolitan community' (Berchervaise and Associates, 1990:10). The character of the area is reflected in its turn of the century timber and iron cottages, the manner in which the cottages fit into the landscape and follow the contours, elements of the buildings

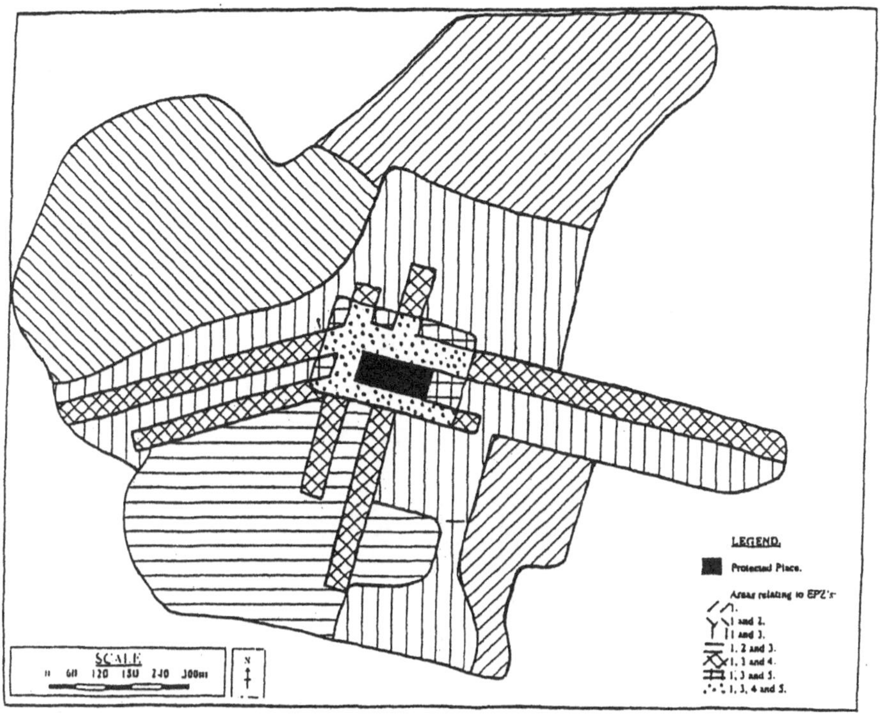

Figure 10.8 Final areal delineation of land use measures (Kozlowski and Vass-Bowen, 1977)

themselves such as their verandahs, window-hoods and roof forms that follow the line of the contours, and the traditional subdivision pattern with vegetation retained mid-block.

Current management practices attempt to retain the traditional character and amenity of housing within the Special Character Area, but they do not address the threat of external negative influences associated with the typical form of redevelopment and activities in the surrounding residential development areas. The same applies to the current management plan for the area which again does not include strategies to protect cultural heritage values within the Special Character Area from externally originating environmental threats.

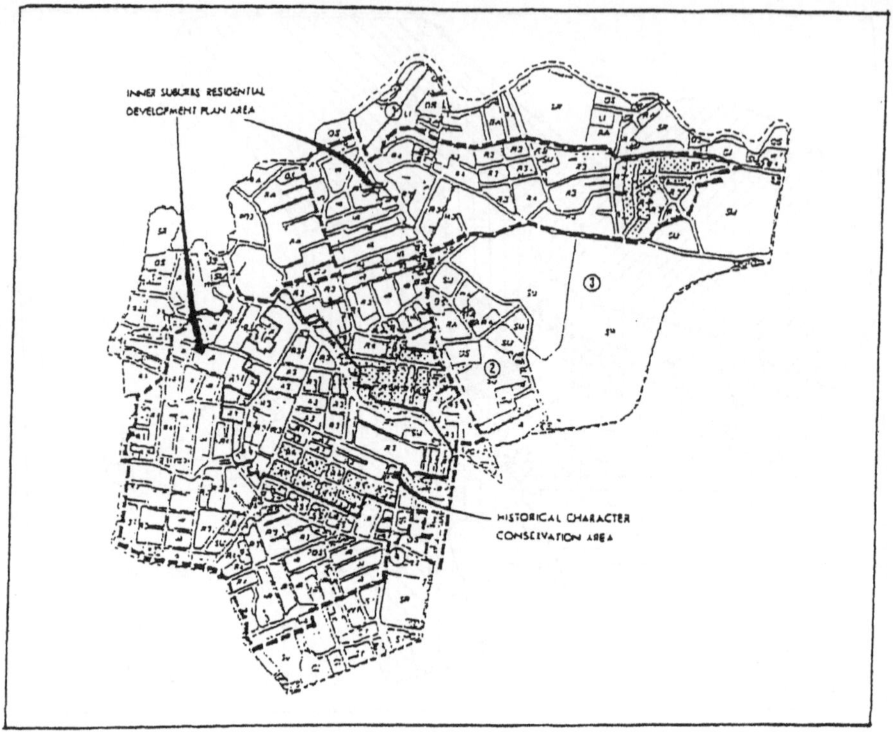

Figure 10.9 The Red Hill-Paddington Special Character Area, Brisbane, Australia (Izatt, 1995)

In Izatt's (1995) study the BZP application was based on a simplification of the original BZP approach as described Peterson (1991), and consists of five main steps. The results of the application are briefly summarised below.

Step 1. Identification of the main cultural heritage values within the Red Hill-Paddington Special Character Area

This step attempted to bridge the gap in current approaches to cultural heritage conservation planning by identifying and describing the 'heterogeneity' of values that are affected, or are likely to be affected, by negative influences emanating from outside the boundaries of Special

Character Area. The values were categorised in the same classes as those in the Ipswich application, that is, visual, structural and functional values. Only sensory values, due to lack of data on noise levels, had to be excluded from this application.

The main attention was directed towards the visual values (both 'association' and 'character') and, particularly, to the identification of landmarks of visual association cultural significance. The following questions were addressed to facilitate this task:

- what are the elements which afford the landmark significance by association with elements or features of its setting and surrounds?
- at what distance are these elements *essential* or *enhancing* to the visual significance within the streetscape ('micro' level) or townscape ('macro' level)?
- which landmark values have visual significance only from micro viewpoints, or from micro and macro viewpoints?

Five buildings, all listed by the Australian Heritage Commission, emerged: the Ithaca Fire Station; Ithaca Tramway Sub-station; Ithaca War Memorial; the 'Terrace Houses'; and St. Brigid's Church. Table 10.4 presents a summary of the visual significance of these heritage elements.

The definition of visual character values was based primarily on evaluation of skyline values (from various viewpoints) at the 'macro' level, and on building form, architectural detail, alignment and colour at the 'micro' level. The information on structural values was not readily available and Izatt (1995) identified only two buildings having distinct structural heritage value that may have been threatened either by heavy trucks or by pile driving vibration occurring in their vicinity. As the functional values of the area were related mainly to its commercial and business establishments, predominantly located along the ridge roads, functional values were examined under 'skyline' and considered together with macro visual character values. As a consequence, the following six specific cultural heritage values were identified within the main classes (Table 10.5).

Step 2. Identification of the main features, land uses and activities surrounding the Red Hill-Paddington Special Character Area

A thorough examination of the heterogeneous features, land uses and activities in the external environment was undertaken, including an examination of: zonings; administrative arrangements; Vegetation Protection Ordinances; conservation studies; and potential redevelopment sites. Information was derived from: *The Brisbane Town Plan* (QSG, 1987) and Development Control Plans for surrounding areas; a Conservation Study of the Inner Suburbs (City of Brisbane, 1989); discussions with the Brisbane City Council's planners; and extensive field work within and surrounding the Red Hill-Paddington Special Character Area.

Table 10.4 Summary of the visual significance of heritage elements of view (in)/landmark values from macro and micro observation points

	Ithaca sub-station		Ithaca Fire Station		Ithaca War Memorial		St. Brigid's Church		Terrace houses	
	MI	MA	MI	MA	MI	MA	MI	MA	MI	MA
Alighment/setback	Enhancing		Essential		Essential			Essential		Enhancing
Archit'ural detail	Enhancing		Enhancing		Essential		Essential		Enhancing	
Building form	Essential		Essential		Essential		Enhancing	Essential	Essential	Essential
Colour			Essential				Enhancing	Essential	Enhancing	Essential
Horisontal order	Essential									
landscaping					Essential					
Lot size/pattern	Enhancing				Essential					
Materials	Enhancing		Enhancing		Essential					
Orientation/shape			Essential		Essential			Essential		Enhancing
Plot ratio	Enhancing				Enhancing					Enhancing
Floor form								Enhancing	Enhancing	Essential
Scale (Height)	Enhancing		Essential					Essential	Essential	Essential
Scale (Mass/Bulk)	Essential		Essential				Essential	Essential	Essential	Essential
Signage			Essential		Essential					
Texture	Enhancing		Enhancing		Enhancing		Enhancing	Essential	Enhancing	
Vertical ordering	Enhancing						Essential	Enhancing		
TOTAL	11	0	10	0	10	0	7	10	7	8

(Where: MI – from identified Micro Observation Points [A,B,C,D,E and/or F]; and MA – from identified Macro Observation Points [A,B,C1,C2,D,E and/or F])

Visual Significance:
Essential   Enhancing   Not significant

The City's forward planning policy statements (BCC, 1994) highlight a need to identify and maintain a particular sense of place for Brisbane, preserve housing and streetscape character, and promote the contribution of significant buildings and structures to the built form of Brisbane. Neither Red Hill nor Paddington was targeted as potential redevelopment sites within the city's strategic plan. The redevelopment of these areas was not considered by Izatt (1995) as a potential threat to the cultural heritage values within the Red Hill-Paddington Special Character Area.

Table 10.5 Specific values examined in the Special Character Area

Class of values	Specific values
Visual association (micro and macro)	Landmarks
Macro visual character	Skyline
Micro visual character	Building Form (including architectural detail) Alignment Colour
Structural	Structural integrity

The information obtained in this step made it possible to undertake, in Step 3 Part A, the determination of existing and potential impacts from the surroundings and, in Part B, to develop the detailed land use principles for preliminary formulation of APZs.

Step 3. Part A: Identification of the interrelationships between the cultural heritage values of the Red Hill-Paddington Special Character Area and determination of existing and potential impacts

The objective of Part A of this step was to outline the nature of the interrelationship between the identified cultural heritage values and the nature of the features, land uses and activities outside the heritage area by examining how each cultural heritage value interacted with the surrounding environment. Synthesis of theses results was to be used to identify existing and potential threats to each value.

As each cultural heritage value is subject to a number of threats, and as each threat has a diversity of causes, the procedure adopted in Part A was to firstly identify the threatening element, establish its causes, and determine the resulting (existing and potential) impacts on each value. A general summary of these interrelationships is shown on Table 10.6, adapted from Vass-Bowen (1994: 74).

Table 10.6 Interrelationship between cultural heritage values of the Red Hill-Paddington Special Character Area and the surrounding environment

Cultural Heritage Value (Element/Feature)	**Interrelationship with the Surrounds**
Macro visual association values View (in)/ landmarks	Visual linkage related to viewing channels created by urban and built form from identified macro observation points outside the Red Hill-Paddington Special Character Area.
Macro visual association values View (in)/ landmarks	Visual prominence related to height, form and location at a micro scale (micro observation points), and manifest in dominance of the streetscape through the scale of a landmark building or site within the Red Hill-Paddington Special Character Area.
Macro visual character value Skyline	Visual contrast between roof and built form within the Red Hill-Paddington Special Character Area, and roof and built forms in the surrounds, at a macro level of observation.
Micro visual character values	
Building form (includes architectural detail)	Perceived character values related to the height and width of buildings, visually related to the surrounds and setting of the Red Hill-Paddington Special Character Area.
Architectural detail	Visual relationship between the ornamentation of heritage streetscapes within the Red Hill-Paddington Special Character Area, and the detailing of buildings within the setting and surrounding these streetscapes.
Alignment	Visual feature related to continuity of the street's built wall within/outside the Red Hill-Paddington Special Character Area.
Colour (includes vegetation)	Visual dominance or mix of colour in an area of proximity to heritage streetscapes within the Red Hill-Paddington Special Character Area eg. between buildings and different sections of a heritage streetscape.
Structural values Structural integrity	Physically interrelated to the surrounds with aspects such as geology, soils, water table and air quality, which are significant to maintaining the value of Ithaca Tramway Sub-station and Fire Station within the Red Hill-Paddington Special Character Area.

Step 3. Part B: Preliminary formulation of the criteria for demarcating Analytical Protection Zones and initial definition of the principles of land use within each zone

The objective of Part B was to establish APZs by determining areas from which particular threats originate and to devise, within each APZ control measures, supported by a range of specific land use policies, aimed at eliminating or, at least, mitigating the impacts of these threats. The process used is illustrated below on an example of 'landmark heritage values' identified in step 1.

Major existing and potential threats to landmark values, at both macro and micro levels, were: building scale (height and mass); signage; landscaping, materials and colour; plot ratio (site coverage); amalgamation; solar glare; and roof form. An example of the process to address threats from 'building scale' and 'signage' is illustrated in Table 10.7.

Table 10.7 Landmark Analytical Protection Zone B (sample only)

Threats (existing and potential)	**Control Measures**	**Land Use Policies**
Building scale (height and mass)	Control over the scale of new developments and building extensions by maintaining current zoning within view corridors to micro and macro landmark values	Include existing visual resources and significant characteristics in assessments of cultural heritage significance Intent of height control guidelines to include a clause to maintain the visual importance of major landmarks
	Control over building height within the view catchment of macro landmark values	Encourage the creation of a maximum building height Environment Impact Assessments to be initiated
Signage	Control over signage within view corridors to macro and micro landmark values	Prohibit the siting of signage above the prevailing height of the built form Restrict the siting of horizontal projecting signs to the extent of the pavement width or verandah line Restrict the usage of street-banner signage. Encourage the siting of pylon and free-standing street signage outside major view corridors

All the existing and potential threats to 'landmark values' with corresponding control measures were mapped and then overlapped to jointly delineate the Landmark APZ, which was to protect the visual association value of landmark buildings from the negative influence of developments and/or activities surrounding the Red Hill-Paddington Special Character Area. The zone and the areas where the measures regarding seven of such threats were to be applied are shown on Figure 10.10.

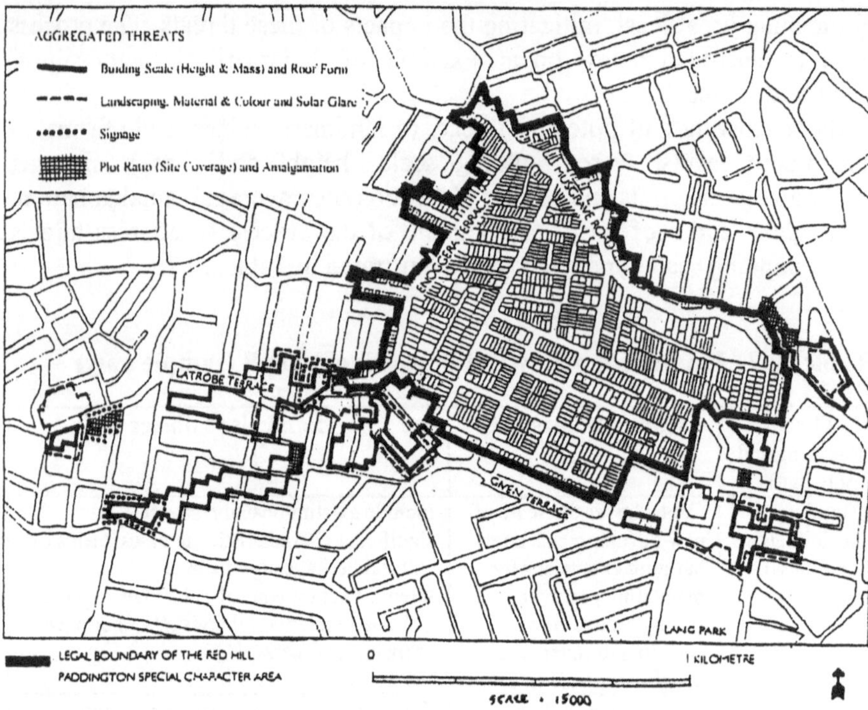

Figure 10.10 Macro and Micro View (in)/Landmark Analytical Protection Zone (aggregated) (Izatt, 1995)

The extent of the measures was mainly limited to areas within view corridors and view catchments. Those regarding 'building scale', for instance, attempted to minimise or, at least, mitigate against view obstruction caused by tall buildings within view corridors to St. Brigid's Church and the 'Terrace Houses', as viewed from eleven macro observation points. Thus, they go well beyond mere 'architectural cosmetics' and address specific effects of proposed developments. In total twelve such control measures against existing or potential impacts to the 'landmark values' were determined. As already demonstrated, to ensure their proper implementation, each of them included a range of specific land use policies.

In addition to 'Landmark Values', APZs were defined for the remaining five cultural heritage values, that is for 'Skyline', 'Building Form' 'Alignment', 'Colour' and 'Structural Integrity'. Interrelationships between these values and the relative threats (existing and potential) were then summarised in a matrix form.

Step 4. Part A: Synthesis of the criteria for demarcating the Analytical Protection Zones

The objective of Part A of this step was to synthesise the results from Step 3 by listing all the threats to cultural heritage values of Red Hill-Paddington area and the proposed control measures required to eliminate or, at least, mitigate each of these threats. The matrix highlighted eleven existing and potential threats and provided the basis for the definition of EPZs.

The process for formulating EPZs is illustrated with reference to the threat of 'Building Scale' (height and mass), which threatens the cultural heritage values of 'Landmarks' (macro and micro), 'Skyline' and 'Building Form'. From the relevant criteria used to determine respective APZs, all areas affected by building scale were combined along with the corresponding control measures and land use policies. Table 10.8 summarises the recommended policies to be implemented to minimise or eliminate the impacts of building scale on the heritage values of the Red Hill-Paddington Special Character Area.

Step 4. Part B: Formulation of Elementary Protection Zones based on a synthesis of the criteria for demarcating the Analytical Protection Zones and definition of the principles of land use within these zones

Table 10.8 Building scale Elementary Protection Zone policies (sample only)

Purpose	To minimise the impact of building scale on the macro and micro views, skyline and building form.
Controls and Land use policies	Policy 1 Control over the scale of new developments and building extensions by maintaining current zonings within view corridors to micro and macro landmark values Inclusion of existing visual resources and significant site characteristics in assessments of cultural heritage significance Environmental Impact Assessments to be initiated Intent of height control guidelines to include a clause to maintain the visual importance of major landmarks Policy 2 Control over the scale of new developments and building extensions by encouraging a rezoning of certain areas within view corridors to micro and macro landmark values

The objective of Part B was to spatially define, for each of the eleven threats, an EPZ within which the identified control measures and land use policies are to be applied to eliminate or mitigate the particular threat. Figure 10.11 illustrates the 'Building Scale' EPZ, the area where specific control measures and land use policies were proposed to address the threat of tall and bulky buildings to several cultural heritage values in the Red Hill-Paddington Special Character Area.

Step 5. Part A: Synthesis of Elementary Protection Zones leading to the delineation of a Heritage and Development Integration Zone surrounding the Red Hill-Paddington Special Character Area

The objective of Part A was to define the total extent of all threats to the cultural heritage values of the Red Hill-Paddington Special Character Area and, as a consequence, of the proposed control measures and land use policies previously devised to eliminate or mitigate these threats. Such an area has been seen as the final 'buffer protection zone' in all applications of BZP to date. However, according to Vass-Bowen (1994), this may imply

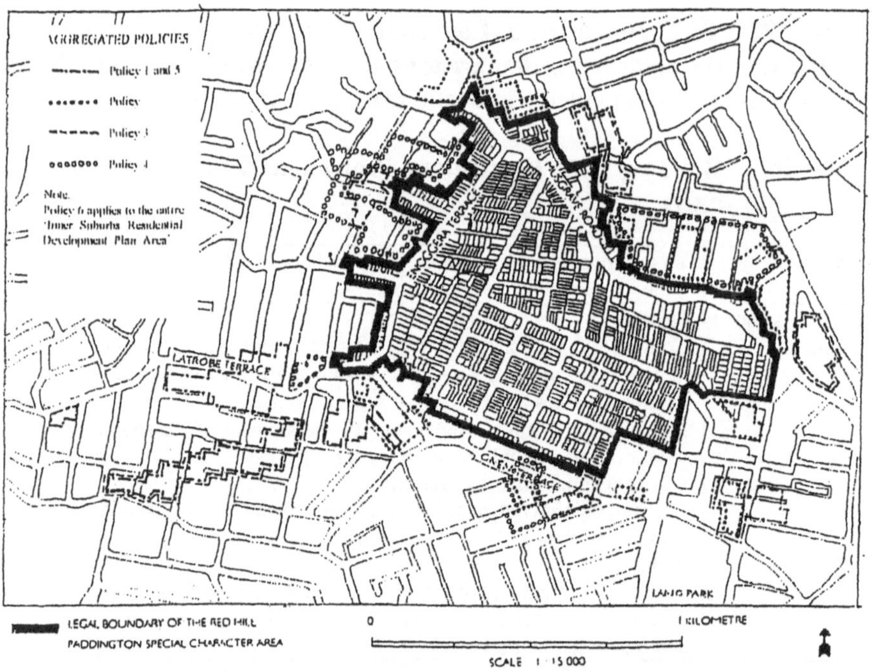

Figure 10.11 Building scale Elementary Protection Zone (Izatt, 1995)

that heritage precincts are protected by *isolation* from their surroundings and as a consequence, the term 'buffer zone' may be seen as non-representative of the intent of the BZP methodology. He therefore suggested that this area be called the 'heritage and development integration zone'. Although Vass-Bowen did not use the term in his own, previously described, application of BZP to Ipswich, the term was adopted by Izatt for the Red Hill-Paddington Special Character Area.

The eleven EPZs, defined in step 4, were now overlapped and areas affected by the same 'combination' of threats and, as a consequence, of same control measures and land use policies were grouped into a variety of sub-zones which produced the final heritage and development integration zone (Figure 10.12).

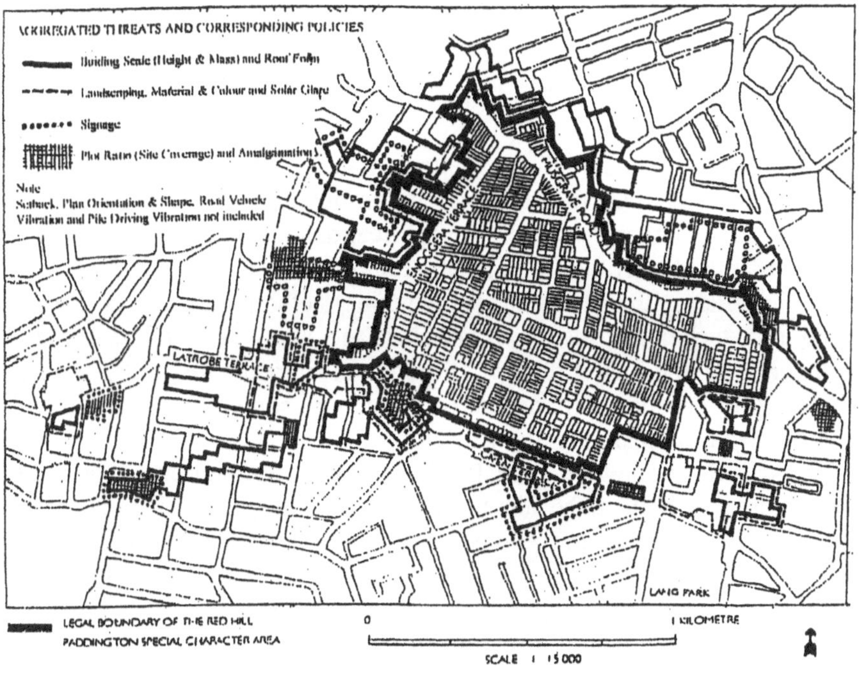

Figure 10.12 Red Hill-Paddington Special Character Area Heritage and Development Integration Zone (Izatt, 1995)

Step 5. Part B Formulation of guidelines for the management of land uses and activities within the Heritage and Development Integration Zone

The objective of Part B was in short the implementation of the Heritage and Development Integration Zone and this was, primarily achieved by synthesising all formerly developed control measures and land use policies on one map giving, thereby, precise information on where they should be applied. Other potential forms of development control, within the framework of the current and new City of Brisbane Town Plan, were also suggested. However, this aspect was not developed in more detail as the general aim of the case study was to examine the general validity and applicability of the BZP to cultural heritage conservation planning and not to dwell on the legal aspects relating to the implementation of control measures and land use policies identified during the exercise.

Conclusion

Although quite limited, both applications highlighted a number of strengths exhibited by the BZP methodology in formulating buffer zones for the conservation of urban heritage. The applications have confirmed that BZP can be adapted to heritage conservation and that its strengths may make it useful in bridging the gap in current planning methodology in this field. By using this technique, the important heritage place is not conserved in isolation and the many forms of potential impact are addressed and integrated to produce comprehensive protection, something which is often lacking in current planning approaches.

The controls applied are spatially defined and allocated only where they are required to protect the identified values of the place. This approach, rather than conservation at all cost and blanket exclusion of development over a wide area is, thereby, closely associated with the concept of sustainable development as it effectively integrates conservation and development by allowing development to continue in areas where it does not degrade the heritage place. In the applications undertaken, in no area was development excluded, rather, through BZP control and guidance, the preferred form of development was established aerially with respect to the place being protected. Even though implementation of these wider buffers may be difficult in reality, the technique increases the awareness of such concerns, which may add weight to the call for the development of regional systems to addresses this problem in the future. Finally, the technique is highly flexible in its scope, being applicable to a single building or a much larger area. This is important in heritage conservation as the size of an identified culturally significant place varies greatly.

Due to the limitations of both the Ipswich and the Red Hill-Paddington pilot studies and their intent to primarily provide a preliminary testing of BZP in the urban context, it is recognised that further research needs to be undertaken in the form of several comprehensive practical applications of the technique to a range of case study areas within which the following specific issues should be addressed (Kozlowski and Vass-Bowen, 1997):

- The approach to define and assess values of a heritage place need further development and refinement. This should include detailed analysis of all the most commonly known heritage values, the refinement of a check list of values, and better understanding of the relationships between the values of the heritage place and its surrounds.
- The process of establishing impacts, specifically non visual ones, and of how the potential impact on the values of a heritage place can be determined aerially, must be further elaborated.
- In assessing visual impacts, techniques that are more objective in analysing visual elements and linkages should be identified and incorporated within the technique, preferably by drawing from the wide experience of landscape planning.
- Principles for developing detailed measures that can be applied to ameliorate the degrading impacts originating from development and activities in the surrounds and setting should be established to guide practical applications of the BZP methodology.
- Further applications should be undertaken in a *multi-disciplinary* manner with input from various professionals, particularly urban designers, architects and engineers, whose insights would be highly valuable in identifying potential impacts on the protected place's values.

This chapter presents a new avenue for BZP by addressing and offering a solution to the problem of 'buffering external threats' to heritage places. So far, the approach cannot be backed by substantial research, as it is still in its initial phase. Yet the problem can be seen as one of the gaps in the field of planning and protection of heritage. The BZP methodology allows planners to avoid making totally arbitrary decisions, by providing a rational base for decision making. The methodology, although not perfect, nor fully tested, provides an avenue for planners to experiment with it and refine it.

Through further research and testing, the potential contribution of the BZP technique to urban heritage conservation, perhaps in the form of Vass-Bowen's (1994) suggested 'Heritage and Development Integrated Zone

Planning' can be further refined to provide an additional planning tool to address what is an important and continuing heritage conservation problem faced by urban planners, engineers, designers and architects.

References

Berchervaise and Associates (1990), *Red Hill-Paddington Character Housing Study*, Berchervaise and Associates, Adelaide.

Brisbane City Council (1994), *The Liveable City for the Future*, Brisbane City Council, Brisbane.

Brolin, B.C. (1980), *Architecture in Context: Fitting New Buildings with Old*, Van Nostrand Reinhold Company, New York, Cincinnati, Toronto, London and Melbourne.

Carlhian, J.P. (1980), 'Guides, Guideposts and Guidelines', in *Old Architecture: Design Relationships*, The Preservation Press, Washington D.C.

City of Brisbane (1989), *Conservation Study of Area 3 – Part of the Inner Suburbs Action Plan*, Brisbane City Council, Brisbane.

Davison, G. and McConville, C. (eds) (1991), *A Heritage Handbook*, Allen and Unwin Australia Pty Ltd, Sydney.

Feilden, B.M. (1982), *Conservation of Historic Buildings*, Butterworths Scientific, London.

Hedman, R. and Jaszewski, A. (1984), *Fundamentals of Urban Design*, Planners Press, American Planning Association, Washington D.C.

IUCN, UNEP and WWF (1991), *Caring for the Earth: A Strategy for Sustainable Living*, World Conservation Union, United Nations Environmental Programme and World Wildlife Fund, Gland.

Izatt, C.S. (1995), *Planning for Protection Cultural. Heritage Precincts*, BRTP Thesis, Department of Geographical Sciences and Planning, The University of Queensland, Brisbane.

Kerr, J.S. (1985), *The Conservation Plan: A Guide to the Preparation of Conservation Plans for Places of European Cultural Significance*, National Trust of Australia (N.S.W.), Sydney.

Kozlowski, J. and Vass-Bowen, N.W. (1997), 'Buffering External Threats to Heritage Conservation Areas: A Planning Approach', *Landscape and Urban Planning*, **37**, pp. 245-267.

Marquis-Kyle, P. and Walker, M. (1992), *The Illustrated Burra Charter*, Australian ICOMOS, Prestige Litho, Brisbane.

Peterson, A.E. (1991), *Buffer Zone Planning for Protected Areas: Cooloola National Park*, Master of Urban and Regional Planning Thesis, Department of Geographical Sciences and Planning, The University of Queensland, Brisbane.

Queensland State Government (QSG) (1992), *Queensland Heritage Act No. 9*, QSG, Brisbane.

Roughan, J. (1986), *Planning for Buffer Zones – An Application of Protection Zone Planning to Nicoll Rainforest*, Bachelor of Regional and Town Planning Thesis,

Department of Geographical Sciences and Planning, The University of Queensland, Brisbane.

Regional Planning Advisory Group (RPAG) (1993), *Creating Out Future: Towards a Framework for Growth Management in South-East Queensland*, State Government Printers, Brisbane.

Satterthwaite, L. (1992), *Ipswich Heritage Study*, Vol. 1, Final report Ipswich City Council, Queensland.

Smith, J.W. (1988), *Vibration of Structures: Applications in Civil Engineering Design*, Chapman and Hall, London.

Vass-Bowen, N.W. (1994), *A Role for Buffer Zone Planning in Urban Heritage Conservation?* Bachelor of Regional and Town Planning Thesis, Department of Geographical Sciences and Planning, The University of Queensland, Brisbane.

Watkins, L.H. (1981), *Environmental Impact of Roads and Traffic*, Applied Science Publishers, London.

PART FOUR

INTEGRATED BUFFER PLANNING (IBP) MODEL

The concluding part presents, in Chapter 11, the Integrated Buffer Planning (IBP) model as a recommended 'best practice' approach to buffering environmentally sensitive areas and in Chapter 12 reviews the progress of buffer planning, discusses the strengths and weaknesses of the IBP approach and provides direction for the ongoing development of buffers for important natural and cultural sites. A step-by-step Guide follows part four as a stand alone document giving practitioners a ready to apply, user-friendly planning tool.

Chapter 11

Integrated Buffer Planning Model

Introduction

One of the problems preventing the widespread implementation of the generally recognised concept of buffers for the protection of important natural areas has been the absence, both in the literature and in a practical sense, of an effective model process to draw upon (perhaps with the exception of the Buffer Zone Planning [BZP] model), and the failure of many approaches to place the design of the buffer within a broader planning context. In this chapter, the general process of the new integrated buffer planning (IBP) model is briefly elaborated, including a summary overview of the rationale for its development and promotion, its general principles and structure.

In addition, however, a detailed, step-by-step Guide is presented at the end of the book as a user-friendly document designated for practitioners, who want to plan, implement and monitor an integrated buffer for any particular core area of natural or cultural significance and who are not necessarily interested in studying why and how the approach has been developed. The Guide is, therefore, a 'stand alone' document for all those who already have a sound knowledge of environmental and planning issues.

A pluralistic approach has been taken throughout this book to explore the scientific basis for good buffer design, to understand the principles of good planning and to evaluate current buffer approaches. This has enabled a science-based approach to be merged with a practice-based approach to arrive at a workable and innovative methodology for buffering environmentally sensitive area (ESAs). In doing this it is implicitly assumed that the loss and fragmentation of natural areas, which are associated with increasing human populations and their associated high levels of resource consumption, has a long-term impact on biodiversity, and that past approaches, which have focused on establishing formally protected areas, need to be augmented by approaches that integrate, link and buffer ESAs from a range of threatening processes, both internal and external.

Underlying Concepts

The IBP model is based upon the following:

- key principles of sustainable development (Chapter 2);
- key issues of integrated landscape/ecosystem planning and management (Chapter 3);
- key principles of good planning (Chapter 2); and
- effective aspects of existing practice-based approaches to buffer planning (described and evaluated in Chapters 4 to 10).

Sustainable Development Principles

Several principles of sustainable development underpin the development and implementation of the IBP. They emphasise the need to:

- conserve biodiversity;
- assess values;
- use resources sustainably;
- integrate planning, economic, socio-cultural, institutional, political and environmental goals;
- plan within ecological boundaries;
- incorporate the effective involvement of the local community and key stakeholders;
- ensure inter-generational and intra-generational equity;
- use precaution;
- take a global view;
- incorporate effective implementation; and
- plan within realistic time frames.

Integrated Landscape/Ecosystem Planning and Management – Key Issues

The creation of a comprehensive, adequate and representative reserve system is a cornerstone of many conservation strategies. However, protected areas and ESAs may become isolated and embedded in a matrix of incompatible land uses, unless effective strategies are implemented to ensure their integration with the often complex social, cultural, economic and environmental framework of their surroundings. It is in this context that planning, as a discipline, has an important role to play in both developing and implementing effective methodologies to ensure improved outcomes for biodiversity conservation.

Habitat fragmentation may subject remnant patches to several chance events, including demographic, environmental and genetic stochasticity and natural catastrophes. It frequently impacts on both the biotic and abiotic components of ecosystems. To minimise the effects of habitat fragmentation, such as altered water and nutrient cycles, wind regimes, radiation balance and species composition, structure and function, planners need to consider landscape spatial structure, including the number of remnant patches, their total area and effective core area, and their spatial relationships, in relation to the availability of dispersers and their dispersal ability, the aim being to enhance the movement ability of wildlife to enable dispersal of wildlife to remnant patches and the recolonisation of habitat patches following local extinctions. Although larger remnants may contribute to higher species diversity and abundance, both large and small habitat patches, if sufficiently interconnected, have an important role to play in conserving biodiversity at the local and regional scale. As the viability of remnant patches is dependent on the maintenance of essential ecosystem processes, the minimum viable area which planners should incorporate into their plans is dependent on the relevant ecosystem and its component species. As data may not always be available, there is an overriding need for planners and resource managers to be precautionary in developing networks of interlinked habitat fragments.

The retention of high integrity habitat is also important. However, it may be equally important to retain poorer quality sink habitats in order to maintain a larger overall metapopulation size, a larger size of source sub-populations and greater genetic diversity. There is some evidence (Smyth, 1996) that species will track suitable environmental conditions, over limited distances and colonise where conditions improve. Thus planners need to ensure that populations do not become too fragmented and that strategies are in place to enhance wildlife movement among patches. Hence planners should consider both the extent of habitat cover and its configuration simultaneously. There is also broad agreement (Sattler and Williams, 1999) that approximately 30 per cent habitat retention, both at the landscape and small patch size, is necessary to help ensure the viability of habitat fragments.

This book has also identified that although external threats have traditionally been the focus of approaches to conserve the biodiversity values of ESAs, it is equally important to consider the ways in which the ESA itself interacts with its surroundings, for processes which originate within the ESA may be seen as threats by adjacent communities and if unmanaged may result in a loss of biodiversity. An important role for planners is in helping to develop strategies to shape a positive community attitude to conservation of ESAs and their component biodiversity. Individual threats to ESAs also have the potential to interact to create cumulative effects, which occur as interactions between activities, between activities and the environment and between components of the environment along pathways. Thus planners need to consider the pathways, which indicate processes that combine impacts synergistically or through a compounding effect.

Principles of Good Planning

Several problems are inherent in planning including: the often complex and interdisciplinary focus of planning; the frequently unknown futures that planners need to consider in plan development (the more distant the planning horizon, the less certain the future); the conflicts that may arise concerning the use of land by competing activities; the lack of data that are available to develop effective plans; the complex organisational structures within which planning occurs; and the uncertainty and complexity surrounding the decision-making processes that take place within various overlapping political frameworks that are often in conflict with each other.

As planning aims to ensure sustainable outcomes, several principles of sustainable development are important in guiding good plan development and should underpin the development of the IBP, including the need to incorporate:

- a multi-objective approach to help minimise or eliminate the conflict between and among competing land use activities;
- effective participation from all stakeholders, including the local community, early in the planning process and in an ongoing way;
- the promotion of a planning approach that is interdisciplinary, transsectoral and transboundary;
- precaution;
- a strategic focus;
- dynamic processes;
- restrictive and promotional planning frameworks; and
- monitoring, research and evaluation.

In particular the IBP involves the development of a physical buffer plan that identifies the spatial and temporal relationships between environmental resources and development activities as well as an institutional plan (incorporated within the Guide) that identifies the key players and their roles in the planning process and the available resources to solve the identified problem.

The IBP model also reflects the need for promoting a strategic and interactive approach due to the anticipated complex and uncertain nature of the problems that the process must address and the need to involve stakeholders fully in the process to better define the planning problem and the approaches to be undertaken to achieve the identified aims and objectives and to implement the plan. Adaptive processes are also important to accommodate the changing circumstances in which the plan is developed.

Integrated Buffer Planning: The Main Assumptions

There are six main assumptions (based on the criteria identified in Chapter 6) that underpin the IBP model. These include the following:

- *The buffer addresses the heterogeneous external threats (both existing and potential) to the core area, and its boundary is based on a synthesis of the source areas of all major identified threats.*
  The buffer should identify all external threats, both existing and future, to the core area. Hence it should be designed without prior consideration of the suitability or tenure of land as a basis for the final buffer boundary. Although the suitability of land is an important consideration in the selection of the 'right mix' of buffer policies and implementation strategies, it is important to integrate conservation with development on all lands, regardless of tenure. Thus where buffers are designed to minimise a range of externally occurring threats, the buffer should be delimited to encompass all necessary lands, regardless of tenure, with a variety of buffer policies applied to accommodate the different tenure arrangements. The complexity of land tenure arrangements in many countries necessitates that the buffer strategy should consider traditional land tenure systems and ensure the continuance of these traditional practices so long as these are sustainable in the long term. In situations where settlements are encroaching onto protected areas or the land surrounding these areas, the provision of secure tenure to occupants within a buffer zone may enhance biodiversity outcomes. Open access to resources, without secure tenure is likely to contribute to over-exploitation of resources and impact negatively on ESAs. In addition, buffers need to be based not only on a consideration of spatial elements, but also include

consideration of the temporal nature of the identified threats, thus ensuring that resultant policies reflect spatial and temporal parameters.

- *The buffer addresses the heterogeneous internal threats (both existing and potential) of the core on its surrounding local communities, land uses and activities. Resultant policies, which are relevant to each identified threat, are developed to help ensure the conservation of the core's values and the minimisation of its impacts on surrounding lands and communities.*

  Consideration of the present and potential impacts of an ESAs resources and activities on its surrounding environment, calls for recognition of the two-way interactions between them as a fundamental criterion of the IBP process, one which so far has not been incorporated adequately into most existing buffer strategies. Resultant policies, which are relevant to each identified threat, can then be developed to help ensure the conservation of the core's values and the minimisation of its impacts on adjacent lands, with their resources and communities.

- *The buffer incorporates sound ecological principles.*

  Rather than defining buffer boundaries by using arbitrary (and thereby, indefensible) distance measures, it is important that the buffer methodology is based on the best available information, and where data are limited, generally accepted scientific principles, rational inference and the consensus of scientific opinion. Where possible, IBPs need to be planned within a bioregional or landscape context, rather than on an individual basis, by ensuring that important natural areas are of sufficient size, are interconnected, and then buffered from threatening processes by including appropriate policies, which are negotiated with the relevant stakeholders. Where necessary, such an approach may become transboundary in scope, requiring joint development with neighbouring shires, states and even countries. In addition, the final positioning of the buffer boundary should coincide with logical boundaries. Such a strategy is useful to aid the clear definition of the buffer boundary on the ground and thus alert management staff and the community to the required actions to be conducted within the buffer.

- *The buffer planning process effectively involves the local community and other key stakeholders.*

  Socio-cultural factors are as relevant in developing effective IBPs as are the biological and economic dimensions. Effective community participation and partnerships are likely to create ownership of the plan, a more cooperative relationship between the managers of the core and local people, enhanced compliance and more acceptable enforcement. It is also important to identify clear links between development and conservation such that any benefits to the local community that are gained from their conservation efforts are recognised by the local community, through negotiation, not imposition, as being dependent on the sustainable development of their natural resources.

- *The buffer is dynamic and responsive to changing circumstances.*
  The buffer process, as in any planning process, is cyclical and must include continual review of plans based on effective monitoring and interrelated research programs. There is also a need to take a long-term view to ensure continuing commitment, support and financing from relevant levels of government, industry, the community and other stakeholders.

- *Buffer planning has an effective methodology that allows effective application.*
  The benefits of a model process include logical progression, efficiency, rational data gathering, plan consistency, transparency, community involvement and improved implementation and daily management. The buffer should also be developed taking into consideration the social, economic, political, institutional/organisational and regulatory framework within which the buffer is to be developed. The development of the physical buffer plan should incorporate, where appropriate, an associated performance evaluation and monitoring program, a research plan, education strategy, implementation plan, enforcement program and review process.

In conclusion, it may also be beneficial to carefully consider an emblem or symbol for the IBP to help promote the buffer strategy and to gain greater community awareness and acceptance of the plan.

The Integrated Buffer Planning Methodology

The Approach

The recommended IBP model process (Figure 11.1) is founded on generic principles of planning, which were presented in Chapter 2 and its overall methodology adapts the elements of current best practice buffer design with important principles of landscape/ecosystem management, good planning practice and sustainable development, which were discussed at length in earlier chapters. The IBP should always be considered in the context of the wider planning framework, ensuring that the development of a physical buffer plan takes place within a socio-economic, legislative, administrative and political framework. The model process is transparent and together with the attached 'working' process, in the form of a Guide, is ready for immediate practical application by planners and other land managers.

Generic planning is a continuous activity that proceeds in cycles. Each cycle commonly includes three essential components, all of which are incorporated into the IPB model process. These include: problem identification; problem solving; and implementation (including monitoring and evaluation). The first two components are further subdivided into phases, which identify the main questions to be answered (Table 11.1).

Structural Arrangements in the IPB Process

The main components, phases and questions are briefly described in this section.

- Component 1: Problem identification

  Phase 1: Setting the framework (Is IBP intervention needed?)
  In this phase an examination is undertaken to determine whether, in the existing socio-economic, institutional and political context, the environmental problems generated by external and internal threats can be eliminated or minimised with the assistance of the IBP process. If the answer is positive, a range of pertinent tasks is identified.

  Phase 2: Setting the planning process (How is intervention to proceed?)
  In this phase the relevant planning problems are defined, aims,

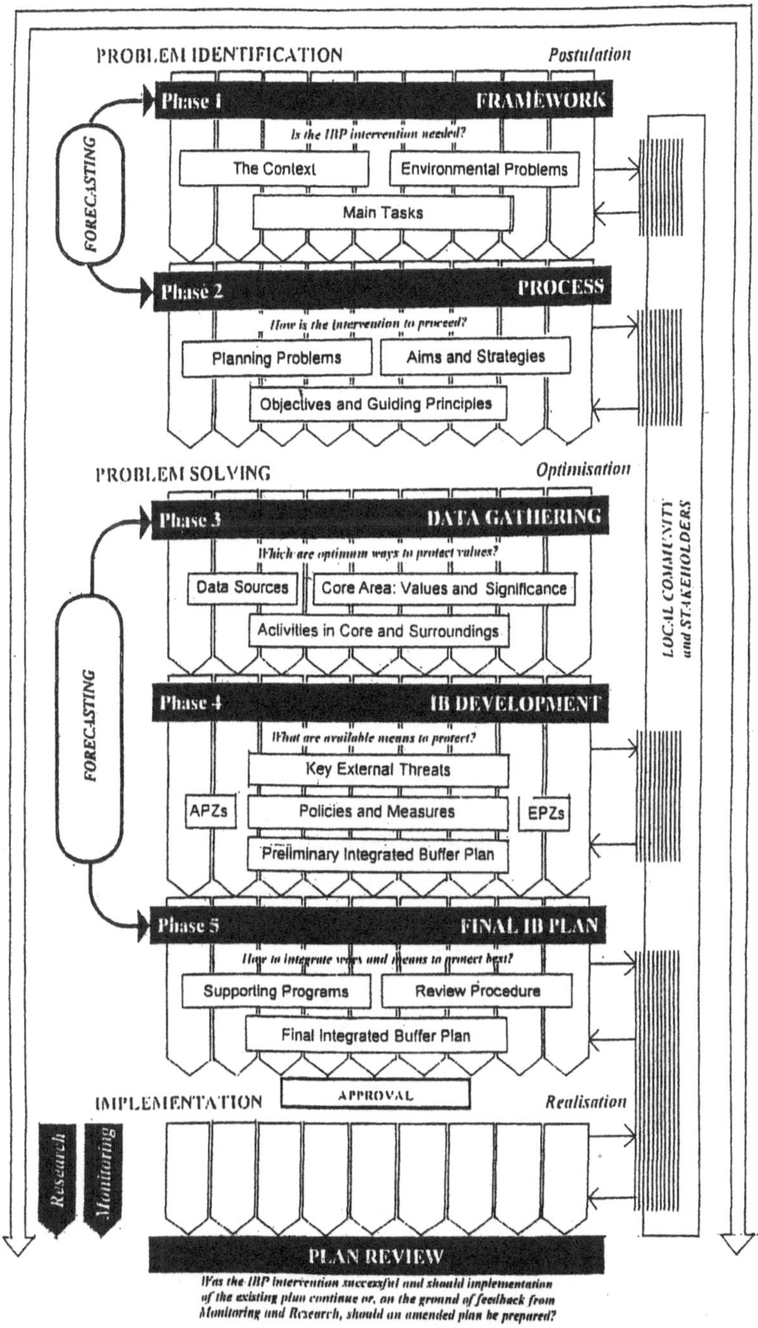

Figure 11.1 The recommended integrated buffer planning model

Table 11.1 Summary of the main components, phases and questions of the integrated buffer planning model

Component	Phase	Question
1.Problem Identification	1. Setting the framework	Is IBP intervention needed?
	2. Setting the planning process	How is intervention to proceed?
2. Problem Solving	3. Data gathering	How are the values to be protected?
	4. Data analysis and draft IBP	What are the means to protect the values?
	5. Final IBP	How can the ways and means be integrated to protect the values?
3. Implementation (including monitoring and evaluation)		What is the most effective mix of strategies?
4. Plan Review		Should the plan be continued, amended or discontinued?

strategies, objectives and guiding principles identified and considered (in relation to the main assumptions of the IBP). The planning team decides how the model is to be adapted and begins to analyse data needs.

- Component 2: Problem solving

Phase 3: Data gathering (How are the values to be protected?)
Data sources to determine the location and nature of the buffer (phase 4) are identified and the data assembled, a focus being the identification of relevant values and activities in the core and its surroundings. The most critical are biological data (e.g. regional ecosystem types and target species) that help to delimit core areas, and specific data relating to the land and resident communities surrounding the core.

Phase 4: Data analysis and draft integrated buffer plan (What are the means to protect the values?)
In this phase key external threats are determined followed by successive definitions of Analytical Protection Zones (APZs) and

Elementary Protection Zones (EPZs) that provide the basis for establishing policies and measures to eliminate or mitigate these threats. At the end of this phase a preliminary IBP is prepared with associated draft policies and land use measures.

Phase 5: Final Integrated Buffer Plan (How can the ways and means be integrated to protect the values?)
The production of a final IBP is a major outcome of this phase. It also incorporates, however, the formulation of recommended programs that support the plan as well as strategies for enforcing and monitoring its implementation.

The approval of the final IBP, which normally follows Phase 5, represents an 'external' input into the planning process. Approval is essential as only a formally approved, and thereby, legally binding plan can ensure its most effective implementation.

- Component 3: Implementation (including monitoring and evaluation)

Effective implementation helps to ensure that the plan takes the form and shape envisioned by its creators, rather than assuming a different structure as a result of changes introduced by those who implement the plan. To proceed with this component, detailed knowledge of government bureaucracies, including their culture and structure, is needed (Clark, 1992). The final IBP identifies limitations within the planning process that may affect the final form of the buffer plan (e.g. limitations on time, resources and data), and highlights potential opportunities that may enhance the implementation process (e.g. effective community structures).

Monitoring, evaluation and review are both closely interrelated with implementation. Through built in feedback loops the IBP model incorporates the results of monitoring and evaluation to establish whether the aims of the plan are being met during implementation. Thus, in situations where limitations are experienced improved strategies may be cooperatively developed. Monitoring also seeks answers to questions about what policies are effective, how performance can be measured and evaluated and what criteria or indicators are needed to measure performance. Monitoring is expected to be interconnected by a feed-back loop with research activities that should permanently accompany the implementation process, feeding it with new or refined data or information.

This process is then followed by Plan Review, which usually heralds the beginning of a new planning cycle. It is an inherent element of a dynamic model of continuing, cyclical planning that is responsive to changing circumstances. The Plan Review is, primarily, to answer the question whether the IBP intervention was successful and whether implementation of the existing plan should continue, or whether on the grounds of monitoring and research, an amended plan should be prepared.

In addition, the IBP process includes forecasting and the involvement of the local community and other key stakeholders as two external elements that have a major influence on the development of the IBP and its implementation.

In conclusion, the policy process by which the IBP is developed consists of a dynamic and reciprocal set of interrelationships between different groups of stakeholders, rather than a top down hierarchical approach. The process aids reconciliation between the often-conflicting interests of the protected area/ESA managers and surrounding communities. It encourages cross-sectoral cooperation and information sharing among state/provincial and local government agencies, statutory authorities and non-government organisations (NGOs). Importantly, the process, although complex, is continuing and cyclical in character, enabling adaptation to many situations.

The Place of IBP in the 'Model' Planning Process

The model planning process (refer to Figure 2.1) identified several stages in plan development, including problem identification, problem solving and implementation. The place and role of the recommended IBP in the context of the model planning process can now be demonstrated (Figure 11.2). It is contained within '*strategies*' and '*means*', which address questions related to '*how*', and '*by what means*' the aims can best be achieved. However, the IBP is also influenced by the '*task*', where the question '*why*' is posed and by '*forecasting*', which strongly assists in defining expected threats. Furthermore, the results of the IPB also have a definite impact on '*implementation*' (including monitoring and research). Finally, the IBP requires an ongoing involvement in the process by the surrounding communities and major stakeholders.

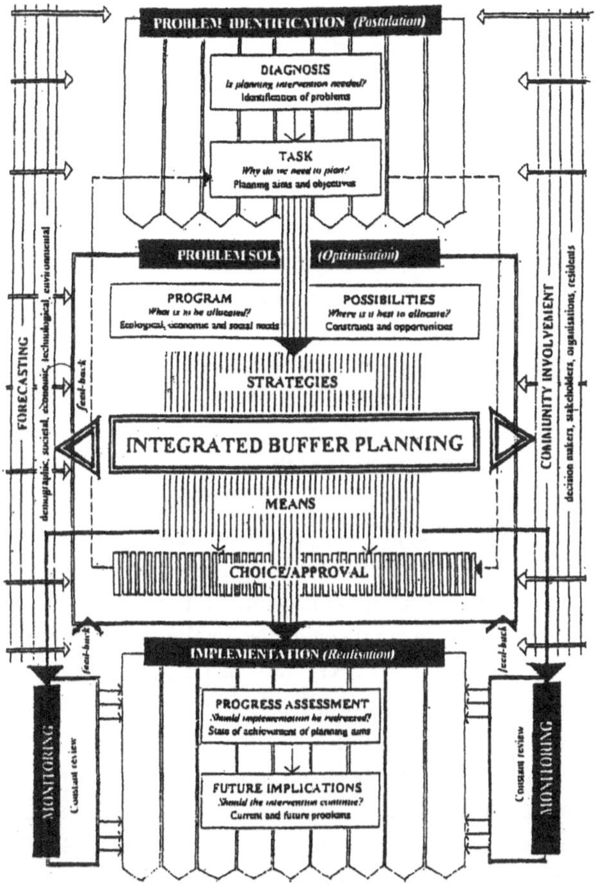

Figure 11.2 The location of the integrated buffer planning model within the framework of the 'model' planning process

Conclusion

The process of design always uses models... (Lynch, 1989: 278).

A model may attempt to represent reality in a simplified way by identifying the most important features of a real life situation. Although model development is a common approach used in the development of planning theory, models are not always easy to apply due to the nature of the planning problem.

Previous buffer planning approaches have largely been *ad hoc*. There has been limited used of a model structure or framework and many approaches have failed to place the design of the buffer within a wider planning framework. The IBP model intends to fill the current gap in buffer planning by incorporating a 'science-based' approach in combination with effective 'practice-based' approaches, resulting in an 'innovative' model to plan for the conservation of significant natural areas. Although many individual elements of the IBP model are not new, it is the step-by-step logical structure, multi-disciplinary/cross-sectoral focus, ecological base and participatory processes that are the main innovations and contributions of the model to sustainable land use planning and management in relation to buffers.

The IBP model enables the delineation of integrated buffers for ESAs in a consistent manner using a number of criteria, thus assisting in the creation of scientifically defensible buffers and a more transparent planning approach. The stepped nature of the methodology aims to facilitate consensus building amongst stakeholders at various critical points in the process. The methodology is systematic and comprehensive in its consideration of the major components of the ecosystem and the nature of interrelationships with existing and proposed human activities and land uses. It aims to provide a logical structure for the development of a physical buffer plan and an administrative and procedural guide for the plan's preparation within a planning framework that incorporates consideration of wider planning issues. The model thus provides links between the physical buffer plan and the planning environment. It also incorporates aspects of timing and emphasises the need for an educational and research strategy, as well as enforcement, monitoring and review processes. Importantly the model is a flexible approach, allowing individual solutions to identified problems, based on the cooperative involvement of stakeholders. The model simplifies a potentially complex planning task, minimises data gathering and can be wholly or partially applied, dependent on local circumstances and needs.

References

Clark, T.W. (1992), 'Practicing Natural Resource Management with a Policy Orientation', *Environmental Management*, **16**(4), pp. 423-33.

Lynch, K. (1989), *Good City Form*, The MIT Press, London.

Sattler, P. and Williams, R. (eds) (1999), *The Conservation Status of Queensland's Bioregional Ecosystems*, Environmental Protection Agency, Brisbane.

Smyth, A.K. (1996), 'The significance of ecological tolerance to fragmentation: foraging and nesting by birds in rainforest', PhD thesis, The University of Queensland, Brisbane.

Chapter 12

The Way Forward

Introduction

For over two decades it has been widely accepted that protected areas, which are linked in many ways with their surrounding regions, have been experiencing increasing levels of ecosystem stress due to impacts from incompatible and often unsustainable land use practices occurring outside their boundaries. The Third Congress on National Parks and Protected Areas, held on Bali in 1982, called on governments to initiate and foster sustainable social and economic development to relieve the pressures of local populations on protected areas, and to reinforce measures to reduce the external threats to protected areas (McNeely and Miller, 1984). There was clear concern that unstainable land use practices on the lands surrounding protected areas might critically endanger the security of those areas, if not their very existence. It was also rightly stressed that effective resource management could not occur when conservation planning and development planning proceeded in isolation (McNeely and Miller, 1984). These views were reinforced at the Fourth Congress on National Parks and Protected Areas, in the Caracas Declaration (IUCN, 1992), which called on governments to take urgent action to consolidate and enlarge national systems of well-managed protected areas by including buffer zones and corridors. The World Resources Institute (1992:129-133) similarly stressed the importance of the management of resources surrounding protected areas and stated that 'the concept of "buffer zones" or "transition zones" is an essential complement to protected area design'. The 2003 World Parks Congress (IUCN, 2003) reiterated these concerns and again called for the integration of protected areas with their surroundings and the implementation of buffer and corridor systems.

Although there is world wide interest and support for buffer zones to minimise threats to protected areas and other environmentally sensitive areas (ESAs), there is as yet little agreement as to how this should be achieved. Roughan (1986), Peterson (1991) and Hruza (1993) examined buffer zone planning around the world, although with particular emphasis on Australia, and concluded that buffer zones were produced largely on an intuitive basis. Wells and Brandon's (1993:25) examination of Integrated Conservation and Development Programs

(ICDPs), which incorporated buffer zones, in Africa, Asia and Latin America concluded that '...Despite their intuitive appeal...buffer zones have not been adequately defined, and there are few working examples...'.

More recently, Peterson (2002) examined a variety of planning and management approaches that have incorporated buffer zone strategies. This examination included biosphere reserves, the multiple use module concept (Harris 1984; Noss and Harris, 1986), core-buffer-multiple use zones as applied to protected areas in India, ICDPs in several countries, various wildlife specific buffer zones, and others (refer to Chapter 4). She concluded that a practical and effective methodology to aid in the delimitation of buffer zones and the incorporation of effective buffer strategies was lacking and that practitioners were searching for guiding principles and a more structured approach to integrated buffer planning.

Recent Research

In Chapter 5, practical experiences of buffer zone planning in Australia were summarised, following contact with over 70 agencies involved in the planning and management of land, both in and around protected areas and other environmentally sensitive areas. Approximately 70 per cent of respondents saw a definite need for buffer zones, particularly where protected areas were surrounded by land intensively developed for tourism and residential uses, and 44 per cent had attempted to produce what the respondents called 'buffer zones'. Where buffer zones had been formally established the majority were single purpose zones designed specifically to deal with only one external threat, for example, fire hazard or watercourse protection and usually these zones were simple in structure being defined by a prescriptive distance rule. The examination revealed that not much had changed over time in relation to the design of buffers, and it highlighted the lack of an ecological base to buffer planning, with buffers being designed largely on an *ad hoc* basis with criteria such as land suitability, logical boundaries, shape, location and prescriptive distances being the dominant concerns. Buffers, in general, were without legal definition, being implemented on an informal basis and lacked a planning methodology to assist in their design.

This summary historical overview confirmed that over the last two decades, there has been no evident breakthrough in the field of buffering natural areas, including protected areas, and that a methodology for defining buffer zones has remained a blank spot in the field of physical planning. Thus, the objective of this book, in putting forward some ideas and thoughts on buffering natural areas, for the purpose of discussion and further testing, appears to the authors to be justified.

The Integrated Buffer Planning (IBP) methodology, which is highlighted as a recommended approach, does not purport to be perfect or to be a thoroughly tested planning tool. However, one of the most common dilemmas in planning practice is the often pressing need to make decisions, which once implemented, may be irreversible. This, in turn, frequently forces planners to make a choice – should such decisions be totally arbitrary, or should some attempt be made to provide a rational base for them, as far as possible within existing experience, available time and knowledge? This book clearly supports the latter approach by offering not only theoretical deliberations, but also a comprehensive Guide (illustrated by several concrete examples), to enable planners and land managers to apply the IBP method in real-life situations. Such a decision may well be challenged as it means that ideas, concepts and methods, that are as yet not fully prepared and matured, have to be formulated and thrown into the field of practice for further testing. This may lead to controversy and criticism, but it can also help to expose any errors and shortcomings in the approach and enable refinements to be made more quickly. This approach would appear to be essential in our current situation, where time to find effective solutions to the definition and management of buffer zones, is definitely running out.

Furthermore, it should be emphasised that the IBP method relies heavily on the Australian and Polish applications of the BZP methodology (Kozlowski and Ptaszycka-Jackowska, 1981). These case studies confirmed the practical validity of the BZP method, which ensured the development of a comprehensive buffer, one that recognised the heterogeneous nature of the environment and devised land use policies in relation to the differing needs for protection of the various park resources in relation to the nature of the external threat concerned. The methodology recognised the importance of ecological principles in the determination of the buffer and through the identification of potential threats, the buffer played a very important role as a form of proactive planning, ensuring that future threats did not eventuate, or at least were minimised in their impact. The methodology also has a distinct multi-disciplinary character, as the approach requires the involvement of, or consultation with, specialists from many disciplines, before planing synthesis can be reached.

The Integrated Buffer Planning (IBP) Method

An important aspect of the IBP methodology that is strongly recommended for actioning is the implementation of the IBP through physical development plans, which can provide the main basis for the steering and control of human activities within the areas concerned. A source of potential difficulty for the future implementation of the recommended land use policies may be that the final buffer zones are bound to cover relatively extensive areas. This problem need not necessarily become serious as one of the main objectives of the method is to limit the policies and land use measures to only those that are indispensable. Where effective community consultation and involvement has occurred, the buffer policies are more likely to be acceptable to surrounding communities and to be supported by them.

Certainly, the boundaries of buffer zones and their land use policies cannot be of an absolute character and must be continually verified and accommodated to the ever changing reality. The development of knowledge, as well as continuous input from the monitoring of interrelations between protected areas and their surroundings, would also imply the need for periodic changes to criteria and methods of environmental protection and management. Therefore, it seems necessary to closely link further research on buffer zones and their introduction and management with economic and physical planning systems in such a way that fresh input is always provided at the beginning of a new phase in a cyclical planning process.

Strengths of the IBP Model Process

The IBP model fills a significant gap in current planning practice and can be used as a basis for further refinement following on-going application. Some of the important strengths of the model process are grouped and summarised below and relate to its structure, ecological basis, planning process, implementation and application.

Structure

- the IBP combines sound ecological and sustainable development principles with best practice aspects of current buffer strategies, resulting in an innovative approach to the conservation of ESAs;
- the IBP simplifies a potentially complex planning task;
- the five phases of the model contain several sequential steps and associated actions, providing a rational and transparent basis for decision making;
- the use of decision points, which pose a series of choices concerning the progression of the planning process, allows the planning team to tailor the process to their specific requirements, enabling the elimination of unnecessary steps or actions, based on resource and data availability, technical expertise, statutory requirements, timing and expediency;
- the model minimises data gathering;

- it can be wholly or partially applied, dependent on local circumstances and needs;
- contextual analysis, which is an important component of Phase 1, ensures broad identification of the factors that may influence policy development and implementation.

Ecological Basis

- The IBP process is based upon a comprehensive incorporation of the main elements/features and values of the core area and its surrounding landscape and the consideration of a wide range of threatening processes, both internal and external to the core. This approach helps to ensure the following:
-
- the interactions between the core and its surrounding landscape are considered to better understanding a wide range of possible pathways through which threats operate;
- the cumulative effects of several minor threats are considered in the design process;
- the source areas of the threats are identified and incorporated into the process, ensuring that buffer policies are specific to the area of operation of the threat and that the resultant buffer is heterogeneous in relation to its structure and functioning;
- external threatening processes that affect the identified core, as well as internal threats that originate in the core and impact on the surrounding lands are identified and minimised;
- the development of policies and land use measures that ensure sustainable development within the core and buffer;
- the buffer is not restricted to being located on state/council owned land, but may encompass privately owned land (although the final buffer boundary will relate to identifiable features in the landscape); and
- the methodology is systematic and comprehensive in its consideration of the major components of the ecosystem and the nature of interrelationships with existing and proposed human activities and land uses.

Planning Process

Use of an accepted methodology, based on relevant criteria, enables IBPs for ESAs to be developed in a consistent manner, thus assisting in the creation of scientifically defensible buffers and a more transparent planning approach. Particular strengths relate to participation, plan making, policy and other issues.

Participation:

- the methodology incorporates and stresses the importance of appropriate participatory strategies being implemented early in the planning process and throughout the development of the IBP and continuing through to its implementation and review;
- the model encourages the development of participation strategies that are compatible with the social and cultural context of the planning area;
- the stepped nature of the methodology aims to facilitate consensus building amongst stakeholders at various critical points in the process; and
- the model process recommends an interdisciplinary/cross-sectoral focus.

Plan making:

- the model provides a logical structure for the development of a physical buffer plan and an administrative and procedural guide for the plan's preparation within a planning framework that incorporates consideration of wider planning issues. The model thus links the preparation of the physical buffer plan with how its preparation is to be achieved within the organisation(s) responsible for preparation and administration of elements of the plan;
- the IBP incorporates aspects of timing in the 'Action Plan';
- the model is a flexible approach allowing individual solutions to identified problems, based on the cooperative involvement of stakeholders;
- the model is dynamic as it recommends the inclusion of feedback loops throughout the plan making process and during the review, implementation and monitoring phases. This enables a re-examination of issues with each iteration of the planning process and helps to ensure internal consistency within the final plan while allowing maximum flexibility in its development and implementation;
- the model encourages a proactive and anticipatory approach to the conservation of ESAs, where consideration is given not only to examining existing use rights over land, but also to any future use rights that may exist, as identified in strategic plans and other planning documents and which may pose a threat to the conservation of core areas and their buffers;
- the model encourages the identification of limitations that may affect the final form of the buffer plan and highlights potential opportunities that may assist the implementation of the plan;
- it recommends the inclusion of performance indicators to aid in the evaluation of the plan's success in meeting stated objectives in relation to

the state of the environment, the pressures or threats in operation and the responses to these pressures; and

- the methodology is capable of implementation by 'non-experts' and can be completed at several levels of complexity, depending on data availability and resources.

Policy:

- the policy process encourages interaction between different groups of stakeholders, rather than a top down hierarchical approach. This approach facilitates the reconciliation of conflicting interests and encourage cross-sectoral cooperation and information sharing; and
- IBP policies are directly related to the area of operation of specific threatening processes and hence the overall management framework, while potentially complex, is neither excessive, nor incomplete in spatial coverage.

Other:

- the model emphasises the need for an educational and research strategy, as well as enforcement, monitoring and review processes.

Implementation

- the model encourages the involvement of those who implement the plan in all phases of the process to ensure that the aims and objectives of the plan are not inadvertently diverted by these people.

Application

- the model can be applied to a variety of core areas including protected areas and similar reserves, state forests, the habitat of particular species, regional ecosystems or other ESAs. It is not necessary for the core to consist of a formally protected area with clearly identified boundaries and secure tenure;
- the model can be applied at the broad landscape scale to develop a network of interconnected cores, corridors and buffers, or it can be applied at a smaller scale to better conserve the biodiversity values of individual cores;
- the model can be applied to provide a framework for development by identifying areas where development should be constrained or is permitted

subject to the implementation of land uses and activities that are ecologically sustainable and compatible with the conservation of the area's biodiversity;

- the flexible nature of the policy process enables the model to be easily adapted to a variety of planning and resource management situations;
- the model may be applied proactively before biodiversity issues become problematic or it can be used reactively to respond to particular existing biodiversity issues; and
- the model can be applied not only to the protection of the natural environment in protected areas, nature reserves and water reservoirs, but also to such protected objects as, for instance, buildings of historic or architectural value.

Limitations of the Model

As with any model, new or well-tested, there are likely to be limitations to its effectiveness, and for the IPB model these include the following:

Participation

- comprehensive identification and inclusion of stakeholders may be difficult. Where there is inadequate identification of stakeholder and user groups this may result in the omission of important aspects of the planning problem, reduced compliance and inefficient use of time and resources; and
- stakeholders may be unwilling to compromise, thus limiting the effectiveness of the IBP's goals and objective. It is thus important to ensure the effective participation of representatives of all major groups, and to allow sufficient time for consensus decision making.

Planning Framework

- the lack of a thorough understanding of how ecosystems and their components function is a significant limitation of the model;
- development and implementation of a comprehensive and effective application of the IBP methodology may require a long-term commitment of time and resources;
- the development and implementation of an IBP may require comprehensive education strategies and a range of voluntary incentive

schemes to encourage effective participation and understanding of the process;

- difficulties may arise where cross-boundary buffers are developed (e.g. international, interstate and inter-provincial). There may be varying levels of commitment in human and financial terms, different levels and styles of enforcement and variable monitoring effort. Such a situation will require comprehensive consultation and negotiation with all authorities and the possible sharing of resources;
- resource constraints, related to staffing and budgetary issues may limit the outcomes of the IBP; and
- a failure to integrate the buffer into the statutory planning framework may weaken the outcomes of the integrated buffer.

The Way Forward

The existing nature of buffers and current approaches to buffer planning indicated a need for a better understanding of buffer types and for the development of a model approach that can be applied in several planning and resource management contexts (e.g. developed and developing countries, urban areas, lands under increasing resource use pressure and those in more isolated and wilderness contexts). The aim of this book was to outline a new buffer model. The resulting recommended IBP is based on principles of good planning practice, sustainable development and elements of good buffer practice. Although the model incorporates many existing elements of current buffer practice and good planning strategies, it is the comprehensive framework that is presented and the logical procedures that are established to enable planners to select the relevant phases, steps and actions in the development of an IBP, that are the significant contributions of this model approach.

Application, critical review and further adaptation of the model are important steps in the way forward.

References

Harris, L.D. (1984), *The Fragmented Forest: Island Biogeography Theory and the Preservation of Biotic Diversity*, University of Chicago Press, Chicago.

Hruza, K.A. (1993), 'Buffer Zone Planning. A Possible Management tool for Fraser Island', Bachelor of Regional and Town Planning thesis, Department of Geographical Sciences and Planning, The University of Queensland, Brisbane.

International Union for conservation of Nature and Natural Resources (IUCN) (1992), '*Parks for Life: The Caracas Action Plan*', Draft 1Vth World Congress on National parks and Protected Areas, Caracas, 10-21 Feb.

IUCN (2003), Recommendations of the Vth IUCN World Parks Congress, 9-17 Sept. Available at:
http://www.iucn.org/themes/wcpa/wpc2003/english/outputs/recommendations.htm.

Kozlowski, J. and Ptaszycka-Jackowska, D. (1981), 'Planning for Buffer Zones', in P. Day (ed.), Queensland Planning Papers, The University of Queensland, Brisbane, pp. 244-38.

McNeely, J. and Miller, K. (1984), 'National Parks, Conservation and Development. Proceedings of the World Congress on National Parks', Bali, 1982, Smithsonian Institution Press, Washington.

Noss, R.F. and Harris, L.D. (1986), 'Nodes, Networks and MUM's: Preserving Diversity at All Scales', *Environmental Management*, 10(3), pp. 299-309.

Peterson, A. (1991), 'Buffer Zone Planning for Protected Areas: Cooloola National Park', Master of Urban and Regional Planning thesis, Department of Geographical Sciences and Planning, The University of Queensland, Brisbane.

Roughan, J. (1986), 'Planning for Buffer Zones – An Application of Protection Zone Planning to Nicoll Rainforest', Bachelor of Regional and Town Planning Thesis, The University of Queensland, Brisbane.

Wells, M.P. and Brandon, K.E. (1993), *People and Parks. Linking Protected Area Management with Local Communities*, World Bank, World Wildlife Fund, U.S. Agency for International Development, Washington D.C.

World Resources Institute, IUCN, UNEP (1992), 'Global Biodiversity Strategy. Guidelines for Action to Save, Study, and Use Earth's Biotic Wealth Sustainably and Equitably', WRI, IUCN, UNEP, np.

Index

INTEGRATED BUFFER PLANNING GUIDE

Ann Peterson

Who is the Guide intended for?

The Guide provides detailed and practical information to assist planners in the development, implementation and on-going monitoring and review of buffers for environmentally sensitive areas. It aims to provide a step by step, adaptive planning process that will result in the effective conservation of core natural areas and at the same time allow for sustainable development within buffer areas.

Foreword

An innovative working process for the integrated buffer planning (IBP) model is outlined in the Guide, which is designed for those who are willing to test and apply the model in real-life, or use it for teaching purposes. The Guide accompanies the book *Integrated Buffer Planning: Towards Sustainable Development* (Kozlowski and Peterson, 2005), which should be read by those wanting to gain a more complete understanding of the theoretical and scientific basis of the IBP model. For those who are not interested in that type of background support, and who have a sound knowledge of environmental issues and planning, the Guide can be used by itself without getting involved in studying why and how the approach was developed.

The Guide responds to the pressing need to streamline and improve the development and implementation of effective buffers for environmentally sensitive areas. It reflects fundamental and commonly recognised objectives relating to sustainable development and the need to protect biodiversity through maintaining essential ecological processes and life support systems. It is based upon key principles of sustainable development, integrated landscape/ecosystem management and good planning and builds upon existing effective practice-based approaches to buffer planning.

The IBP model incorporates five main phases: setting the framework; setting the planning process; data gathering; data analysis and preliminary integrated buffer plan development; and final integrated buffer plan and implementation. It includes important issues that should be considered in buffer design, such as implementation, monitoring and research and recommends practical ways of dealing with these. Its application is expected to result in a more consistent and environmentally sound approach to buffering natural areas.

The Guide provides many examples of the practical aspects of buffer planning and gives ‘suggestions’ relating to how to implement compatible developments or activities in areas of biodiversity importance. Use of the Guide is an important means of ensuring the long-term viability of significant natural areas and the biodiversity they contain and will answer calls by the International Union for the Conservation of Nature (2003) to develop tools which facilitate the integration of protected areas with their surrounds.

Contents

List of Figures

List of Tables

List of Abbreviations

a	Action
APZ	Analytical Protection Zone
BZP	Buffer Zone Planning
CCCS	Common Nature Conservation Classification System
EPZ	Elementary Protection Zone
ESA	Environmentally sensitive area
IBP	Intgegrated buffer plan
IUCN	International Union for the Conservation of Nature
KCA	Koala Conservation Area
NCA	Nature Conservation Act 1992 (Queensland, Australia).
NGO	Non-government organisation
OMH	Other Major Habitat
P	Phase
s	Step

Background

National parks and other types of protected areas and reserves play an important part in the conservation of biodiversity. However, many species are poorly represented in this system, with their habitat being located in less secure landscapes. The existing network of reserves will not afford effective long-term conservation of biodiversity due to the small size of many reserves, the increasing fragmentation and isolation of the reserves and the high level of threats to individual areas. Similarly, the habitat outside this system is subject to a wide range of human related impacts that threaten the long-term survival of many species.

Management of biodiversity must continue to focus on integrating the biodiversity in protected areas and other natural ecosystems with their surrounding landscapes and on reducing or eliminating threats that impinge on these areas. It is increasingly likely that what happens to the biodiversity within a reserve or remnant patch may be less important than what happens in the surrounding landscape matrix.

Conventional Planning and Development

Planning aims to develop a use of land that will ensure the sustainable development of all resources within an area or region. Previous approaches to incorporating environmental concerns, and wildlife issues in particular, into planning have concentrated on identifying protected areas and reserves where development is constrained. Such an approach frequently results in isolated reserves that may not adequately represent and conserve an area's biodiversity. The reserves are also frequently surrounded by landscapes containing valuable patches of wildlife habitat, but in which there is competition with land uses such as housing or industry. In the past wildlife has suffered by having its habitat destroyed or degraded by the impacts of these competing land uses. Effective planning must ensure that all resources, including biodiversity, are effectively considered in the planning process.

Integrated Planning and Management

No longer is it possible to assume that biodiversity in protected areas will be conserved in perpetuity, for the most significant threats to such areas and their biodiversity are from land uses and activities in surrounding lands. Similarly remaining patches of habitat outside of protected areas are being placed under increasing development pressure. Growing fragmentation and isolation of these remnant patches threatens their viability and may result in the local extinction of many species. The conservation of biodiversity requires an integrated planning and management approach, one which is based on identifying important regional ecosystems and corridors, and integrating the management of these lands with surrounding land uses through the development of buffers. In such a way development can be directed to the least sensitive areas, threatening processes can be identified and effective management strategies implemented to ensure more sustainable outcomes.

How to Develop an Integrated Buffer Plan (IBP)

The purpose of this Guide is to ensure effective outcomes for the conservation of natural areas by providing end users with a step-by-step practical approach to the design of buffers for environmentally sensitive areas (ESAs). The IBP process that is described in this Guide is a physical planning tool that can be used to identify core areas and integrated buffers.

- *Core or important natural areas* represents areas that may contain significant regional ecosystems, the habitat of rare or threatened species, habitat of high integrity and wildlife movement corridors, such areas being crucial to the conservation of biodiversity. At the local scale, core areas may be defined for specific target wildlife and may include remnant habitat that is important for food, shelter, breeding and social interaction.

- *Integrated buffer for core habitat* includes lands that may be the source of existing and/or potential threats to the core and which need to be managed, through effective planning and design strategies, to minimise or eliminate the impact of each threat on the biodiversity of the core. The integrated buffer plan incorporates consideration of the ways in which the core may also impinge on the surrounding areas and communities, and includes strategies and policies to be applied within the important habitat to minimise these negative impacts. By encouraging land uses and activities that are compatible with biodiversity conservation, land does not need to be fenced off and totally protected, but rather important habitat and its buffer may include a wide range of uses and activities, so long as these are sustainable.

The identification of core areas and buffers is an effective integrated resource management strategy to ensure the conservation of biodiversity and the perpetuation of functioning ecosystems.

Advantages of Integrated Buffers

Lifestyle: In rapidly urbanising areas, people may continue to live and work in natural settings and achieve the associated lifestyle benefits of large lot living, plenty of open space, peace and quiet, and a relaxed atmosphere, while at the same time being secure in the knowledge that their use of the land will have little or no impact on biodiversity, in a local and regional context.

Sustainable communities: In many developing countries, pressures from human use within and around protected areas can cause significant deterioration in their environmental values. Integrated buffers that are designed on sound ecological principles and which include the community can result in the implementation of a range of sustainable devlopment projects, which benefit the local community within the buffer and at the same time enhance local and regional conservation outcomes.

Environment: A network of significant natural areas, interconnecting corridors and buffers will help to ensure the conservation of important biodiversity values. Integrated resource management focuses not only on species conservation but also on the protection of genetic diversity and the maintenance of ecosystem processes and functions across the landscape. This also includes strategies to enhance the hydrologic values of an area, with issues such as run-off and erosion, stream bank stability, riparian vegetation, in-stream aquatic fauna and water quality being important issues for management.

Recreation: Provision of a network of natural areas that maintain viable populations of wildlife will provide valuable active and passive recreational opportunities.

Education and scientific: Natural areas that are managed to retain their ecosystem values are an important education and scientific resource in a local community. Such sites may be used as outdoor education facilities and to provide baseline data on the natural values of the region.

Amenity: The maintenance of a network of natural areas with viable populations of wildlife will enhance the 'liveability' of neighbourhoods and communities.

Intergenerational equity: By the effective conservation of viable patches of remnant habitat the present generation may fulfil part of its responsibility to pass on to future generations, natural resources that are capable of fulfilling their needs.

Economic: Functioning environmental resources provide free ecosystem services and may reduce costs associated with management of degraded systems such as soil and hydrologic systems and reduce spending on land degradation projects. Productive ecosystems may also provide a range of valuable products thus enhancing the economic benefits to local communities.

Others: Implementation of an integrated approach to environmental management helps meet international, national, state, regional and local policy requirements.

Potential Pitfalls of the Integrated Buffers

Every planning strategy is likely to encompass a number of potential shortcomings. However, through ongoing community consultation and involvement, such problems will be minimised. It is recommended that the Guide be implemented cooperatively with all relevant stakeholders.

Design of an Integrated Buffer Planning (IBP) Model

The process of design always uses models... (Lynch, 1989:278).

A model may attempt to represent reality in a simplified way by identifying the most important features of a real life situation. Although model development is a common approach used in the development of planning theory, models are not always easy to apply due to the nature of the planning problem.

Previous buffer planning approaches have largely been ad hoc. There has been limited used of a model structure or framework and many approaches have failed to place the design of the buffer within a wider planning framework. The IBP approach outlined in this Guide is based on principles of sustainable development, good planning and sound science, and the incorporation of elements of 'best practice' buffer design that are evident in several existing buffer approaches.

Although managers of natural areas may have extensive knowledge of particular target species or habitat, planning for the conservation of biodiversity needs to encompass more than the biological, and include cultural, social, ethical, policy, legal and political aspects that may bear on the survival of particular species or ecosystems (Clark and Kellert, 1998). As a consequence, the IBP model's sequential structure is set within a broad planning framework, utilising an interdisciplinary approach. Thus, the IBP model does not simply identify stages in data collection, but rather includes procedural matters that relate to how the buffer model should be developed and implemented, who should be involved in the process and when, and the nature of the existing administrative and other structures that influence both design and implementation. Planning in general terms has several important purposes, all of which will be incorporated into the IBP model. These include the following:

- *Problem identification*
  The IBP model provides a framework to identify and solve some biodiversity planning problems, and gives a rational basis for decision making. It includes a process for determining whether and where a biodiversity problem exists and for identifying the precise nature of the problem to be addressed by the IBP process. It permits a range of possible policy responses to be identified and evaluated.

- *Problem solving*
  The IBP process is a technique to solve biodiversity related problems. It provides a structure to investigate locational and temporal issues related to where the most suitable places for development are in areas of nature conservation significance, and when various activities may take place. The model provides a mechanism for addressing conflicting viewpoints that may exist in relation to biodiversity and provides a means for identifying forms of development, policies and land use measures that are compatible with the long-term conservation of biodiversity. The model however, is not suited to solving all biodiversity planning problems and may need to be incorporated with other techniques, where necessary.

- *Implementation*

  This is a distinctive phase of the policy planning process and involves the execution of the IBP to achieve its aims and objectives. Effective implementation helps to ensure that the plan takes the form and shape envisioned by its creators, rather than assuming a different structure as a result of changes introduced by those who implement the plan. This stage requires detailed knowledge of government bureaucracies, including their culture and structure (Clark, 1992). The IBP identifies limitations within the planning process that may affect the final form of the buffer plan (e.g. limitations on time, resources and data), and highlights potential opportunities that may assist the implementation of the plan (e.g. effective community structures). The implementation process however, should encourage the involvement of plan implementers, who may provide feedback on the effectiveness of implementation, so that in situations where limitations are experienced improved strategies may be cooperatively developed.

- *Monitoring, evaluation and review*

  Through built in feedback loops the IBP model incorporates the results of monitoring and evaluation to establish whether the aims of the plan are being met during implementation. Evaluation asks questions about what policies are effective, how performance can be measured and evaluated and what criteria or indicators are needed to measure performance. This phase also incorporates, at regular intervals, review of the main components of the plan, including its goals, objectives, policies and outcomes, to produce a dynamic planning model that is responsive to changing circumstances.

The policy process by which the IBP is developed consists of a dynamic and reciprocal set of interrelationships between different groups of stakeholders, rather than a top-down, hierarchical approach. The process aids reconciliation between the often-conflicting interests of the managers of reserved or protected areas and surrounding communities. It encourages cross-sectoral cooperation and information sharing among state/provincial and local government agencies, statutory authorities and non-government organisations (NGOs). Importantly, the process, although complex, has no set beginning or end, enabling adaptation to many situations.

The IBP model includes the processing of information in each of its five major phases (Figure 1), namely information input, processing and output.

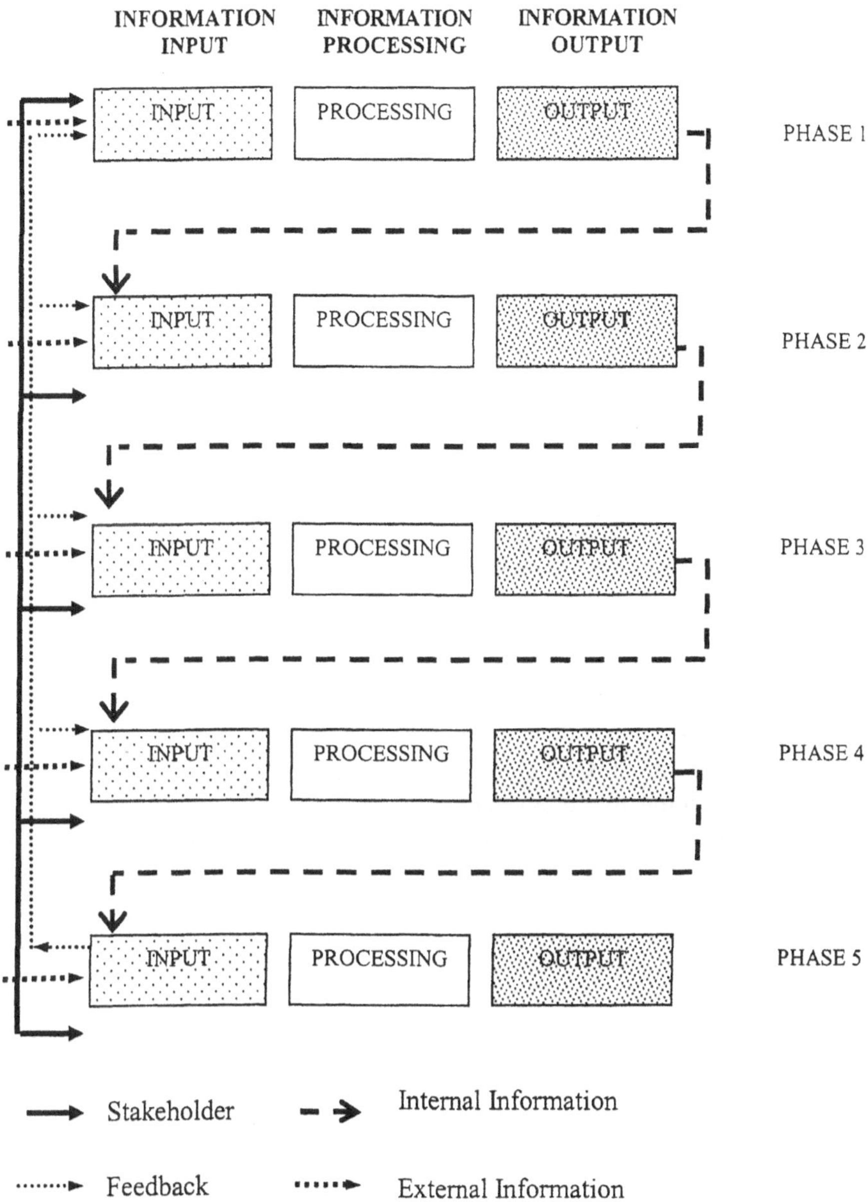

Figure 1 **Phases in the integrated buffer planning model**

Information Input

Information input may consist of the following:

- *external information*: data obtained from external sources to specifically address aspects of the problem;
- *internal information*: the data output from previous phases of the model (refer to information output);
- *stakeholder input*: the early and ongoing participation of all relevant individuals and groups (e.g. planners, decision makers, decision influencers, end users, affected communities etc.); and
- *feedback loops*: information loops that input data gained throughout the process, into relevant previous planning stages, to enable a re-examination of issues in the light of recent developments in the plan's preparation. This helps to produce internal plan consistency, while allowing for maximum flexibility in plan development and implementation.

Information Processing

This consists of a number of sequential steps (Figure 2), each step having a series of actions. Also included throughout the model process are decision points that provide flexibility in plan development by posing a series of choices concerning the progression of the planning process. The use of decision points enables the planning team to eliminate unnecessary steps or actions and to tailor the process to their specific requirements, based on resource and information availability, technical expertise, timing and expediency. There are five phases of information processing within the model (Figure 2).

Phase 1: Setting the framework A preliminary examination of the issues is undertaken, the purpose being to determine whether the IBP process is suitable to use in the existing circumstances. In this phase the institutional/organisational, socio-economic and political context of the task is examined.

Phase 2: Setting the planning process Where stakeholders have identified a biodiversity problem or issue, the precise nature of the problem is detailed in this phase. This important, preliminary part of the model helps to

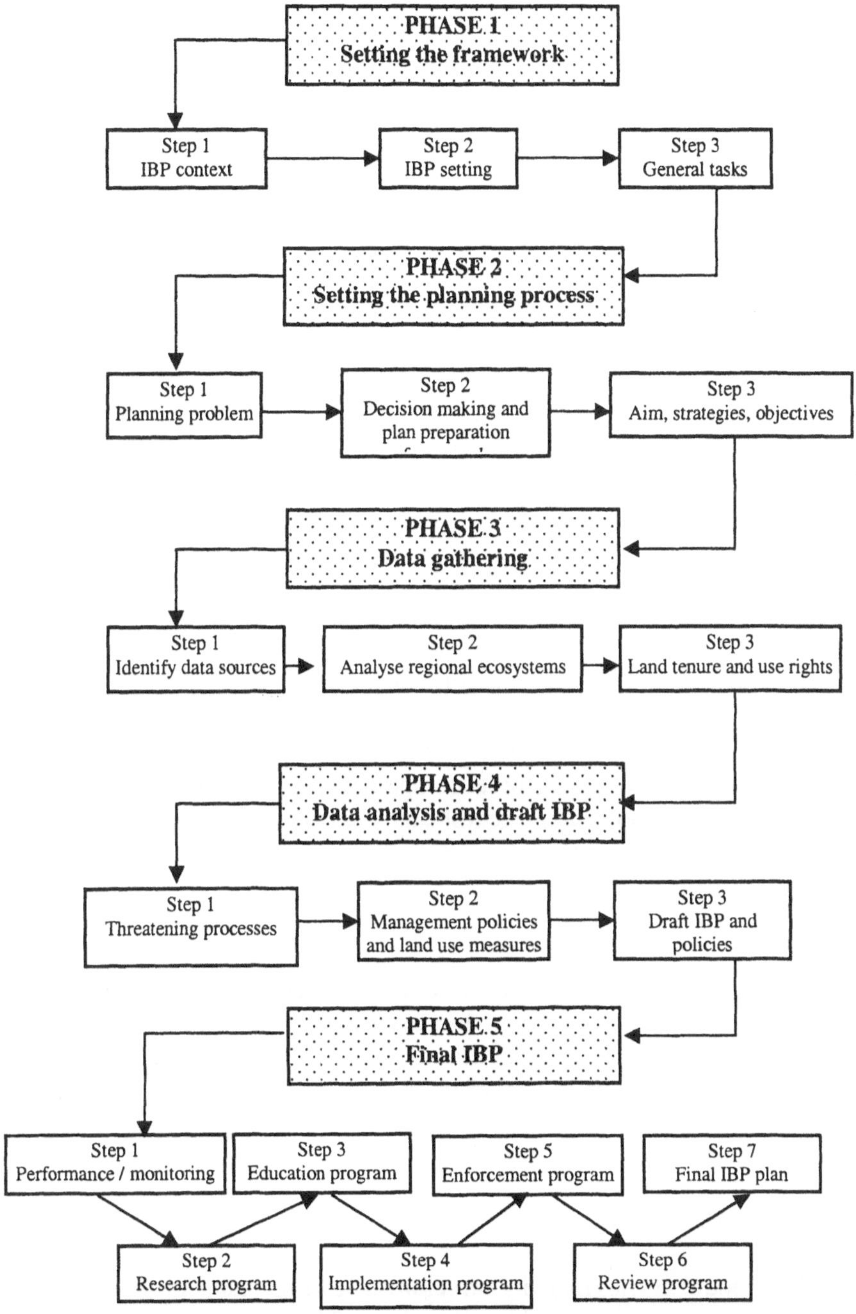

Figure 2 Information processing phases of the integrated buffer plan

identify the necessary prerequisites for data gathering (phase 3) and analysis (phase 4). The phase includes the definition of the planning problem, statement of aims, strategies, objectives and the identification of the main assumptions and specific limitations of the process. The planning team also decides how the IBP model is to be applied in the current circumstances and begins to analyse data needs. It is in this phase that the model is adapted from a theoretical model process to a pragmatic and realistic applied process that is space and time specific.

Phase 3: Data gathering Once the planning task has been established, data to determine the location and nature of the buffer (phase 4) are assembled by the planning team. A major focus is to identify the environmental values of the area (e.g. the habitat requirements of particular species and the landscapes in which they are found). Also important is the gathering of specific data relating to the resident communities, both within and surrounding the core natural area.

Phase 4: Data analysis and draft buffer plan This phase outlines the process of delimiting the IBP and developing policies and land use measures to be implemented within the core and buffer. A draft buffer plan, policies and land use measures are produced.

Phase 5: Final IBP The production of a final plan is an important outcome of this phase. The plan may be accompanied by programs which identify how to implement, enforce, monitor and review the final plan, as well as education and research programs.

Information Output

This part of the IBP process (Figure 1) represents a largely measurable product that can be achieved within a particular phase. Hence, for each phase of the model, it is necessary to identify the output to be achieved, and usually this data will be the main input into the successive phases of the model process.

How to Use the Guide

The IBP model is flexible, its structure allowing planners to select the relevant steps and actions within each phase. The selection may depend on time and resource availability, the size of the area, the cultural setting, data availability, political or ecological expediency, the planning team's expertise, prior knowledge of and experience in the planning area, and the statutory requirements of the organisations involved in plan development. There is no requirement to complete all components of the process, especially where the wider planning issues are well understood by the planning team.

Each phase of the Guide has a consistent format. For example, a flow diagram will illustrate the main steps and actions to be undertaken in each phase. Figure 3 illustrates and explains the main components of the process. The remainder of this Guide is devoted to outlining the five phases of the model IBP process.

PHASE NUMBER
TITLE

Information input
The information required to begin the phase is listed.

Expected outcomes
Specific outcomes to be achieved in the phase are listed.

P1, s1 **Step 1: Title**

A brief description of the purpose of the step is outlined.

P1, s1, a1 **Action 1 Title**

An explanation of the purpose of the action
This provides background and context to clarify and reduce ambiguity.

Planning actions required
Specific tasks are detailed and in some instances possible barriers to effectively accomplishing the tasks are indicated, as well as suggested mechanisms for overcoming these barriers. It may not be necessary to undertake all of the identified actions in every situation. Local circumstances and a range of other factors will influence the selection of actions to be undertaken.

Examples / illustrations / case studies
A range of applications or case studies is provided to help clarify the planning process required. Their inclusion is based on the premise that one of the best ways to learn about IBP design is to observe and improve on what others have done. However, this does not imply that these assessments represent the 'state of the art', for buffer design is continually evolving.

Key players
Each action requires an actor or groups of actors. A range of key players is indicated for each action. However, due to the wide variation in the planning setting, it is not possible to cover all possible actors. Planners should give additional thought to possible individuals, groups or agencies that may assist in implementation.

♦ **DECISION POINT (DP)**
These occur throughout the process and provide direction on the subsequent progression of the model.

Figure 3 Explanation of the format used in each phase of the integrated buffer planning process

PHASE 1

SETTING THE FRAMEWORK

Brief Description

The purpose of this phase is to understand the context or framework within which the IBP process is conducted. It involves understanding the institutional/organisational, socio-economic and political framework (Figure 4).

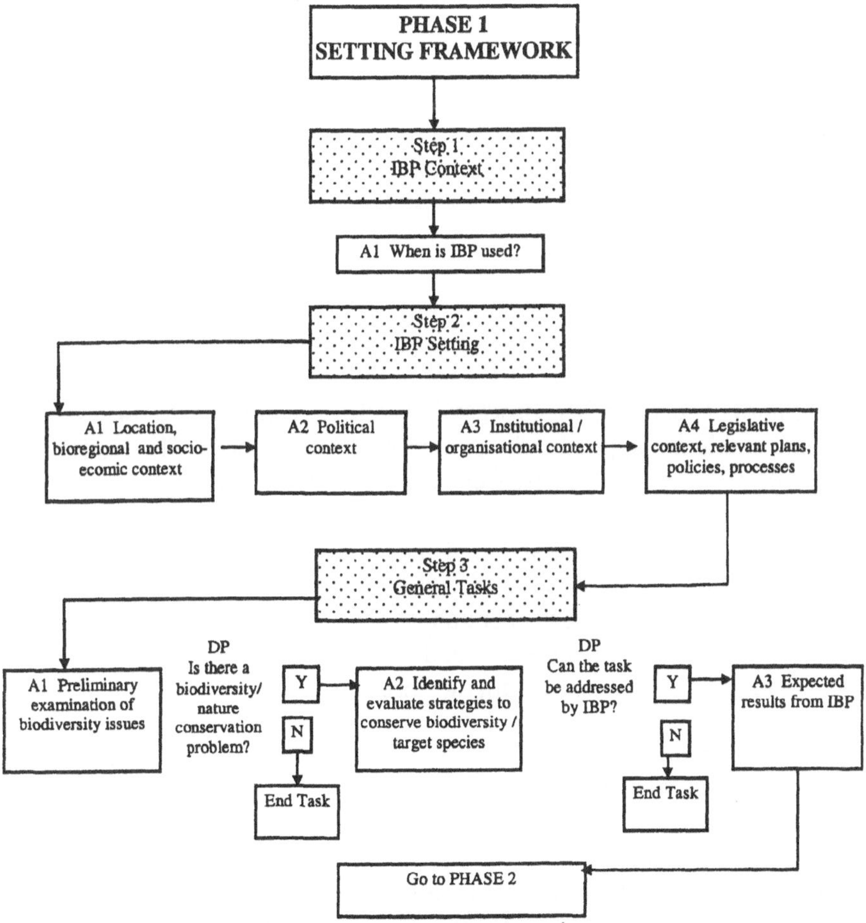

Figure 4 Phase 1 – Setting the framework

Information Input

- external information that indicates there may be a planning problem (e.g loss of biodiversity or poor water quality) needing further consideration.

Expected Outcomes

- a decision on whether the IBP process is appropriate for the task; and an understanding of the broad planning framework.

Pl. sl **Step 1: Understand the context of the IBP planning process**

This step describes the context in which a buffer plan may be a useful planning strategy, identifies the existing issues surrounding the problem and examines the conceptual framework for problem solving, focusing on contextual analysis to provide a broad identification of the factors that may influence policy development and implementation.

Pl. sl. al *Action 1* *Indicate when an IBP process may be used*

It is important that the IBP process is used for a purpose for which it is designed and is not applied in inappropriate circumstances. The model IBP process offers a logical, step-by-step approach leading to:

- the development of an integrated buffer that minimises the impact of external threats on one or several identified core areas and also minimises the impact of the core on its surroundings; and
- the development of policies and land use measures that ensure sustainable development within the core and buffer.

There are a number of possible situations in which the IBP model may provide an effective planning tool:

- *to conserve significant natural areas by reducing or eliminating threats (both existing and future) to these areas*
  Planners need to adopt 'anticipate and prevent' strategies. An IBP may be implemented proactively before ecosystems or target wildlife are threatened, or ecosystem damage and loss of biodiversity become

critical. Although this approach is preferred, the IBP model is more likely to be initiated reactively in response to an existing problem, which may relate to changes in the composition, structure and functioning of biodiversity. For example, the number of individuals in a particular population may be declining, particular habitats and ecosystems may be degrading, or productive systems such as fisheries may be failing. Structurally, the spatial arrangement of a species across a landscape may be changing due to altered land use activities in a region, which in turn may produce functional changes due to altered ecosystem processes. These changes may occur over time scales ranging from days, to years and decades. For example, changes to a population may be relatively rapid, as a result of natural catastrophes (e.g. fire), or more subtle and slowly evolving (e.g. due to changes in the reproductive rate or as a response to the impact of dogs, cats and cars). These changes may be detected on the basis of a noticeable loss of a species from an area, the presence of road kills, animals injured from dog attacks, diseased animals, the absence of young, the loss of wildlife habitat, or the threat of future developments. Potentially threatening land uses include residential development (pest species and polluted water), industry (point source pollution and contaminated storm water runoff), agriculture (nutrients, pesticides, fertilisers and spray drift) and grazing (trampling and fouling of waterways).

- *to indicate areas suitable for new development or an intensification of existing development and which will result in little or no adverse impacts on the biodiversity of an area and on physical processes*
  Where human populations are expanding into natural habitats or areas subject to dynamic processes, such as coastlines and riverbanks, the IBP model can be used to identify sites where development should be avoided and/or permitted subject to the implementation of land uses and associated activities that are compatible with the conservation of the area's biodiversity and the maintenance of ecosystem processes.

- *to indicate policies or guidelines for land use and development in ecologically sensitive areas or areas subject to dynamic processes such as erosion and deposition*
  The IBP process can be used to develop relevant policies and land use measures that can be applied to retain and restore the values of environmentally sensitive areas (ESAs). This will enable landowners, developers, government officers and the community to know the

constraints and opportunities for development in particular areas and to respond appropriately.

The model thus provides flexibility, being dynamic and responsive to a range of biodiversity planning needs and to changing circumstances, and it may be applied in various social, economic, cultural, organizational, political and technological situations. Hence, the process of developing an IBP in a remote habitat of an endangered species may differ from that for a habitat fragment in a more developed setting. Also, the procedures for producing an IBP plan may be varied to reflect the knowledge and expertise of the planning team.

Pl. s2 Step 2: Identify the general setting in which the IBP may be developed

To solve complex problems, a good policy framework requires contextual analysis, allowing the planner/planning team to make sense of large, complex and potentially bewildering information, by understanding how events are interrelated (Clark, 1992). The aim of contextual analysis is to 'make the obvious inescapable' (Torgerson, 1985:250) and to 'remove the ideological blinders' (Lasswell, 1971:220). It aids problem definition as it is a broad and in-depth identification and examination of all the factors that directly or indirectly influence the issue/problem at hand, thus preventing a narrow, fragmented conception of these issues/problems and their context. It also aids the identification of possible solutions and highlights the role of individuals and groups within the policy process.

Developing a social process map produces an explicit conception of the entire policy process and social process. It also identifies the context of the values of individuals and institutions and identifies the expectations of participants. Biodiversity conservation cannot be carried out in isolation from the social, economic, organisational and political environment. An understanding of issues beyond the confines of any physical plan is necessary to ensure a more comprehensive, systematic and rational process of planning. Issues such as the compatibility of the project aim with the broader goals/objectives of the organisation and the social and political context in which the plan will be implemented need to be identified. There may be conflicting needs and conflicting legislated mandates within agencies and between interested parties. As well, the land ethics of a society may produce both constructive and destructive interactions among managers of ESAs and surrounding landholders. This step consists of five actions.

P1, s2, a1 ***Action 1*** ***Establish the location and context (both environmental and socio-economic)***

The broad location of the proposed study area should be identified by setting tentative boundaries within which the plan is to be developed. Where bioregions or ecoregions have been identified, a bioregional framework is recommended for addressing biodiversity problems and specific wildlife issues. Where water quality is a prime concern a catchment or sub-catchment may be an appropriate spatial setting. In coastal areas, the definition of a coastal zone for planning may be more problematic, requiring consideration of the specific planning issues and their spatial extents.

The planning team should also understand the social, cultural and economic context within which the plan may be developed. The implementation of compatible land use strategies within the buffer will require wide community cooperation, support and ownership of the final plan. A failure to understand relevant socio-economic issues may result in poor plan compliance (*Note*: specific issues relating to land use practices, customary rights and responsibilities, and land tenure arrangements are examined more closely in P3, s3). The gathering of sociological information includes the need to understand the characteristics of the human use of the area so that monitoring may indicate any changes in the needs and perceptions of communities, and in their land use activities. Where the planning team is familiar with the socio-economic context, this element may require less investigation.

This action is designed to begin building communication links between the community and the planning team and to encourage the community to begin thinking about their resource use patterns and future needs. The following actions should be considered:

- *Identify the context (e.g. bioregion or catchment)*
  The bioregion/ecoregion or catchment in which the IBP is to be developed should be identified. In areas where there is extensive environmental variation a bioregional framework will enable the identification of priority conservation issues that should be considered in the IBP, thus enabling a more systematic approach to nature conservation.

- *Identify the likely 'end users' of the plan*
  Although the final plan boundaries are unknown, the potential users of the plan should be identified in the initial stages and throughout the planning processes.

- *Identify the values, attitudes and perceptions of the participants in the policy process and the 'end users' of the plan (both existing and future)*
  This requires discussion with participants to gain an understanding of the following:

  - *Current natural resource use and community livelihood practices*: this may include a map of general land use patterns, forests, fisheries and water system use and other issues as relevant to particular areas.
  - *Historical usage of the area*: this may include data on past events, cultural sites, pre-history, history of management (e.g. burning regime). Information sources may include government archives and libraries, local residents and user groups.
  - *Living standard*: issues to consider include health and food security status, mortality and education.
  - *Cultural values and traditions*: this may include base values such as power, enlightenment, wealth, well being and skill (Clark, 1992). A systematic evaluation of the community should be undertaken prior to initiating change. This will serve as a benchmark for measuring any subsequent changes in knowledge, attitudes and behaviours. Knowledge of such issues will enable the planning team to design an effective public relations campaign to improve support for the plan and to educate the community on relevant issues. (*Note*: in this part only a general understanding of values and traditions is required, for a comprehensive survey of values may be undertaken in P3, s2, a2).
  - *Community perspectives and needs*: identify conflicting interests among the community to identify issues of most importance to them.
  - *Strategies of action*: through discussions with community members begin to identify acceptable strategies for implementation of the plan.

The effects of all the above elements on each other and on the decision and policy processes should also be understood (Clark, 1992).

- *Identify resources and support that may be available locally*
  This may include local people who may be employed during the course of plan development, and community groups that may be able to offer support, or assist with data gathering.

- *Identify possible socio-economic barriers and develop contingency plans to overcome these*
  These could include:

  - *Lack of acceptance and/or active support for the IBP plan*, combined with perceptions of risk or uncertainty that may impede the effectiveness of the plan and limit effective negotiation. To minimise community fears, the process may need to include a community education strategy and ensure effective consultation and involvement in developing, implementing and monitoring the plan. Potential opponents of the IBP plan should be included in the planning process in the early stages and their concerns addressed as they arise.
  - *Lack of baseline information on the needs and desires of local communities*. Community surveys may be a useful strategy to provide this information.
  - *Regionally declining resources*, or increasing economic pressure for consumptive use of natural resources, or rapidly changing socio-economic system including high turnovers in land ownership. These issues should be considered in the design of appropriate implementation strategies.
  - *Lack of finance* to ensure effective plan preparation and implementation. The planning team or organising committee should maximise its contacts with community, business and government agencies and ensure that realistic budget provisions and an action plan are in place. The IBP planning process must be tailored to the financial resources available and where possible should utilise partnership agreements that encompass resource sharing.

Examples:

Golden Lion Tamarin Conservation Project (Dietz et al., 1994) All three lion tamarin projects in Brazil included surveys of local knowledge and attitudes regarding wildlife and local protected areas, and previous projects. Communities surrounding the reserve were targeted, as well as residents in the cities of Rio de Janeiro and Sao Paulo, who were illegally purchasing wild animals. Also surveyed were government bureaucrats and politicians in Rio de Janeiro and Brasilia and the general public. The information gained served as a basis for planning strategies and capitalising on local interests that were compatible with the project's conservation objectives.

Mountain Gorilla Project in Volcanoes National Park, Rwanda (Wells and Brandon, 1993) As part of the project, community surveys were conducted before and after plan implementation to provide insight into the community's views of wildlife and conservation issues.

Port Stephens Koala Management Plan (Lunney et al., 1997) The plan was based, in part, on a local community survey to ascertain community attitudes to particular threatening processes and their views on possible conservation strategies.

Pl. s2. a2 ***Action 2*** ***Establish the political context***

The political context, which is a central part of planning, relates to the prevailing authority or power structures and the interplay of organisations and laws. Where many stakeholders are involved, issues of authority and power need to be understood, because many biodiversity/species management programs are as much political as biological/scientific. Factors that may influence this political context include land tenure patterns, access to and control over resources, property relations and rights, social stratification and traditional authority (Reading et al., 1991). Biodiversity management issues may produce conflicts, for example, between agriculturists, conservationists and mining interests, as well as those fearful of losing traditional rights or power, and employment.

The prevailing political organisation will influence many aspects of buffer design and implementation, because decisions are frequently taken in the political context and may be outside the control of the planning team. The type of strategies developed, whether they are formal or informal, regulatory or non-regulatory, or incentive based will depend on the political context. An understanding of this context is crucial to the establishment and ultimate effectiveness of integrated buffers. In general, the strategy should be aimed at the political level which is strongest e.g. where local government is weak, local communities may be a target. Generating support from all levels of government will help to ensure long-term conservation within shifting political arenas.

Integrated buffer planning takes time – usually more than the life of one government. A commitment (of time and personnel) of around 10 to 15

Table 10.3 Matrix of existing and potential treats to heritage values

	Macro Visual Character				Micro Visual Character				Macro Visual Association				Micro Visual Association				Structure			
Values	A		B		C		D		E		F		G		H		I		J	
Impact	E	P	E	P	E	P	E	P	E	P	E	P	E	P	E	P	E	P	E	P
Height/ mass	Major	Major	Significant	Significant	Significant	Significant	Significant						Major	Major	Major	Major	Major			
Roof form	Low	Significant																		
Landscaping			Significant	Significant			Significant	Significant												
Signage													Significant	Significant		Significant				
Traffic vibrat'n																				Significant
Pile driving																				Significant

(Where: Values include A. Skyline; B. Colour; C. Scale; D. Colour; E. Architectural detail; F. Alignment; G. View towards protected place [landmark]; H. View towards protected place [landmark]; I. View away from protected place; and J. Structure integrity; and threats include E. Existing threat; and P. Potential threat).

Major Impact	Significant Impact	Low Impact

historic nature of the study area in contrast to the flat rooved modern development in the foreground.

The landmark value of the Post Office Tower is reduced by the larger mass and height of the building in the foreground (located on the corner of Brisbane and East Streets). Vass-Bowen (1994) considered this a major impact, in terms of building height and mass, one which degraded the landmark value of the Post Office Tower. Similarly, the skyline of the city centre is also dominated by the same building from viewing point M1, although its horizontal dimensions are not sufficiently extensive to completely obscure the skyline. The Grammar School grounds behind are visible creating a contrasting backdrop to the skyline. The impact of height and mass on the skyline value was therefore considered significant.

Due to time and resource limitations, the next five steps (3 to 7) were undertaken for only the most significant values. They were the macro character values of skyline and colour, the macro association value of views towards the protected place (landmark), and the structure value of structural integrity. In a full application of BZP in an urban context, the extent of threats would be highly variable in form and the relationship between the place and its surrounds highly complex. Given this constraint Vass-Bowen decided to use the variant B of the BZP methodology, in which the derivation of APZs precedes that of the EPZs.

Pl. s2. a3 Action 3 Establish the institutional/organizational context

The IBP process must link what is to be prepared with how its preparation is to be achieved within the organization(s) responsible for preparation and administration of elements of the plan. Ultimately, the success of the IBP will depend on the commitment from all levels, especially the highest levels of management within the organisation. By understanding the institutional or organisational context, the IBP is more likely to 'fit' into the existing national, regional or local system of planning.

The institutional context may be a product of a number of factors including:

- the historical circumstances surrounding the emergence of concern for natural systems and environmental quality;
- the mix and relative weights of national objectives (e.g. economic development, income distribution, regional development and environmental quality);
- the nature of the cultural traditions and values of a society;
- the nature of the socio-economic system;
- the nature of the political and governmental system; and
- the nature and relative importance of international economic links (e.g. multilateral and bilateral aid agencies).

Understanding this institutional framework is crucial in developing appropriate environmental strategies and for monitoring ecological and social changes. Ideally, the IBP process needs an organisational structure that will permit a variety of management approaches (regulatory, non-regulatory and incentive based). Planning approaches that are practical and suit the institutional context are most desirable. Actions to be taken include:

- *Identify planning bodies and other organisations that may help to develop the IBP*
  One, or several agencies or departments may be responsible for developing and implementing a buffer plan, e.g. national, state or local governments, regional planning bodies, catchment management and Landcare groups, and conservation NGOs. It is important to identify and understand the administrative and institutional structures within which the plan is to be developed.

- *Identify the expected barriers*
  Possible barriers may include:

  - *The organisational context* This includes the structure, culture, goal orientation, personnel characteristics and agency hierarchy (Clark, 1992). The effectiveness of the buffer strategy may be related to the level of commitment from those developing, administering and implementing the plan. Success may be limited if the organisation is resistant to, or lacks commitment to the buffer strategy, or there are conflicting organisational goals.
  - *The decision making processes* This may include policy formulation and implementation processes, management philosophy, operating procedures, degree of organisational conservativeness, relations with constituents and the public and the level of administrative discretion. For example, Reading et al. (1991) state that the organisations that dominate endangered species programs are conservative, government bureaucracies with fixed standard operating procedures. In some cases power differentials and states' rights can dominate the kinds and frequency of interaction among the organisational stakeholders and this has major implications for the conservation of biodiversity.
  - *External pressures* An organisation may be subject to external pressures from controllers, clientele groups, constituencies, allies and adversaries, the media, the legislature and judiciary (Clark, 1992). Thus an IBP may fail to eventuate if sufficient negative pressures are applied to the responsible agency.
  - *Inter and intra-agency relations* There may be a lack of communication with other organisations and the wider community.
  - *Lack of expertise in conflict management* The resolution of conflicts requires effective education, communication and carefully designed mechanisms for planning, cooperation, and coordination.
  - *Cross boundary problems* Where a buffer crosses administrative boundaries, cooperation among all stakeholders is crucial for the success of the plan.
  - *Resource constraints* These may relate to staffing and budgetary issues.

- *Develop contingency plans to overcome likely organisational barriers*
  This may include:

- A Memorandum of Agreement Where organisational responsibility for the buffer is shared, a contractual memorandum of agreement may be necessary to formalise the responsibilities of each party.
- Workshops and training session This may be useful for administrators of the plan.
- Development of informal structures Members of the planning team should be represented in relevant planning forums within the region to enhance plan consistency and to develop informal networks.

P1. s2. a4 Action 4 Identify the legislative context and relevant plans, policies and processes in the wider area (Government and NGOs)

- *Identify the legislative/administrative boundaries in the region*
These may include the boundaries of the local, state/territory, regional planning areas, and water catchments etc. Where the IBP may cross a number of legislative/administrative areas, all relevant players should be included in the planning process e.g. local and state governments and NGOs.

- *Identify any laws, policies or plans relevant to the wider area*
Those responsible for plan preparation should understand the legislative framework and work closely with the responsible agencies. Existing laws and policies may contain general objectives/intents or conservation objectives specific to particular species and habitats and restrictions concerning land uses/human activities that may affect the ecological and social contexts within which the buffer plan is being prepared. A thorough understanding of the legislative basis is crucial for the development of effective buffer policies, especially where numerous Acts and regulations are present. These laws may deal with the purposes for which the land can be used, management of natural and cultural values, soil and water quality, feral animal control and fire. A failure to understand this context may result in inappropriate buffer policies. Hence, statutory responsibilities and obligations that are stated in legislation and other policies should be identified. Consideration should be given to:

  - international agreements/conventions;
  - national laws, policies and strategies;
  - state laws, regulations, policies, codes and guidelines;

- regional plans, policies and guidelines;
- local initiatives, plans, policies and strategies e.g. proposed acquisitions, voluntary conservation agreements, utilities planning (e.g. access routes and electricity/water supply routes), development proposals under consideration, strategic plans, planning schemes, catchment management plans, fire prevention plans, or specific natural resource management plans such as open space plans, corridor plans, landscape/streetscape plans, and vegetation management plans); and
- legislative instruments (e.g. covenants) that strengthen customary laws and which may provide a basis for community stewardship and a vehicle for conflict resolution and arbitration.

At all stages of plan preparation and beyond, planners also should be aware of changes to relevant legislation, so that these can be incorporated into the buffer plan. The planning team should be in continuing contact with relevant agencies during plan preparation. Prime responsibility for these actions rests with the planning team.

- *Identify local and regional planning bodies and existing programs that may be able to assist in implementation of buffer policies*
  It is not desirable to establish additional programs and structures if existing ones can be adapted to the needs of the IBP. This may help to minimise the conflict between the buffer plan and existing programs, and established groups may be able to give additional expertise and support to the developing plan. Coordinated natural resource management may be enhanced if representatives of the IBP's working groups are appointed to the relevant committees or advisory groups working on associated conservation strategies.

- *Identify relevant research being undertaken*
  The planning team should identify the research currently being undertaken in relation to the natural environment or the local area, as the outcomes of this research may affect the outcome of the buffer plan.

P1. s3 Step 3: Describe the general tasks

The purpose of this step is to gain a better understanding of the broad nature conservation issues within the proposed plan area and to assess the suitability of applying the IBP methodology.

P1. s3. a1 ***Action 1*** ***Undertake a preliminary examination of the nature conservation issues***

The process of problem definition is one of search, creation and initial examination of ideas for solution. Hence planning intervention may be necessary where a gap is identified between the existing 'state' and the community's 'desired future state' and it is necessary to prevent an undesirable future state. Alternatively, even where no problems are currently evident, planning intervention may be warranted to proactively plan for the continuing conservation of the area's values. In either situation a knowledge of the general task or issues in relation to biodiversity generally, or specific species/populations is essential before the precise nature of the planning problem can be identified. Problem recognition is a function of both the information received and the ability to interpret it (Brewer and deLeon, 1983) and thus needs to consider both the individual and the organisation.

- *Identify broad community goals or philosophy in relation to nature conservation*
  Recognition of a problem implies a certain philosophy or set of values and goals on the part of the individual and community. Many nations have signed international agreements and conventions in relation to nature conservation, and these have been translated into legislation at the national, state/provincial and local government levels. Some communities have established planning goals in relation to biodiversity and some local governments have stated a commitment to conserve biodiversity. Where such broad agreements are in place and particular goals have been identified, they should be clearly specified.

- *Access information sources through which the conservation status of ecosystems and wildlife may be identified*
  For a problem to be recognised, information must become available. However, data relating to structural, functional and compositional changes in ecosystems, which are the result of landscape fragmentation, isolation and degradation may be difficult to obtain. In some situations, species diversity may remain unchanged yet the community structure may alter due to the introduction of more generalist and noxious species. These qualitative changes may represent a significant loss of biodiversity, in particular that of specialist species.

  Problems may be identified informally by officers of government departments, statutory authorities and other agencies, scientists, academics, community groups, individuals, businesses, the media and politicians. More formal processes may include:

- *Formal legislative requirements* This may include State of the Environment Reports, which are a legislative requirement of many countries, states/provinces, regional planning groups and local administrative bodies; and 'recovery plans', which may incorporate regular monitoring and evaluation time frames. Such legal requirements are an avenue for identifying issues of importance for nature conservation.
- *Advisory committees* Government departments may have formal procedures established, whereby biodiversity values may be monitored. This may include a range advisory committees established to provide guidance to decision makers. Such advisory committees are an avenue to address biodiversity issues of concern.
- *Formal administrative structures* Structures may exist within governments, agencies and organisations to regularly obtain data on a range of biodiversity values. This may include vegetation monitoring at a landscape scale or activities conduced by local governments and community groups on a smaller scale to monitor the status of selected target species.
- *Workshops and conferences* These enable issues to be highlighted by a wide range of individuals and organisations and may be a valuable source of data on a range of biodiversity values and the nature of threatening processes.
- *Scientific literature* Scientific research may provide a source of data relevant to species' conservation and management.

• *Bring the problem to the attention of a responsible organisation*
Once an individual or group is satisfied that a problem exists it must be brought to the attention of an appropriate organisation. However, several barriers may prevent this.

• *Identify barriers to problem identification*
Barriers may exist at two levels:

- *Individual* An individual's perception of whether there may be a nature conservation problem is conditioned by the individual's personal background (e.g. psychological, cultural and educational) and institutional/professional background (Brewer and deLeon, 1983). Individuals may choose to suppress or ignore some information (e.g. due to cognitive dissonance, information overload, recognition of the costs involved in potential remedies, feelings of powerlessness and futility, and the fear of invoking negative community reactions, etc.); and

- *Organisational* An organisation can facilitate or impede information inflow and recognition of problems based on the organization's: distribution of authority (pyramidal or horizontal); lines of communication; size and age; demands already made on it; composition and attitude of the workforce (Brewer and deLeon, 1983). Organisations may be reluctant to recognize problems and initiate policy changes due to: the costs involved in replacing or revising existing policies; conservatism and a preference for the institutional status quo; and an inability to perceive problems.

♦ DECISION POINT

Is there a general biodiversity or nature conservation problem?

At this point, a preliminary assessment of whether there is a biodiversity/nature conservation problem or issue that requires further planning action is needed. The decision will be made, usually at a political level in either the informal or formal review processes indicated above. Perceiving a problem does not mean that new policy alternatives will be generated, however.

If 'yes' state precisely the nature of the task to be investigated and **proceed to Phase 1, Step 3, Action 2** (P1, s3, a2) within the model IBP process.

If 'no' the IBP **process is halted**, perhaps temporarily. Such a decision may be based on lack of verifiable information, or an evaluation that the problem currently is not significant, or a response is premature.

However, the 'no' decision may initiate action to overcome some of the barriers. A lack of information may suggest the need for more research or perhaps the use of a Delphi type process to gain a consensus of 'expert' opinion.

P1. s3. a2 ***Action 2 Identify and briefly evaluate general strategies to conserve biodiversity and/or target species***

Once a problem is recognised, possible strategies to alleviate or resolve it should be explored and the risks, advantages and disadvantages of each examined, before a decision to proceed with the IBP plan is made. This action emphasises empirical and scientific issues to determine the consequences of the suggested strategies or options and an assessment of the desirability of these outcomes. The strategies may include both statutory and non-statutory approaches and will relate to the specifics of each planning situation. The range of feasible alternatives should include a 'do nothing' option as well as more 'radical' options. The alternatives should be sufficiently varied to permit the decision makers to choose between different capabilities, technologies, resources and policy levers (Brewer and deLeon, 1983). This action is thus a creative one where many ill-defined suggestions may be put forward. This action aids the definition or re-definition of the problem to get a sense of it in terms of its possible importance and whether it merits further attention and expenditure of resources. The choice of which strategy to pursue should be based on suitable criteria, such as: effective conservation; low risk of failure; cost effectiveness; and social and political acceptability. Costing of alternatives over the life of each proposed plan is important. Possible evaluation techniques include: decision analysis; cost-benefit analysis; safe minimum standard; and multi-criteria analysis (refer to Crosthwaite, 1995).

♦ **DECISION POINT**

Can the task be addressed, 'in full' or 'in part', using the model IBP process?

If 'Yes' then **proceed to Phase 1, Step 3, Action 3** (P1, s3, a3) in the model IBP process.

If 'No' then IBP **process is halted**. Such a decision may be based on: lack of verifiable information; an assessment that the risks/uncertainties of doing something outweigh those of doing nothing; lack of resources; an assessment that other more suitable strategies may be applicable; lack of political will; or lack of community support.

However, the 'no' decision may initiate action to overcome some of the barriers. For example, a lack of community support may suggest the need to inform and educate the public on broad nature conservation issues and thus lay the groundwork for future integrated management approaches. Lack of political will may be a temporary situation, especially where strong public support gathers impetus, or where a change in government occurs. A lack of information may suggest the need for more research or perhaps the use of a Delphi type process to gain a consensus of 'expert' opinion.

P1. s3. a3 ***Action 3 Specify the expected results from the IBP analysis and the form in which the results are to be presented***

Identify where the main emphasis of the IBP analysis should lie and what specific outcomes are required. For example, the desired planning outcome may be the production of a map indicating important habitat and integrated buffer areas. Alternatively the buffer map may be accompanied by a comprehensive development of appropriate policies and land use measures to conserve biodiversity.

PHASE 2

SETTING THE PLANNING PROCESS

Brief Description

In this phase, the planning problem, aims, strategies, objectives, main assumptions and any specific difficulties are defined. The planning team also decides how the IBP model is to be applied in the current circumstances and begins to analyse data needs (Figure 5).

Information Input

- knowledge of the planning framework;
- a decision to apply the IBP planning model to address the identified problems; and
- a preliminary outline of the expected outcomes of the process.

Expected Outcomes

- statement of needs and issues relevant to key stakeholder groups;
- clear statement of the problem;
- statement of goals, strategies and objectives; and
- knowledge of the decision making framework.

P2. s1 Step 1: Identify the planning problem

This step aims to identify all relevant stakeholders to ensure effective identification of the planning problem. It consists of three actions.

P2. s1. a1 *Action 1* *Identify key stakeholders*

A clear statement of the precise nature of the problem is needed to focus the remaining phases of the model. Stakeholder participation will play a role in achieving this outcome. However, participation processes can be either bureaucratic and elitist (Maywald, 1989), based on the premise that elected officials and other elites can best serve the community if left largely

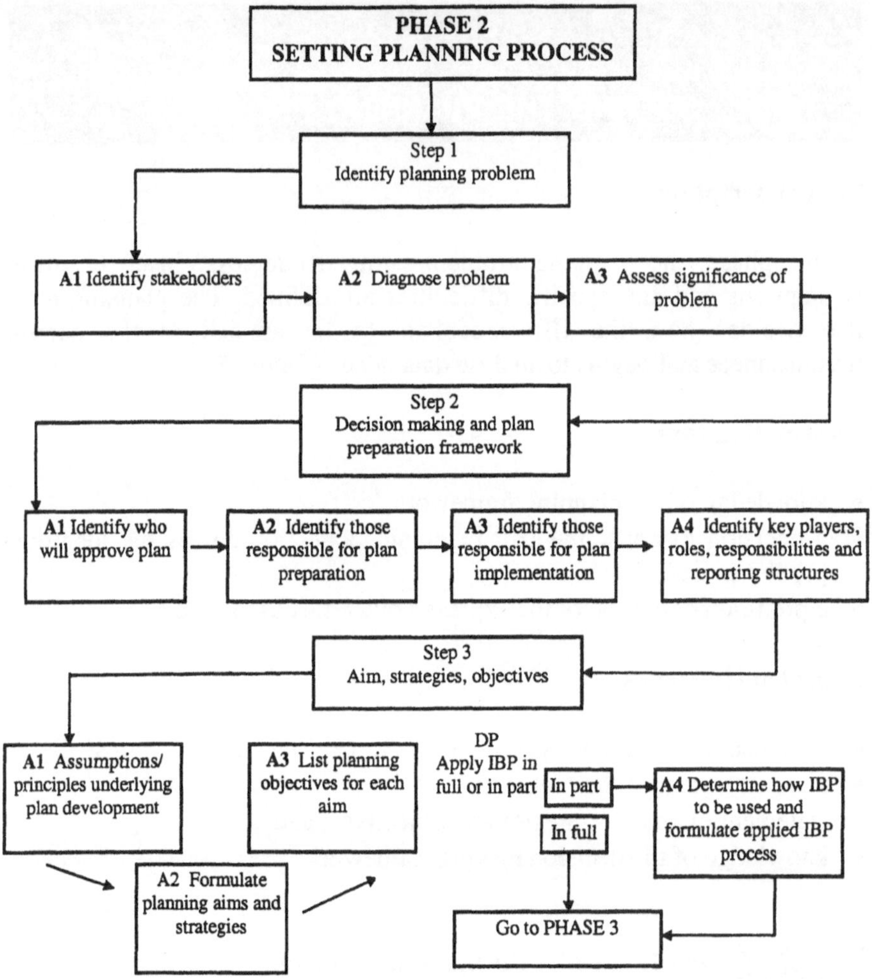

Figure 5 **Phase 2 – Setting the planning process**

to their own devices, or they can be more pluralist and social democratic, based on the premises that those who have a stake in the issues should be consulted and involved in the decisions that affect them, and that they are often able to make important contributions to the process. As a buffer plan may impact on many individuals and groups, wide community input is essential. Bringing all stakeholders or their representatives together will promote a joint understanding of the relevant issues, aid problem definition and enhance acceptance of the final plan. For example, a planning policy to prevent further vegetation clearance as a means of protecting habitat, may appear rational to a biologist, but may impose restrictions and possible financial loss on others in the community. Understanding the goals of all groups in the community and examining a range of possible actions is critical to effective policy making. Such understanding of the problems or issues experienced or perceived by others may encourage exploration of new, innovative 'win-win' solutions. Inadequate identification of stakeholder and user groups may result in the omission of an important aspect of the particular planning problem, reduced compliance with the final plan, as well as inefficient use of time and resources. A further political implication is that the resultant plan may conflict with other policies. The implication is that the biologist's 'rational policy' may not be a workable solution.

Stakeholders will vary depending on the planning context, the level of development within and around the core areas, tenure arrangements and the like. Identifying stakeholders is a difficult part of the IBP process. However, it is essential that consultation is ongoing and includes all stakeholders early in the planning process before a course of action has largely been decided. Stakeholder representation must be based on social equity considerations. Broad participation must be across gender to ensure participation of women, across generations, social strata and ethnic groupings. The process of involving stakeholders must ensure that they are effectively consulted, perhaps through a process of consensus decision making, and not merely given 'final' drafts that have been developed by the responsible agency. Participation may be oral or written and may occur at many venues. It should also include information feedback to those who contribute to the plan.

The organising or sponsoring agency should have high level input into ensuring adequate stakeholder representation. Groups to consider for inclusion in the IBP process include those who:

- *make decisions*
  This may include politicians and community representatives;
- *develop the plan*
  As a decision on who will develop the final plan has not been made at this stage, it is important to include a range of possible contributors e.g. officers from relevant local governments and state or provincial government departments;

- *use and interpret the plan and manage the land*
  This may include government and semi-government agencies, NGOs such as conservation agencies or the water boards, the business community, industry, academics, the media, private landholders, lessees and the general public.
- *implement and enforce the plan*
  This group may not be clearly evident at the start of the IBP process and continuing consideration should be given, throughout the planning process, to who are the appropriate stakeholders. Failure to include those responsible for implementation and enforcement may result in: lack of ownership of the plan, especially if it is non-statutory; a lack of obligation to use the plan; and no means for follow-up to evaluate whether the plan is meeting its aims and objectives. Those responsible for implementation should also assist in the identification of stakeholders (King Cullen, 1993).
- *will be affected by the plan*
  Representation of those likely to be affected by the plan is critical. Buffer zones may require changes in existing land use practices to more ecologically sustainable ones and hence it is important to foster the community's interest and involvement in resource use through effective participation in the planning process. Local communities may be important partners in helping to conserve valuable resources. Cooperation will be enhanced if the end users feel that have been effectively consulted and included in the planning process. This may include landowners (freehold and leasehold), visitors/users, members of the public who wish to retain a use option, local government (councilors and staff), state and commonwealth instrumentalities, agencies with authority over a particular site (e.g. road reserves) or those with statutory rights to use particular sites (e.g. utilities' agencies).
- *have an interest in the plan*
  This may include public and private interest groups and neighbouring landholders.

The number of stakeholders involved and the method of consultation (e.g. meetings, workshops and distribution of material for comment) has resource, funding and timing implications that should be taken into account.

P2. s1. a2 *Action 2* *Diagnose the problem(s)*

Recognising certain problems in relation to a particular ecosystem or species and their habitats is only a first step. The next crucial part is to accurately assess the situation and precisely define the problem. This critical step of problem definition affects the direction of all succeeding phases. It is important to consider both existing and possible future conditions, the underlying causes of the problem, rather than the symptoms, and the need to obtain relevant data. All facets of the problem, as represented by the various stakeholder groups, need to be understood clearly before the following steps and later phases of the model are attempted, for wrong problem definition may produce wrong decisions, an undesired outcome, a waste of time and resources, loss of credibility and alienation of stakeholders. All too frequently a narrow focus on wildlife conservation at the expense of other valid views on the significance and purpose of particular areas, will minimise the chances of success of the IBP. A list of key existing and potential issues and problems related to the target area's environmental values should be produced. Most preventable program and policy mistakes stem from failure to ask the right questions or to appreciate the answers, with some critical part of the context misconstrued or overlooked (Clark, 1992). The identification of a wide range of stakeholder or interest groups will assist in accurate problem definition.

Primary responsibility for problem definition should rest with all identified stakeholders, in particular those responsible for plan development and implementation. The planning team should act as facilitators not decision makers.

P2. s1. a3 *Action 3* *Assess the significance of the problem(s)*

At this point it is useful to identify the implications of not proceeding with the planning process.

- *Identify whether there are serious risks if the identified problems are not eliminated or minimised*
  Identification of the seriousness of the risks associated with the problem enables the planning team to more clearly understand the problem and the urgency of the planning process, and hence the time frames for action. If there are likely to be serious risks, a means of dealing with the problem(s) need(s) to be identified and addressed quickly. For example,

failure to produce a plan may see a target species or ecosystem decline rapidly or even be lost. If there are not likely to be serious risks, the planning team may proceed more slowly in the development of the IBP, or adopt alternative strategies.

P2. s2 Step 2: Develop a project statement to identify the decision making and plan preparation framework

The decision process is a way to reconcile, solve or manage conflicts among policies through politics (Clark, 1995). Although people pursue different policies that reflect their particular interests, in many situations, people must reconcile their policy differences to secure their common interests. The common interest takes the form of rules, substantive or procedural, and may be formal (e.g. law) or informal (e.g. social norms accepted in a group) (Clark, 1995). To clarify the decision-making and plan preparation framework it is necessary to develop a project statement that outlines the potential scope of the project and the proposed course of action. The project statement will need to be refined as the preliminary goals, objectives, outcomes and structure of the process are more clearly defined. This step identifies who is responsible for approving, developing and implementing the plan and consists of four actions.

P2. s2. a1 *Action 1* *Identify who will approve the IBP*

The plan will be developed and approved in an organisation (e.g. state/local government or NGO) and it is important to understand the decision-making framework of that organisation. The following actions should be taken:

- *Identify laws that define the legal framework and determine who will approve the buffer plan*
  The plan, if legislatively based, will be approved under existing legislation, as an amendment to existing legislation, or new laws may be required. The relevant legislative framework should be identified to clarify who is responsible for approving the IBP. Similarly, where a non-regulatory approach is used to implement the buffer plan, the sponsoring agency and the responsible decision maker(s) should be identified.

- *Obtain executive commitment to the IBP*
  Executive commitment to the project is necessary to ensure that the planning process and resultant plan is integrated into the management and administrative system of the organization. This is consistent with the principles of sustainable development, which involve integrating environmental criteria into economic practice to ensure that the strategic plans for an organization ensure growth and development and also conserve the environment.

- *Determine the level of interest in resolving the issue and develop consensus decision-making process*
  Obtain a preliminary understanding of the level of support that governments, NGOs and the community have for a resolution of the identified problem(s). Where support is evident, strive to develop a consensus decision-making process, which is based on the premise that effective decisions should have the full support of all the interest groups involved in the identified problem. This puts the onus on all stakeholders to commit to a successful outcome. Where support is lacking, individual concerns should be dealt with to aid consensus. Failure to do so may result in individuals/groups displaying a reluctance to fully commit to the outcome and this may weaken the chances of success of the plan. Further such groups/individuals may feel disempowered and alienated if they are simply out-voted by the majority. In situations where agreement looks uncertain it is important to examine the cost of not reaching the final outcome at this time and perhaps seek additional information in order to resolve the issue.

These actions are primarily undertaken by government representatives in consultation with stakeholder groups.

P2. s2. a2 *Action 2* ***Identify those responsible for plan preparation***

Responsibility for preparing the IBP must be decided in the initial stages of planning, so that those responsible are involved in all aspects of the process, including problem identification and the statement of aims, goals and objectives for the plan. The following actions should be taken:

- *Identify necessary skills/knowledge required to design the buffer plan*
  This may include: knowledge from particular sciences e.g. biology, botany, ecology and social sciences; skills in negotiation; extensive local area knowledge; and administrative skills. Highly skilled professionals will be important, particularly in the initial stages of plan development.

- *Appoint a steering committee/advisory group*
  It may be necessary to establish a steering committee or advisory group to guide plan preparation, set directions, allocate resources and to ensure the effective and equitable participation of stakeholders. Members may be representatives from the organization responsible for preparing the plan and from the key stakeholder groups.

- *Establish relevant working groups*
  Development of an IBP strategy is a complex, multifaceted task reliant on a planning team, or working group(s) with an interdisciplinary structure. Where a planning team is used, members should be carefully selected to ensure effective coverage of the areas of skill/knowledge required. The representatives from the different disciplines should work as a team to produce the final plan, rather than work independently to produce separate reports that are then edited by a team leader. This structure will enhance communication among all members of the team. Alternatively, a series of working groups may be established and report to a steering committee. The working groups may address the following:

  - community consultation and information delivery – to gather and assess information from the community and to keep the community informed about the progress of the IBP;
  - planning – to assess data, compile progressive drafts of the IBP and provide general support to the steering committee and other working groups; and
  - technical working groups – these may be established as the need arises and would be responsible for providing expert advice to the steering committee on technical/scientific issues.

- *Determine whether formal approvals are necessary, when, how, and what the lead time is for obtaining approvals*
  Clearly identify where the primary responsibility for plan development lies and outline the strategic framework for producing the IBP (e.g. Action Plan and Gantt chart). This action is important where a

regulatory approach is taken and formal approvals are necessary. However, it is also important to follow such a plan of action where a non-regulatory approach is used and endorsement of the IBP is required, either by the sponsoring agency or an independent body, within specific time frames.

Responsibility for this action lies with the stakeholders, particularly representatives of government.

P2. s2. a3 *Action 3* ***Identify those responsible for plan implementation***

An important aspect of the planning process is implementation, which is 'the social activity that follows upon, and is stimulated by, an authoritatively adopted policy mandate' (Brewer and deLeon, 1983:256). This definition stresses the process aspect where implementation 'transforms conceived policies into actual programs' (Brewer and deLeon, 1983:156). In this phase the main aim is to decide who will implement the plan. Such individuals or organisations should be actively involved in the remaining processes of buffer plan development and assist in devising realistic integrated buffer boundaries and policies. In many instances the roles of plan maker and plan implementer will be combined.

Administrators usually have broad discretion in implementing legislation or policies and may modify them to fit the bureaucracy's particular interests or personal requirements, thus undermining or altering the purpose of the original plan (Brewer and deLeon, 1983). 'One should never underestimate the power of even a small, apparently politically disadvantaged group to redefine the policy intentions of those with formally constituted authority' (Brewer and deLeon, 1983:255). Inclusion of implementers in plan making enhances their familiarity with the goals, strategies and objectives of the plan and provides a greater chance of effective plan implementation.

Where legislative arrangements for the buffer are difficult to achieve, and/or where the planning process indicates a non-regulatory approach is preferable, informal structures are an important means to ensure the success of the buffer. These may include creation of: a 'buffer implementation unit'; a community advisory group with a real, two-way sharing of information between the community and managers of the core (e.g. government department or NGO); or agreements with community councils or regional organisations of councils, individual landowners or business groups etc. Such structures may enable the administration, or operational responsibility for the buffer to be removed from government control. Cooperative, informal arrangements, with a high level of community involvement are an important mechanism for improving stewardship within the buffer.

Identification of those responsible for implementation should be undertaken by all stakeholder representatives, or the advisory group and steering committee, where these are established.

P2. s2. a4 ***Action 4*** ***Identify when key players are to be involved, their roles and responsibilities and the necessary reporting structure***

To ensure an efficient and effective planning process, the roles and responsibilities of the key players (e.g. those responsible for plan preparation and implementation) should be agreed to by all stakeholders and clearly stated. It is important to ensure that each group (and each individual) understands the nature of the participation expected, the group's (and each individual's) responsibilities in the planning process, and to whom individuals and groups report. Defining these particular roles, responsibilities and reporting structures will be on-going and will require frequent clarification as the plan develops.

P2. s3 **Step 3:** **Formulate planning aims, strategies and objectives**

Based on a clear understanding of the problem by all stakeholders, the planning framework for solving the problem must be set. All major stakeholder groups should be involved, for if the overall aims and specific objectives are not supported, limited success is the likely outcome. This step consists of four actions.

P2. s3. a1 ***Action 1*** ***List the assumptions/guiding principles underlying plan development***

An important objective of the IBP process is to minimise 'black box' planning in which assumptions underpinning the plan are not made explicit. Although it may be necessary to withhold sensitive information, it is important that the basic premises or assumptions of the planning process are identified early, as they may affect succeeding steps in the process. Stakeholder groups should identify their aspirations and state any assumptions relevant to the group they represent. Conflict in policy and decision processes may result from differences in basic premises that participants take as given, and operate from (Clark, 1995). These basic premises (or myths) can be viewed as a hierarchy of elements (Brunner, 1995) that are accepted as a matter of faith:

- the doctrine e.g. aims, expectations of the community;
- the formula e.g. basic law, or environmental principles; and
- the miranda (symbols) e.g. the koala is a symbol recognised by many in Australia (Clark, 1995).

The main actions to be taken include the following:

- *List the assumptions of the major stakeholder groups and identify any limitations, in relation to possible courses of action, that the planning team or decision-takers may formulate*
  Organizations and individuals have goals or underlying assumptions on which they base decisions. Universal goals/assumptions are unlikely to exist. The assumptions may relate to a range of issues, including: the level of protection to be given to a target species; the planning time frame; resource (including cost) considerations; the amount of compensation payable; the application of the 'precautionary principle'; current options for existing and future uses; the level of public acceptability of the IBP method; and the necessity of meeting human needs. The assumptions that underpin the buffer plan will affect the gathering of information (Phase 3). Changes in the assumptions may necessitate a redefinition of the problem as well as a re-examination of all succeeding stages.

- *Redefine the problem, where necessary*
  This is a feedback loop to ensure that the definition of the problem is relevant to the guiding principles.

Example:

Australia's 'National Koala Conservation Strategy' (ANZECC, 1998) This strategy identifies guiding principles/assumptions that include: the principles of sustainable development; the 'precautionary principle'; a recognition that community input and involvement are crucial to koala conservation; that existing local approaches should be expanded upon; and that the process of decision making should be efficient and transparent and provide for public participation (Table 1). Identification of underlying principles or assumptions is an important basis for the strategy's development.

Table 1 Assumptions of the National Koala Conservation Strategy

Acknowledging that The koala is a national symbol of considerable cultural importance to all Australians. The koala is an important part of Australia's natural heritage. Koala populations have suffered significant declines at a regional level. Koala habitat is poorly protected through much of its range. The community has a significant role to play…
Recognising that Much koala habitat and many koalas occur on private land. The information base on which management must be based is imperfect... There is a substantial tourist industry based on koalas in wildlife parks.
It is agreed that The conservation of koalas depends on the conservation of their habitat… There is a need to better understand the conservation biology of koalas... The Commonwealth and all States and Territories will cooperate to conserve koalas.
Therefore the primary aim of this national strategy for the conservation of the koala is: To conserve koalas by retaining viable populations in the wild throughout their natural range.

(*Source*: adapted from the 'National Koala Conservation Strategy' [ANZECC, 1998:7])

P2. s3. a2 ***Action 2*** ***Formulate the planning aims and strategies***

Once the problem is clearly identified the following actions should be undertaken:

- *Define the aims of the plan*
  The aims should indicate what the plan sets out to achieve. They should be derived from an analysis of the desires, aspirations, values and needs of the present and forecast future community and should, where possible, satisfy their main requirements and interests. Ever present is the difficulty of reconciling differing perspectives. However, agreement on the aims must be reached if the wider society is to begin to accept the policy as legitimate. With integrated planning there is also a need to ensure that the aims cover the full range of possibilities and not only those with which participants are familiar. The project aim may relate to defining the features which need protection, such as a species, population or ecosystem, e.g. 'to ensure the sustainable use of land of importance for the target species'; or 'to ensure the long-term conservation of the target species'. The aims should also incorporate the general principles of sustainable development, including the need to conserve biodiversity, to include wide community participation, to work towards social justice and an improved quality of life.

  The degree of consensus on the final buffer will depend upon agreement reached in this initial step, regarding how much protection the buffer should afford. Strong stakeholder support for the aims is necessary before the design phase begins. The statement of aims should be clear and concise to focus forthcoming steps in the model. The initial statement of aims is preliminary only, as it may be redefined as subsequent steps in the IBP process are undertaken.

  All stakeholder representatives should be involved in this action, especially those responsible for plan development and plan implementation.

- *Identify planning strategies relevant to the planning aims*
  Identify a broad range of possible strategies to achieve theaims. Gain an understanding of each strategy's potential and decide which strategy may be relevant for the specific task. Also gain an appreciation of how they are linked to each other and select a package of strategies.

Example:

The National Koala Conservation Strategy (ANZECC, 1998) The strategy outlines six aims (called objectives in the strategy), including:

- to conserve koalas in their existing habitat (ANZECC, 1998:9);
- to rehabilitate and restore koala habitat and populations (ANZECC, 1998:11);
- to develop a better understanding of the conservation biology of koalas (ANZECC, 1998:12); and
- to ensure that the community has access to factual information about the distribution, conservation and management of koalas at a national State and local scale (ANZECC, 1998:13).

P2. s3. a3 *Action 3* ***List the planning objectives for each aim***

The objectives should refine the scope of the identified problem. They will determine what priorities are assigned and what policies are selected, provide guidelines for the implementation of the chosen programs and determine the criteria for program evaluation (Brewer and deLeon, 1983). The determination of the IBP's objectives is fundamental to the task. Involvement of a wide range of stakeholders creates a complex process for the plan maker. Where the values of some of the stakeholders are not congruent with those held by the decision maker, a policy conflict may occur, and as Brewer and de Leon (1983: 51) state 'rationality does not necessarily dominate'. They stress that the question of, 'what values' and 'whose values' ultimately are to count, is a question that may be answered through political process, not rational analysis. Legitimate political power may reject certain value judgements or may even deny that there is a problem that needs any solution.

Where possible, measurable objectives for each goal should be stated. For regional ecosystems these may relate to the percentage of each ecosystem that is to be conserved within a bioregion, and for individual species, they may relate to numbers of individuals, populations, aspects of habitat quality, and metapopulation characteristics required for long-term conservation of the species. Identification of measurable objectives will assist with policy development and the definition of performance indicators and will help to ensure compatibility with the terms of reference for required scientific studies and with post-implementation monitoring, review and evaluation. Where possible, an indication of those who are responsible for implementation of the objective should be stated. All major stakeholders, including those responsible for plan development and implementation should be involved in this action.

Example:

National Koala Conservation Strategy (ANZECC, 1998) The aim of the strategy, which is to conserve koalas in their existing habitat, is to be achieved by listing seven measurable objectives. Each specific objective lists those with 'Primary Responsibility' for the action.

♦ **DECISION POINT**

Should the IBP process be applied 'in full' or 'in part'?

Based on the plan's proposed aims, strategies and objectives, decide which steps in the model process need to be undertaken. This will reflect the resource availability, expertise of the planning team and data availability.

If the model is applied 'in full' **proceed directly to Phase 3.**

If the model is applied 'in part' **undertake Phase 2, Step 3, Action 4** (P2, s3, a4).

P2, s3, a4 ***Action 4*** ***Determine how the IBP process will be used in the existing decision-making framework and formulate the applied IBP process***

- *Develop a modified IBP flow chart by redefining the scope of the project* The planning team, in consultation with stakeholders, should develop a modified flow chart to indicate the scope of the project, i.e. the phases, steps and actions that are to be undertaken in the applied process. This requires a thorough understanding of the complete IBP process and an assessment of the resources, both financial and human, that are available to undertake the planning process. It is a crucial action that brings into focus the important relationships between issues and areas of application at regional, local and site levels.

PHASE 3

DATA GATHERING

Brief description

Information gathering should not be an end in itself, but rather be directed to achieving the aims and objectives of the plan and assisting in a redefinition of the problem, where necessary. In this phase the data sources are identified and appropriate data collected for later analysis (Figure 6).

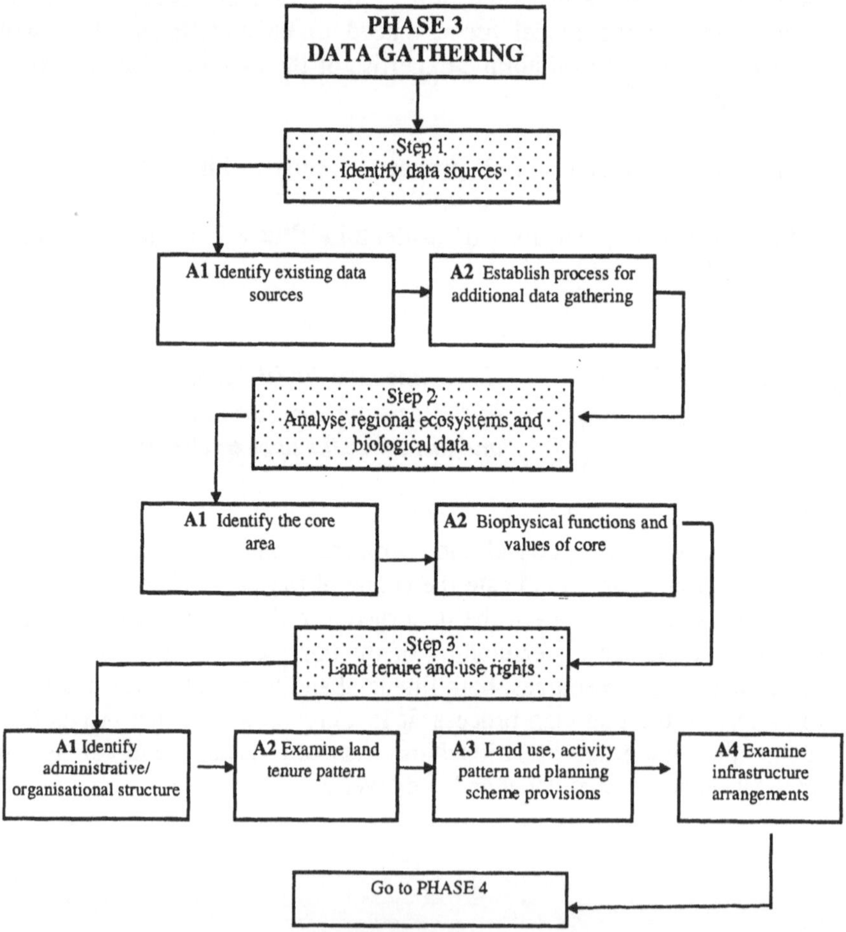

Figure 6 **Phase 3 – Data gathering**

Information Input

- clear statement of the problem, aims, strategies and objectives; and
- knowledge of the decision-making and plan preparation framework.

Expected Outcomes

- map and description of core area; and
- map and description of land use, land tenure and infrastructure arrangements within the plan area.

P3. s1 Step 1: Identify the data sources

Reliable and relevant data are essential to resolve the stated problem. In this step existing data sources and any additional information required to develop the plan are identified. The planning team should begin to identify and sustain strong contact with existing information sources, for although data acquisition will be intensive in the initial phases, it must continue through the life of the project. One of the ever-present difficulties in the area of nature conservation is the lack of data on which to base effective planning and management. Despite this, the planning team must strive to gain a detailed understanding of the environmental values and the social, economic and administrative framework in the plan area. This step consists of two actions.

P3. s1. a1 *Action 1* *Identify existing data sources*

All relevant existing data should be collected, organized into a useable form and evaluated. This review phase may be time consuming especially if the data is not held in a manner that is easy to interpret and collate, or if it is constrained financially. Joint management of the IBP with other agencies may improve the availability of information and facilitate data exchange.

Types of data may include:

- *biodiversity data sets* e.g. regional ecosystems, remnant vegetation communities, flora and fauna site records (especially rare or threatened species and disjunct species populations); historical data; indigenous biodiversity values; and other physical data sets including topography, geology, soil, hydrology, aspect, climate and land capability;

- *planning studies and maps* e.g. strategic plans, zoning maps and local area plans;
- *administrative datasets* e.g. land tenure, land use, land ownership and transport networks;
- *social datasets obtained from recent census information* e.g. population distribution and growth, age structure, dwelling types, number of occupants per dwelling, etc.

Sources of data may include:

- *individuals and groups:*

  - individuals with local knowledge of the resources of the area, including indigenous traditional owner knowledge and information from former and present landholders and long-time residents, resource users and managers, and researchers working in the area;
  - special interest groups such as historical societies, conservation organizations, Landcare, wildlife preservation societies and integrated catchment management groups;
  - industry groups that have a stake in the planning area can provide information on resources, their needs and perceptions. Such groups might include tourism operators, land developers, transport operators, mining and power companies;
  - 'expert panels' that may provide knowledge or the consensus of scientific opinion on particular issues; and
  - land managers (e.g. parks and wildlife service and forestry agencies).

- *academic and research organizations* e.g. libraries, natural history or cultural resources collections or museums, field stations, faculties in history and natural and social sciences;

- *national and state/provincial government:*

  - wetlands of importance e.g. as identified by the Convention on Wetlands of International Importance (Ramsar Convention);
  - heritage listed sites, including World Heritage listed sites;
  - bioclimatic modelling;
  - titles office for information on land tenure; and
  - parks and wildlife service, environmental agencies and museums for information on plant species and fauna site data;
  - regional and local government:

- regional strategies or frameworks for growth management may identify important natural areas and provide direction on future planned growth corridors and areas suitable for development;
- local governments frequently undertake vegetation mapping and assessment of conservation significance to aid in the preparation of their planning schemes and strategic plans; and

- *other collections e.g. private, city state or commonwealth collections.*

Use of such a wide variety of source data helps to gain multiple perspectives on the problem and its context and may result in a more comprehensive understanding of the issues in the plan area.
The format of information may include:

- *satellite imagery* e.g. examination of satellite photographs (1:500 000) to produce a broad scale map of the remnant vegetation and potential corridor links;
- *photographs and aerial photographs* e.g. to identify more precisely the boundary of the remnant patches and existing corridor links;
- *data in reports/journals* e.g. environmental impact studies, local government strategic plans, development control plans and development applications that have fauna and flora survey data included; and
- *GIS/maps* e.g. regional ecosystems, remnant vegetation, land systems, fauna distribution, geologic and soil data. The maps may provide information at a particular point in time or evidence of changes in particular attributes over time. For example, remnant vegetation maps may show:

  - distribution of vegetation (forest, woodland and grassland etc);
  - types of vegetation (e.g. based on taxonomic groups of the dominant plants, or its significance in providing habitat for the target species);
  - quality or integrity of the vegetation (intact, disturbed and cleared); and
  - presence of rare or threatened species, or priority species and ecosystems.

The scale of data collection should be considered at this point. A broad brush approach may be used based on regional ecosystem data, or on habitat modelling to indicate probable areas of nature conservation significance. The loss of information arising from decreasing the scale needs to be balanced against the cost of producing the necessary maps.

Detailed studies of smaller areas of habitat may provide more accurate information, but take a long time to obtain the necessary data, prolong the implementation of the IBP and be more costly.

Decisions on the selected scale will be influenced by the size of the plan area, the nature and extent of the identified threatening processes and the available data. A tentative delimitation of this area should be made at this stage and a map scale selected. For large areas (catchments or bioregions) map scales of 1:100 000 or 1:250 000 are suitable. For local government areas, or parts of such areas, map scales of from 1:25:000 to 1:10 000 are satisfactory. If other similar projects have been undertaken elsewhere, or standard data assessments are available, consider producing compatible datasets that may eventually be synthesised or overlapped in a sieve mapping or GIS process. This will assist in developing a regional or state/province-wide IBP.

Consideration, at this stage, should also be given to:

- joint data collection with other local governments, higher levels of government, other agencies and advisory bodies (e.g. regional organizations of local governments) to minimise costs; and
- how the data will be stored and maintained – consideration should be given to storing raw data from survey sites, so that it can be manipulated later, as necessary, and to using GIS. For example, biodiversity related data bases in government departments/agencies and other sources may be collected, digitised and overlaid in a GIS. Changes in the mapped attributes need to be included in the datasets (e.g. recent vegetation clearing; and sightings of any target species).

Example:

The Regional Nature Conservation Strategy for South East Queensland (EPA, 2003) (Australia) The Strategy is based on the Common Nature Conservation Classification System (CNCCS) (Chenoweth EPLA, 2000), which 'classifies the significance of mapped remnant vegetation units for nature conservation purposes, with standardised criteria and levels of data collection, that can be consistently applied' (Chenoweth EPLA, 2000: 5). The CNCCS, which was developed collaboratively by several local government councils and representatives from state government and community groups, was applied by the state's Environmental Protection Agency to identify areas of nature conservation significance in the south eastern corner of the state (EPA, 2003). The data assessment process involved the collection and collation of several data sets from various sources (e.g. Herbarium, local government, reports and journals etc) as well as the incorporation of 'expert opinion' and Indigenous Traditional Owner biodiversity values. Information management, including the ongoing collation of data, as well as the dissemination of the information to all stakeholders, is an important component of this process. The output of this methodology provides a sound basis for decision making within the region.

Prime responsibility for this action lies with those responsible for plan development.

P3. s1. a2 *Action 2* ***Establish the process for additional data gathering***

Where data are lacking, or are not satisfactory for the plan's purposes (e.g. scale limitations, level of detail or incompatible databases), additional 'new' data may have to be gained or the existing data enhanced. For example, where a particular species requires a certain habitat type and the precise distribution of the species is unknown, it may be useful to map remaining suitable habitat and changes in this habitat over time. This may be based on predictive modelling of suitable habitat factors (e.g. Samedi, 1994). As well as biological data, information may also be required on the sociological, economic, institutional and political context within which the plan is to be developed. It may also be necessary to consider and incorporate the data needs of any baseline studies that may be utilised to enable comparative assessment of the effectiveness of the buffer plan. The type of additional data collected should be considered in conjunction with the future monitoring program (P5, s4).

Additional data may be gathered by means of:

- ground truthing or further elaborating the data collected e.g. using broad-brush sampling techniques;
- on site field work e.g. radio-tracking, vegetation mapping, or field inspection of lots with potential for inclusion in the species' critical habitat, to check for recent clearing or degradation, and identification of disturbance regimes (e.g. edge effects) that may threaten the critical habitat;
- predictive modelling;
- use of expert panels; and
- a pilot project.

Where both existing and additional data are inadequate, the objectives of the IBP may need to be reconsidered. However, a lack of information need not halt the planning process if estimates from general theory and other similar situations can be made. Where such information is lacking, a recommendation for further research may be made at this point, the plan being adjusted when data become available. Those responsible for plan development should be tasked to obtain the necessary data.

Examples:

Pt. Stephens Koala Management Plan (Callaghan et al., 1994) This plan used a postal questionnaire to obtain data as the planning team recognised that the primary source of information on koalas was local knowledge. A community based koala survey was distributed by post to all residents within the Pt. Stephens local government area to establish locations of principal koala populations, population health status and an indication of community attitudes to identified threats to koalas and potential conservation measures. Residents were also asked for any historical information about koalas or koala habitat in the area. Information about other wildlife species was also included, as much to collect rarely reported sightings, as to increase the response rate of the survey. The survey included a colouring competition, the entry to which could only be accepted if it was accompanied by a completed koala survey. Organisers visited each school in the area to promote completion of the survey by the students' parents. The koala plan also included a systematic review of historical literature including newspapers, historical records, documents and reports. As well, a number of personal interviews were undertaken, where respondents indicated they had historical information. The vegetation of the local government area was mapped to identify potential koala habitat, through mapping vegetation associations with particular reference to koala habitat values. These were based primarily on interpretation of recent colour aerial photographs and Landsat imagery.

Yengo National Park (Curtin and Lunney, 1995) Yengo National Park was used as a study site to compare community-based and field-based surveys for koalas. A community survey was distributed to residents adjoining the reserve. Responses covered a 60 year time period (1935-1995) and gave a range of new information on koala sightings. The survey was used as a guide for field investigation, which produced more specific information about local koala habitat, preferred tree species and vegetation types. Curtin and Lunney (1995) recommended that both community-based and field-based surveys should be undertaken where the planning aim is to better manage koalas in particular areas.

P3. s2 Step 2: Identify and assess the values of the core

The IBP process is usually undertaken to provide additional protection, to a core area, from various threats. This step in the process aims to clearly identify the core area and its values. Effective management of biodiversity, including any target wildlife and their habitat, depends on a sound knowledge of the biological and physical resources of the core and the processes contributing to the persistence of the resources. This step consists of two actions.

P3. s2. a1 *Action 1 Identify the core area*

The core may consist of the following:

- protected area or similar reserve;
- dedicated forestry land;
- critical habitat (e.g. as defined in relevant legislation);
- habitat of a target species or group of species;
- a regional ecosystem e.g. wetland; and
- an area of nature conservation significance.

It is not necessary for the core to consist of a formally protected area with clearly identified boundaries and secure tenure. The IBP process can be applied to conserve the habitat of threatened species and ecosystems and unprotected habitat fragments. It can also be applied at the broader landscape scale to buffer areas of high nature conservation significance that occur on a variety of tenures. This action should be undertaken by those responsible for plan development.

Examples:

Common Nature Conservation Classification System (CCCS) (Chenoweth EPLA, 2000) The CNCCS establishes 'diagnostic' and 'other desirable' criteria to aid in the identification of areas of nature conservation significance (Table 2). The diagnostic criteria use data that are sufficiently consistent, reliable and available in database format to be queried and combined to automatically generate mapped significance classes. The other desirable criteria require expert interpretation of data to modify the rankings produced as a result of the application of the diagnostic criteria. The criteria represent a spread across the three broad 'themes' of conservation significance, namely, rarity values (e.g. endangered species habitat, regional ecosystem value and relative size), general habitat values (e.g. remnant size, integrity, diversity, general species habitat, local biodiversity contribution and geomorphological variation) and ecosystem process values (e.g. connections, context and linkages and other ecosystem values).

Table 2 Diagnostic criteria used in the Common Nature Conservation Classification System to define nature conservation significance

Diagnostic criteria		Other desirable criteria	
Type	Description	Type	Description
A	Essential habitat for at risk spp.	H	Other habitat for 'at risk' species
B	Ecosystem value	I	Habitat for other species
C	Remnant size	J	Localised contribution to biodiversity
D	Relative size of ecosystems	K	Corridor links
E	Integrity	L	Geomorphological variation
F	Community diversity	M	Other ecosystem values
G	Context and connection	N	Indigenous Traditional Owner cultural resource values

(*Source*: Chenoweth EPLA, 2000)

Decision rules are established to identify areas of state, regional and local conservation significance. Each diagnostic criterion is rated as low, medium, high or very high based on descriptive statements and measurable indicators. These ratings are combined through database combination filtering to derive a 'first cut' level of significance for each remnant unit. Thus for example, if an area is ranked as very high in terms of its value as essential habitat for 'at risk' species or is ranked very high for its ecosystem value, the area is designated as having 'state significance'. The system incorporates 19 sequential queries that are applied in a GIS format, resulting in a map of relative conservation significance in the three categories of state, regional and local significance. An advantage of the CNCCS methodology is that the criteria can be assessed at three levels, from a basic level using coarse and/or readily available data with little further investigation, to a more advanced level based on additional studies. The database also allows users to examine the range of values that contribute to the classification. The CNCCS has been applied within South East Queensland and is an important component of the 'Regional Nature Conservation Strategy for South East Queensland' (EPA, 2003).

Port Stephens Koala Management Plan (Lunney et al. 1997; Callaghan et al., 1994) (Australia) The process of identifying koala habitat included the following:

- *Community based survey* (see P3,s1,a2);
- *Vegetation map* – the map was based on aerial photo interpretation and field survey and identified the dominant plant species for each vegetation association. Results of the community survey were overlaid onto the vegetation map, using a GIS, to determine the vegetation types in which koalas were observed;
- *Community-based koala habitat map* – the density of koala records in each vegetation association was calculated and used as a means of ranking the vegetation associations and to group vegetation associations into habitat categories (A to E) which were mapped using a GIS;
- *Koala habitat map based on field survey* – field survey involved plot-based searches for koala faecal pellets combined with statistical analyses. Specific koala tree preferences were identified and used as the basis to define koala habitat throughout the local government area;
- *Overlap of community map and survey map*;
- *Combined koala habitat map* – the final map reflected the results of both surveys and the need to consider which areas of koala habitat require greatest protection;
- *Habitat map used for planning purposes* – links between areas of preferred koala habitat were identified and defined according to the nature of vegetation occurring in the link, e.g. cleared land, supplementary koala habitat.

State Planning Policy 1/97 (Conservation of Koalas in the Koala Coast) (Qld Government, 1997) (Australia) This policy regarding koalas in the Koala Coast area of southeast Queensland, designates land of high importance to koalas as 'Koala Conservation Area' (KCA) and 'Other Major Habitat' (OMH). The KCA comprises significant remnant bushland as well as cleared areas, forming a large integrated and relatively undisturbed area. The OMH does not form a single cohesive area and lacks precise boundaries, the onus being on local government to determine the boundaries in their planning schemes or through development assessment.

'Critical habitat' areas The concept of 'critical habitat' is embodied in various legislation at the state level in Australia. The *Threatened Species Conservation Act 1995* in New South Wales defines critical habitat as habitat critical to the survival of endangered species, populations and ecological communities. The state of Queensland's *Nature Conservation Act 1992* (NCA) defines critical habitat as 'habitat that is essential for the conservation of a viable population of protected wildlife or community of native wildlife, whether or not special management considerations and protection are required' (NCA s13[1]), and may include 'an area of land that is considered essential for the conservation of protected wildlife, even though the area is not presently occupied by the wildlife' (NCA s13[2]).

The criteria used to define critical habitat may need to be specific to the target species, but in general terms should describe an area that is essential for the long-term conservation of the species, i.e. without the identified critical resources, the survival of the species in the wild would be uncertain. It may include: critical breeding areas (e.g. nest sites); food sources (e.g. particular tree species); water bodies; particular vegetation communities; home range territories; and movement corridors, and in general will provide continuous habitat capable of sustaining the target species in its ecosystem.

P3. s2. a2 *Action 2* *Identify the biophysical functions and values of the core*

The core should be viewed as a heterogeneous environment with many interacting functions and values, forming a complex system. An understanding of these interrelationships and how they contribute to the overall importance and perhaps uniqueness of the core is important in developing a management strategy within the buffer. Values arise from ecological significance, sociological significance, management objectives and legislation. The following actions should be undertaken:

- *Identify the important values*
  Although biodiversity may be an important value, it must be considered in relation to the wider values that the natural area provides, especially the values that are important to the local community. Understanding all the existing and potential functions and values of the natural area will provide a strong foundation for developing an appropriate buffer plan, one that meets the needs and aspirations of the local community. Although natural areas fulfill a number of important functions, local communities will place value and importance on some of these

functions and disregard others. In some situations poverty may force a community into unsustainable practices that may threaten important functions and values.

Consult with all stakeholder groups to determine the values attached to the core, and using the identified data sources (P3,s1,a1), obtain preliminary information on these values in the planning area (Table 3 provides a suggested format). It may be necessary to use a standard

Table 3 Biophysical functions and values of core natural areas

Values	Description
Physical	⇒ geologic structure, geomorphic processes, edaphic features, hydrology, climate, microclimate
Biodiversity	⇒ genetic, species, ecosystem and regional diversity
Cultural	⇒ the importance of traditional societies using core natural resources
Aesthetic	⇒ scenic qualities, views, landscape appearance
Ethical	⇒ the existence values of species, ecosystems and geomorphic systems
Sensory	⇒ silence or a minimum of noise disturbance; wilderness values
Scientific	⇒ research; baseline monitoring of environmental change
Educational	⇒ field based centres for education
Recreational	⇒ variety of resource based recreational pursuits e.g. photography, bushwalking, sightseeing
Economic	⇒ tourism, agriculture, forestry, grazing, mining, land development
Human	⇒ natural/clean environment, lifestyle, amenity
Option	⇒ provide for a range of future options

survey method for assessing community values (e.g. randomly select community representatives and rank the values they attach to the core area). The results of this analysis will help to balance the values that are attached to the area by government officers and other experts involved in the planning process.

- *Integrate competing values*

  Most natural areas will possess multiple values. The IBP should aim to integrate these values into the developing plan. However, where there is competition over the use of particular resources it may not be possible to protect all the identified values. The values to be given priority should be determined through consultation and consensus decision making with all stakeholders, based on an assessment of the long-term benefits and costs of alternative strategies to the local and wider community. The community survey of values will play an important role in aiding in the selection of the best combination of uses for the core and its surrounds.

P3. s3 Step 3: Examine the land tenure pattern and associated use rights

The core is enmeshed in its surrounding environment. To determine the main threats or issues, the principal features or elements of this landscape matrix should be analysed. As it is unlikely that all land within the identified core will have a high level of protection (e.g. within the protected area estate) it is necessary to examine the land tenure and associated use rights in the core and its surroundings. This step consists of four actions.

P3. s3. a1 *Action 1* *Identify the administrative/organizational structure*

Examine the location of the core in relation to local, regional and national administrative boundaries. Where more than one administration is involved, buffer design may be complex and this should be considered early in the planning phase. Consultation and negotiation will be required in all administrative areas to ensure the success of buffer strategies. In particular each administrative body should be represented as a major stakeholder.

P3. s3. a2 *Action 2* *Identify the land tenure pattern*

This action has importance in directing the form of management and implementation strategies that may be applied in the buffer. Although some buffer plan strategies are restricted only to 'suitable land' or land in government ownership, the IBP process recommends that all land, irrespective of land tenure should be considered for inclusion in the developing buffer.

P3. s3. a3 ***Action 3*** ***Examine the land use, activity pattern and planning scheme provisions***

The way the land is used may help to identify stress factors and indicate management issues. Even where a high level of protection and limited use rights exist over the core, the conservation of biodiversity may not be guaranteed if the surrounding uses and activities are incompatible. The following specific actions should be undertaken:

- *Identify and map the main current land uses*
  Identify the types, distribution and intensity of key activities/land uses (e.g. urban, rural residential and grazing etc) within the region. This will include:

  - areas that are relatively undeveloped and/or restorable, especially those adjacent to, or near the core e.g. nearby reserves and parks, drainage lines, especially those with links to the core and other habitat, and perhaps connecting easements for sewerage and electricity transmissions;
  - areas currently being used extensively for rural/grazing activities, plantation forestry, military training, water supply catchment purposes, sewage disposal, communications facilities and institutional purposes (e.g. prisons, hospitals and schools, and transport corridors). These areas may have a dual function in providing additional habitat and movement corridors for wildlife; and
  - more intensive types of land use e.g. industry, urban, residential.

By identifying the land uses/activities of the planning area, possible direct, indirect and cumulative threats to the identified values can be identified (Phase 4). Aerial photographs are a useful tool to identify the main current land uses.

- *Identify the potential use rights*
  Anticipatory planning is crucial to the success of the buffer strategy. Although land may be used for a particular purpose at the time of planning, future use rights may have been allocated throughout the planning area, or future intents for use may have been stated. These existing lawful use rights, as well as some outstanding development approvals may pose additional threats to the core, affecting its viability. It is important in this element to examine planning documents (e.g. planning scheme and planning policies etc) that may indicate future use

of the land and possible constraints on the IBP and its policies. The buffer plan must anticipate these influences and incorporate appropriate strategies to minimise their impact on the core area.

- *Identify the trends in land use and activity patterns*
Where possible, trends in the condition of the main components (e.g. ecosystems and human communities) of the area should be identified. This historical context should include important human stress factors and pertinent environmental regulations and standards (refer P1,s2,a4). For example, changing land use patterns may be the result of changing socio-economic factors, with the result that a change in the economic base of the area may threaten the components of the core. Thus in agricultural areas it may be necessary to identify trends in farm practices as well as farm prices that could impact on biodiversity. This is a particularly important aspect in the conservation of the mahogany glider in Australia. The sugarcane industry in the early 1990s was in an expansion phase due to the high profitability of growing cane. As cane development was often at the expense of mahogany glider habitat, conservation of this habitat required consultation with cane growers and industry representatives to develop compromise strategies to ensure the protection of glider habitat, as well as provide certainty to cane growers. Conservation measures also may be difficult to implement when landowners are facing income shortfalls. Conservation strategies have had limited success, particularly in many developing countries where the rights of wildlife are perceived by the local community to be placed ahead of the basic needs of human communities. An understanding of these important economic issues and associated policy directions may be significant in some situations.
A changing population structure may also indicate potential threats to the core. For example, population growth and increased household formation may place pressure on remaining habitat. Similarly, where low density land use patterns give way to higher density residential development, a new range of management issues may come to the fore (Peterson, 1991).

P3, s3, a4 ***Action 4*** ***Examine infrastructure arrangements***

The location and management of existing and proposed future infrastructure may impact on the values of the core. Consideration, within the context of the planning area, should be given to:

- *Transport (road, rail, air, water)*
  Transport lines may fragment habitat and act as barriers to movement. For many species of wildlife, roads are a cause of injury and mortality. Aspects to consider include: the location of transport lines in relation to identified habitat and in particular the movement paths of the target wildlife; density of transport lines; the vegetation pattern and amount of clearing within transport reserves; movement speeds; use intensity; and management practices (e.g. signage, lighting and rumble strips).

- *Water supply*
  Availability of and quality of water may be important issues for target species and in particular may impact on the vegetation mosaics within the area. Activities that pollute or reduce the availability of the water supply may threaten the target species directly or indirectly by changing vegetation composition. Provision of water infrastructure (e.g. pipelines and dams) may fragment habitat, and dams and other water impoundments are likely to result in the loss of habitat due to submergence.

- *Power*
  Power lines have the potential to fragment habitat and increase the level of chemical application to the land. Existing management strategies need to be examined as well as proposals for an expanded network.

- *Waste disposal*
  Aspects to consider may include the types of waste disposal and the impact on surface and groundwater.

PHASE 4

DATA ANALYSIS AND DRAFT INTEGRATED BUFFER PLAN

Brief Description

The purpose of this phase is to understand the relationships between the core and its surrounding environment, to identify threatening processes, both internal and external and to delimit the preliminary integrated buffer boundaries and management policies (Figure 7).

Information Input

- map of core area;
- data sets relating to the core area (e.g. regional ecosystems/target species ecology and habitat requirements);
- map and description of land use and land tenure within the core and surrounding landscape; and
- information on the biophysical functions and values of the core.

Expected Outcomes

- identification of threatening processes (internal and external) and their areas of operation; and
- delimitation of the integrated buffer zone and management policies.

P4, s1 **Step 1: Identify threatening processes, their causes and environmental pathways**

In Step 1 the threatening processes are identified. As knowledge of the ecological requirements of individual plant or animal species may be limited, and in the absence of systematic monitoring of regional ecosystems and specific wildlife, only broad statements on readily observable threats may be possible. There are five important actions:

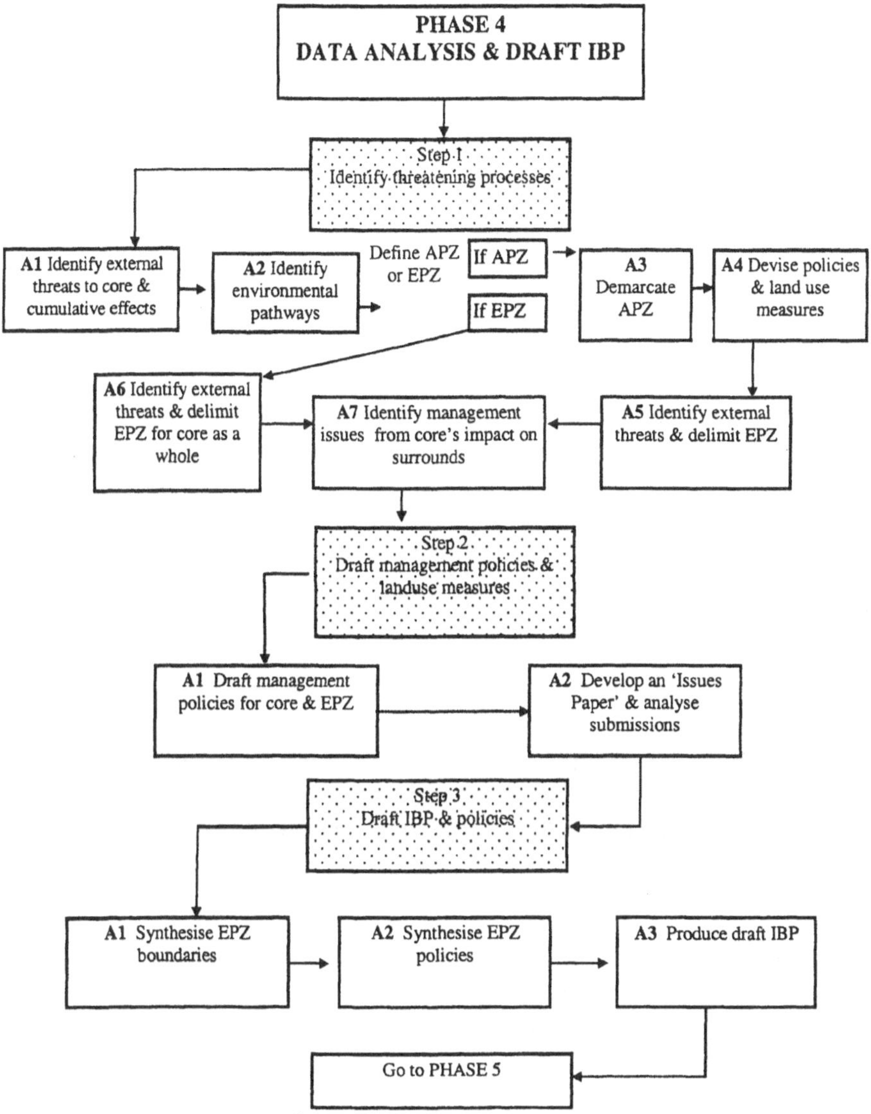

Figure 7 Phase 4 – Data analysis and draft integrated buffer plan

P4. s1. a1 ***Action 1*** ***Identify major external threats to the core and determine if the effect of each threat acts cumulatively***

- *Identify the interrelations between the core (elements and values) and the surrounding area (land uses and activities) to determine the external threats to the core*
  Identify and describe the threatening processes that impact on the components of the core (identified in P3,s2,a3/a4). Provide a clear distinction between a threatening process and its causes. For example, loss of vegetation may threaten the survival of a particular animal. The cause of the vegetation loss may be due to one or more factors operating in the surrounding region, such as rural or urban expansion. The threatening process should be identified as 'vegetation loss' and not for example, residential development, which is one of the causes of vegetation loss. The alternative process of identifying individual sources of threats, rather than the threatening process, may result in an extensive listing of developments or actions taking place in the study area and cause the following steps of the model to be unduly complex.

- *Focus on significant threats*
  For the analysis to inform decision making, the IBP process must be limited, through scoping, to an examination of important issues (e.g. of national, regional or local significance) and effects that can be effectively evaluated. Hence the analysis of threatening processes should be extended to the point at which the resource is no longer significantly affected.

- *Identify cumulative effects*
  Cumulative effects are changes to the environment that are caused by an action in combination with other past, present and future human actions (CEAA, 2000). They occur as interactions between human activities/land uses and the environment, and between components of the environment. Cumulative effects need to be evaluated along with the direct and indirect effects of specific actions. For example, cumulative effects on the wildlife of the core may result from road kills, dog attacks and destruction of habitat, these all being due to increased residential and rural residential living.

  The magnitude of the combined effects along a pathway can be equal to the sum of the individual effects (additive effect) or can be an increased effect (synergistic effect) (CEAA, 2000), or may be countervailing, where the net adverse cumulative effect is less than the sum of the individual effects (Council on Environmental Quality, 1997). Where possible, some attempt should be made to quantify the magnitude of the threat's impact.

Cumulative effects can occur in various ways (CEAA, 2000):

- *Physical-chemical transport*: a physical or chemical constituent is transported away from the land use/activity under consideration, where it then interacts with other land uses/activities, e.g. waste water effluent and sediment.
- *Nibbling loss*: the gradual disturbance and loss of land and habitat, e.g. land clearing and the construction of roads into sensitive natural areas. The cumulative effect may be fragmentation of habitat and a change in landscape structure and function.
- *Spatial and temporal crowding*: when too much is happening within too small an area and in too brief a period of time. A threshold may be exceeded and the environment may be unable to recover to pre-disturbance conditions. Spatial crowding results in an overlap of effects among actions, e.g. close proximity of timber harvesting, wildlife habitat and recreational use in a sensitive natural area. Temporal crowding may occur if effects from different actions overlap or occur before the natural area has had time to recover.
- *Growth-inducing potential*: each new land use/activity can induce further actions to occur, e.g. increased vehicle access into a previously unroaded natural area may add to the cumulative effects already occurring in the vicinity of this area, e.g. road kills and the spread of weeds.

• *Identify spatial boundaries*
Buffer planning has traditionally involved defining arbitrary boundaries around reserved lands and examining the impact of one or a few threats. The IBP process expands those spatial horizons and aims to examine the spatial distribution of all significant threats (e.g. site specific, dispersed and linear). Transboundary effects (e.g. animal migrations) and global-scale effects (e.g. atmospheric effects) should be identified, although the mitigative response may ultimately be beyond the capability of the IBP plan. Thus, the planning team needs to determine at what point to stop identifying impacts, for some constraint on information gathering and analysis is necessary. Accurate and reliable determination of the probabilities of occurrence, and the magnitudes and durations of all potential effects would be costly, time consuming and excessive

(CEAA, 2000). The implication of too small a boundary is that important regional and long-term effects may not be examined (e.g. movements of far-ranging wildlife). Alternatively, the progressive expansion of humans into natural areas suggests the need to assess effects over large geographic areas.

The planning team must determine at what point an impact is trivial or insignificant. The concept of thresholds may be used, but they are often difficult to define. It is recommended that an adaptive approach be used when setting boundaries. The starting boundary should be based on the best professional advice ('an educated guess') and be later changed if new information suggests that a different boundary is required. Boundaries can be assigned on the basis of available data (e.g. available coverage of remotely sensed data, or a well-studied catchment), as the cost and time required to obtain more data may be prohibitive and time consuming (CEAA, 2000).

In determining geographic boundaries it is necessary to consider the distance an effect can travel and its mode of transmission. For example, the area in which water pollution may be a threat to a component of the core would be limited to a catchment or sub-catchment boundary, downstream of the threat's source. For a migratory species, the geographic area may include the species' breeding grounds, migration route, wintering areas, or home range, while for a resident species, the geographic area may include the species' known and predicted habitat or associated regional ecosystem.

In general the boundary should be conservative and incorporate the precautionary principle. It should rely on professional judgement, use an adaptive approach and be based on ecologically defensible criteria. In using the IBP approach boundaries will need to be set for each of the threats that are examined, resulting in the identification of multiple boundaries. Using jurisdictional boundaries, while expedient, ignores the ecological realities of the area. The applicable geographic scope needs to be defined on a case by case basis.

- *Identify temporal boundaries*
  Cumulative effects are caused by the sum of past, current and reasonably foreseeable future actions (CEAA, 2000; Council on Environmental Quality 1997):

  - *past actions*: these are no longer active, yet continue to represent a disturbance to the ESA. Although there may be readily observable effects, significant changes may affect ecological processes (CEAA,

2000). Data describing these past effects may be scarce and hence this analysis may be qualitative.

- *existing actions*: these are currently active actions; and
- *future actions*: these are actions that are yet to occur. Possible future threats may relate to proposed policy and planning changes, population movements, development proposals that may affect land use activities and some natural hazards (e.g. forest fires). It is necessary to classify future actions according to the certainty that the action will proceed. Hence future actions may be:

  - certain (the action will proceed, or there is a high probability the action will proceed);
  - reasonably foreseeable (the action may proceed, but there is some uncertainty); and
  - hypothetical (there is considerable uncertainty whether the action will ever proceed) (CEAA, 2000).

The selection of future actions will be a compromise between under-representing the full extent of future change, and identifying and assessing an unreasonably large number of actions. The planning team should make the selection based on regulatory requirements and professional judgement. Future actions may be identified by investigating the plans of relevant agencies (state/provincial government, local government, private organisations and individuals) (refer to P3, s3, a3). For example, local government strategic plans may provide indicators of future development possibilities, while an examination of approved developments will indicate which areas will be subject to certain development. A useful way to portray this information is to develop a schematic diagram showing the core area and the location of land uses/activities (existing, certain and reasonably foreseeable) that may threaten the core. This may be achieved using a GIS or manual map overlay system. By examining the areas of influence of these land uses/activities it is possible to identify those in which impact zones overlap and affect the components of the core.

P4, s1, a2 *Action 2* *Identify environmental pathways*

Each threat should be described in adequate detail to allow an understanding of how the components of the core are susceptible to its effects. Some threats or their causes may have to be assessed generically because there may be too many to practically characterise them individually, e.g. there may be many small activities suspected of causing minimal effects due to short duration, low magnitude, irregular and unpredictable occurrences, or temporary duration (CEAA, 2000). Such threats should be grouped, where possible. As the environmental stresses or threats may occur at different levels in the biodiversity hierarchy affecting the composition, structure and/or functioning of the ecosystem, all of these levels should be considered. Further, effects at one level may impact on other levels, often in unpredictable ways (Noss, 1990). For example vegetation clearance external to the core may increase edge effects on the core's vegetation, resulting in altered plant and animal communities at the interface, or it may disrupt an important movement corridor between core areas, thus impacting on the biodiversity of the core. An understanding of how vegetation loss is likely to impact on the core in terms of the changes it may bring to the core is critical, particularly in the design of management strategies to eliminate the threat, or at the least to reduce its impact on the core's components and values and also in identifying where these strategies should be targeted.

- *Conceptualise the cause-and-effect relationship*
  It is necessary to characterise the nature of the environmental pathways so that a 'line of enquiry' can be established to identify possible impacts and necessary management responses (Figure 8). Networks and system diagrams are a useful approach to help conceptualise these relationships. Models may identify all pathways and be quite complex, or can be simplified to include only important relationships that can be supported by information.

- *Develop an interaction matrix to illustrate the identified interactions*
  Once the important cause-and-effect pathways are identified the planning team should determine how the components of the core respond to environmental change (i.e. what is the cumulative effect) and the extent to which the core can sustain the impact before changes in condition cannot be reversed. To better understand the interactions, a tabular matrix should be developed. One approach is to identify the environmental components of the core (e.g. fauna and vegetation communities) and the threats to these values (e.g. land clearing and road construction). An interaction matrix is a tabulation of the relationships between the two factors. Matrices are used to identify the likelihood of whether a threat may affect a particular environmental component. They can also be used to identify the intensity of the impacts by highlighting the potentially 'strongest' cause-effect relationships. However, because

they are a simplistic representation of complex relationships, the matrix should be accompanied by detailed explanations of how the interactions and rankings were derived.

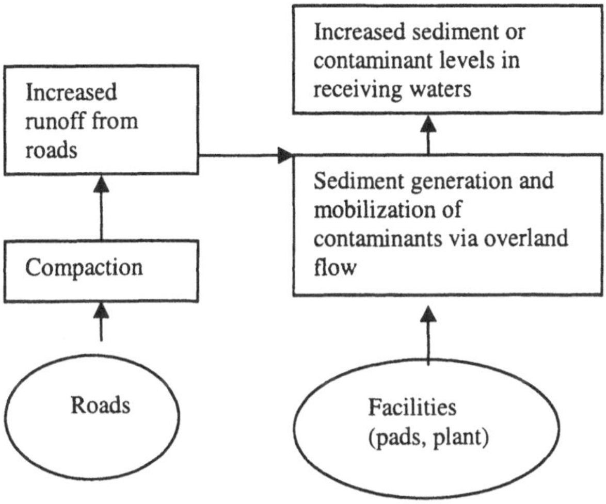

Figure 8 **Pathway diagram illustrating that roads and facilities will result in the generation of sediment and transport of contaminants to receiving waters (adapted from IORL, 1997)**

A useful approach for determining the likely response of a resource or ecosystem to environmental change is to evaluate the historical effects of activities similar to those under consideration. For example, the construction of a new road may be compared to similar roads constructed in the past, or to similar linear features such as the construction of pipelines and power lines (Council on Environmental Quality, 1997). If these relationships cannot be quantified, or if quantification is not necessary, qualitative evaluation procedures can be used. The matrix may indicate the process of interaction and the level of impact through a ranking system (e.g. 1 = low to 5 = high on a 5-point scale, or H = high, M = moderate, L = low on a 3-point scale). It is likely that the planning team will have to rely on qualitative analysis as many cause-and-effect relationships are poorly understood and site-specific data will not be available.

This approach enables the planning team to focus on the most important identified interactions and ensures that data collection is limited to that required to address these issues. This allows a 'coarse filter' approach to data collection, an approach that is necessary due to the probability that the area to be included in the analysis will be quite extensive.

Examples:

Buffer Zone Planning (BZP) Several applications of buffer zone planning (Hruza, 1993; Peterson, 1991; Roughan, 1986) have included the compilation of a matrix to examine threatening processes. For example, by examining the inter-relationships between each of the identified resources of Cooloola National Park and its surrounding environment, 16 existing and/or potential negative impacts were identified. The level of impact ranged from low to medium to major (Peterson, 1991).

Cold Lake Oil Sands Project (IORL, 1997) A ranking of L (Low), M (Moderate), or H (High) was based on the duration, magnitude and extent of an effect in this project (Table 4).

Table 4 Cold Lake Oil Sands project interaction matrix

Duration	and	Magnitude	Extent			
			Local	Regional	Territorial	National/ International
Short-term	and	Low	L	L	M	M
Short-term	and	Moderate or High	L	M	M	M
Medium-term	and	Low	M	M	M	M
Medium-term	and	Moderate or High	M	M	M	H
Long-term	and	Low	M	M	H	H
Medium-term	and	Moderate or High	M	H	H	H

(*Source*: IORL cited in CEAA, 2000:12)

Kluane National Park (Hegmann, 1995) This study assessed the effects of various existing and proposed actions in and around Kluane National Park. Table 5 illustrates some of the effects for grizzly bear. Six types of effects were identified as well as an overall effect that represented the combined influences of all effects from each action on the grizzly bear's habitat. The rankings are defined as: 'blank' = no effect; L = low probability of occurrence, or magnitude of effect, on reproductive capacity of species or productive capacity of habitat, probably acceptable; M = moderate or possibly significant effect; H = high probability of occurrence or magnitude of effect probably unacceptable (e.g. population recovery may never occur or may occur in the long-term). A ranking option for positive effect (+) was also provided.

Table 5 Kluane National Park: Effect's interaction matrix

	Effects						
	Habitat Loss	Fragmentation	Alienation	Destruction	Mortality	Removals	Overall
Existing Actions:							
Backcountry camping	L		M	L	H	H	M
Backcountry hiking			M		M	M	M
Snowmobiling			L				M
Rafting campsites	L		M		H	H	H
Future Actions:							
Goat Head Mtn. Trail			L		M	M	M
Shuttle to Bear Camp			H		M	M	M
Helihiking			H				H

(*Source*: adapted from Hegmann, 1995)

Freshwater Creek Integrated Buffer (Peterson, 2002) The IBP method was applied to an endangered regional ecosystem within the catchment of Freshwater Creek (Australia). Thirteen individual threats (external and internal) to the environmental components of the core and its surrounds were identified. For each threat, its causes were examined and environmental pathway diagrams developed to illustrate the impacts of the threat. For example, a significant threat to the koala, which was a target species in the IBP, was 'barriers to movement'. These barriers included roads, fences, swimming pools, waterways and alien environments. The koala's ability to negotiate roads for example was influenced by the extent of habitat fragmentation, the koala's inherent susceptibility to injury and death, particularly from vehicle collisions on roads, the road design and vehicle speeds. Based on a thorough understanding of this threatening process, an environmental pathway diagram (Figure 9) was developed and used to identify appropriate policies to reduce the impact of 'barrier to movement', both within the core and surrounding landscape. In addition, an interaction matrix was developed to indicate how the components of the core (e.g. fauna and vegetation communities) responded to threatening processes (e.g. land clearing) that originated from outside (i.e. external) the core (refer to Table 6, which illustrates a partial application for vegetation and fauna). An assessment of the duration of the threat (long, medium and short-term) in combination with its magnitude, as reflected in its relative size or amount of effect (low, medium and high) and extent (local, regional and state/national) allowed the threat's impact to be ranked on a four point scale (e.g. H = high, M = medium, L = low, Blank = no effect). The analysis also included consideration of both present and future threats.

♦ DECISION POINT

Either
(a) Are external threats to each resource or element of the core to be mapped to define Analytical Protection Zones (APZs)?
(*Note*: This is the longer version of the model process requiring more thorough examination and mapping of the relationships between the core's resources/elements and its surrounds);

Or
(b) Are external threats to the core as a whole to be mapped to define Elementary Protection Zones (EPZ)?
(*Note*: This shorter version can be more quickly applied and does not require mapping of the specific interactions between each of the elements of the core and its surrounds. It is the recommended procedure where planning resources and time frames are limited).

If (a) go to Phase 4, Step 1, Action 3.

If (b) go to Phase 4, Step 1, Action 6.

P4, s1, a3 *Action 3* *Demarcate analytical protection zones*

The various elements of the core (e.g. geology, hydrology, soil and vegetation) were identified in Phase 3, s2, a2. For this action it is necessary to identify the existing and potential threats (e.g. fire, water pollution and pest plants) to each of these elements and to map the source areas of each threat. This is followed by a synthesis or overlay of the mapped areas. The resultant map for each element is called an analytical protection zone (APZ).

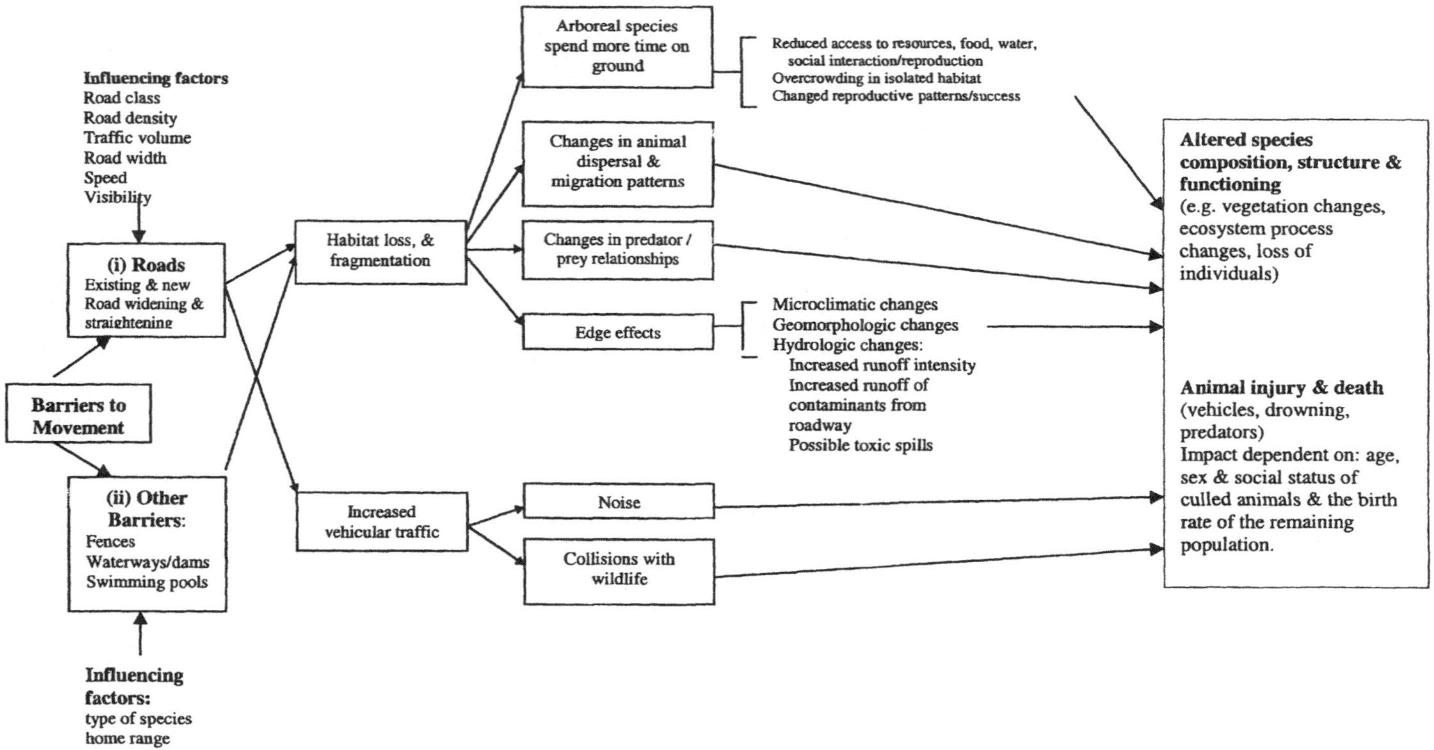

Figure 9 Barriers to movement environmental pathways (Peterson, 2002)

Table 6 Present and future external threats to the vegetation and fauna of the Freshwater Creek core

Legend: High significance; Medium significance; Low significance; No effect

THREAT (EXTERNAL)			Vegetation (Composition & Structure)								Fauna (Distribution & Abundance)							
			Eucalypt open forest		Melaleuca wetlands		Coastal/ estuarine		Koala		Avifauna		Amphibians		Inverteb-rates		Macropods	
		Present/Future Threat	P	F	P	F	P	F	P	F	P	F	P	F	P	F	P	F
THREAT (EXTERNAL)	Veget-ation	Vegetation loss	High	High	High	High	High	High	High	High	High	High	High	High	Medium	Medium	Medium	Medium
		Vegetation degradation1	Medium	High	High	High	High	High	Medium	Medium	Medium	Medium	Medium	Medium	Low	Low	Low	Low
		Vegetation fragmentation	High	High	High	High	High	High	High	High	High	High	Medium	Medium	Low	Low	Medium	Medium
	Barr-iers	Barriers to Movement (i) Roads	Low	Medium	Low	High	Low	High	High	High	Low	Medium	Low	Low	Low	Low	Low	Medium
		Barriers to movement2: (ii) Other	No effect	No effect	No effect	No effect	No effect	No effect	High	High	Low	Low	Low	Low	Low	Low	Low	Low
	Hydrol-ogy	Point sources of water pollution3	Medium	Medium	High	High	High	High	No effect	No effect	No effect	No effect	No effect	No effect	No effect	No effect	No effect	No effect
		Non-point sources of water pollution4	Medium	Medium	High	High	High	High	No effect	No effect	No effect	No effect	No effect	No effect	No effect	No effect	No effect	No effect
		Drainage pattern disruption5	Low	Low	High	High	High	High	High	High	No effect	No effect	High	High	Low	Low	Low	Low
	Other	Fire	Medium	Medium	Low	Low	Low	Low	Medium	Medium	Low	Low	Low	Low	Medium	Medium	Low	Low
		Predators/feral animals e.g. dogs	No effect	No effect	No effect	No effect	No effect	No effect	High	High	Medium	Medium	Low	Low	Medium	Medium	Low	Low
		Recreational activities	Low	Low	Low	Low	Medium	Medium	Medium	Medium	Low	Low	Low	Low	Low	Low	Low	Low
		Noise	No effect	No effect	No effect	No effect	No effect	No effect	Low	Low	Low	Low	Low	Low	Low	Low	Low	Low
		Light	Low	Low	Low	Low	Low	Low	No effect	No effect	No effect	No effect	No effect	No effect	No effect	No effect	No effect	No effect

Where: 1 noxious/exotic plant species; 2. dams, fences, swimming pools; 3 landfill, sewerage treatment plants, construction sites; 4 septic tank drainfields and urban stormwater; 5 dams, infill of low lying areas, canal development and altering stream flow.

Example:

Cooloola National Park (Peterson, 1991) For Cooloola, 10 APZs were mapped (e.g. geomorphology, hydrology, vegetation, fauna, and cultural heritage). Figure 10 illustrates the Geomorphology APZ. The main. geomorphic resource of the park was its sandmass system. External threats to this system were due to interface processes and interactions that occurred mainly at the inter-tidal zone between the marine and terrestrial environment. This was also a zone of contact between the park's natural systems and the influence of humans, providing almost unlimited access from the beaches and providing opportunities for several harmful land use practices.

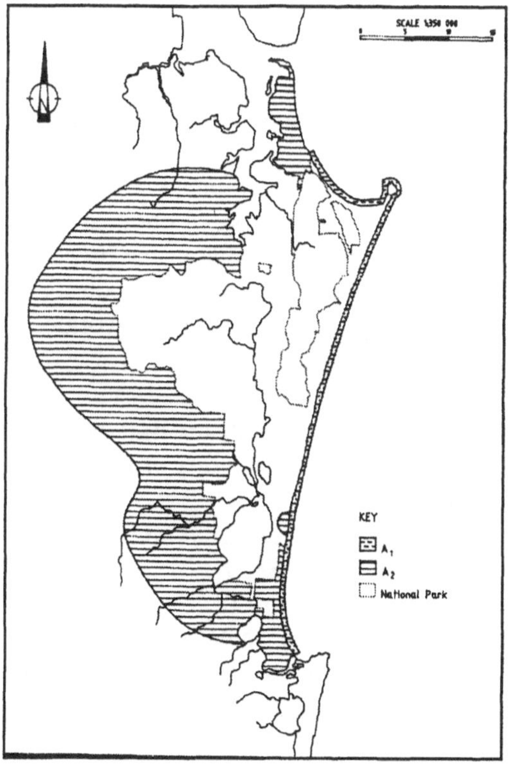

Figure 10 **Geomorphology Analytical Protection Zone, Cooloola National Park**

Where A1 and A2 refer to buffer policy areas. Policies are similar throughout each defined area. (Peterson, 1991: 126)

The Geomorphology APZ was designed to reduce the negative impacts of eight separate threats to the park's geomorphology including: removal of vegetation; recreational activities; fire; noxious and exotic plants; and excessive visitor numbers. The source area of each threat to the park's geomorphology was mapped and the eight maps synthesized. Although the Geomorphology APZ was almost continuous, there was no uniform land use policy throughout the APZ. The fringing beach interface (Zone A1), with seven identified negative impacts, required several management policies while the western landward boundary (Zone A2), with one identified negative impact, required fewer control measures (refer to Peterson [1991] for a complete explanation and presentation of the mapped data).

P4. s1. a4 ***Action 4*** ***Devise policies and land use measures to minimise the impact of the threats on the components of the core***

Develop policies to minimise or eliminate the impact of the identified threats on each element of the core. Policy statements and land use measures should be general at this stage in the process. Specific strategies for implementation may be given preliminary consideration, to reflect the legislative, social, economic and political framework of the study area.

P4. s1. a5 ***Action 5*** ***Identify external threats to core and delimit elementary protection zones***

- *Synthesise the criteria for demarcating the APZs and demarcate elementary protection zones (EPZ)*
  Many of the separate elements of the core (e.g. geology and vegetation) may be affected by the same threatening processes due to the inter-related nature of the ecosystems in the core. For example, the removal of vegetation in Cooloola impacted on six core elements (e.g hydrology, soil, geomorphology, vegetation, fauna, and wilderness). As the purpose of the IBP process is to identify and map threatening processes, it is necessary to synthesise the mapped output for each threat to each element of the core according to the type of negative influence. This results in the formulation of protection zones from the viewpoint of a specific negative influence, such as fire, or water pollution. These zones are called elementary protection zones (EPZ). As EPZs should be

established for each major threat, this longer version of the methodology ensures a comprehensive examination of the identified threats to the core.

Example:

Cooloola National Park (Peterson, 1991) Sixteen threatening processes were identified in Cooloola National Park, resulting in 16 EPZs, each with specific land use policies. For example, the Fire EPZ (Figure 11A) represented the zone to protect the park from fires originating outside its boundaries. The water pollution EPZ boundary (Figure 11B) reflected the source areas of water pollution. Its shape reflected the catchment boundary of the streams flowing into the park and unlike the fire EPZ did not surround the core. In Cooloola a sizeable portion of the western boundary of the park was not affected by water pollution from outside the park. By

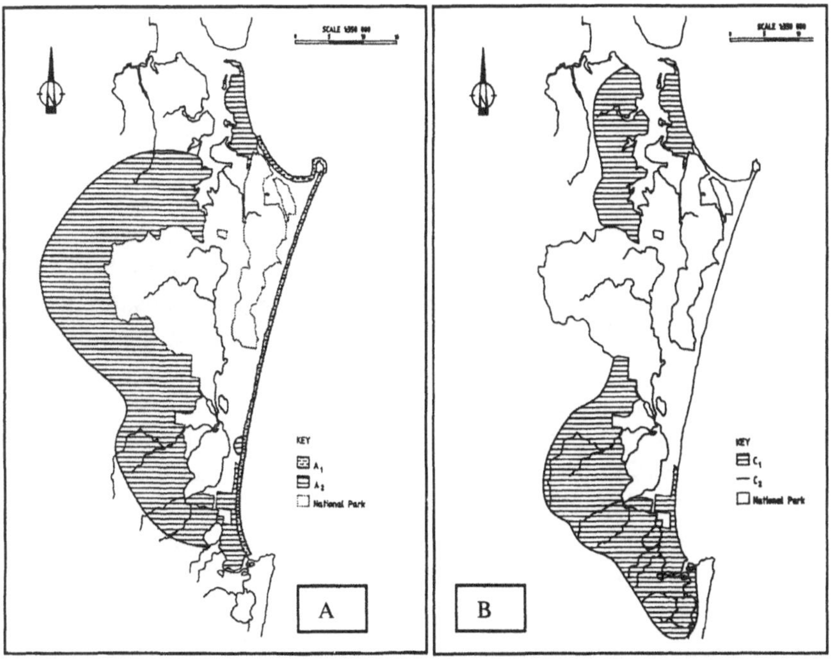

Figure 11 Elementary Protection Zones, Cooloola National Park
(A: Fire EPZ (Peterson, 1991:224) Where A1 and A2 refer to buffer policy areas; and B: Water Pollution EPZ (Peterson, 1991:230). Where C1 and C2 refer to buffer policy areas. Policies are similar throughout each defined area).

identifying the source areas of each significant threat, the buffer plan policies could then be applied only to the areas where the relevant threat was identified. This helped to produce a more realistic final buffer zone that responded to the heterogeneous nature of the surrounding region.

Where APZs are used to define EPZs, proceed to Action 7.

(*Note*: P4, s1, a6 is ONLY undertaken in the shorter version i.e. where APZs are not identified. The progression in this shorter form is to complete P4, s1, a1 and a2 and then to proceed to P4, s1, a6, the decision having been made previously not to undertake the actions a3, a4 and a5 in P4, s1).

P4, s1, a6 ***Action 6*** ***Delineate elementary protection zones (EPZs) for core as a whole***

This action proceeds from P4, s1, a2 and is undertaken where APZs are not formulated. This sequence is a shorter version of the IBP methodology and is recommended where resources of time and money are limited.

- *Map the source of each threatening process*
  Using the data assembled in the matrix (P4, s1, a1 and a2), map the source of each threat (Figure 11). The mapped areas are termed Elementary Protection Zones (EPZ). This procedure ensures that the EPZ will include only those areas where control measures and management strategies are necessary. The boundary of the EPZ is not a prescriptive distance from the core, nor does it necessarily surround the core, nor does it include areas that do not have the threat present. Delimiting EPZs for each identified threat results in a series of potentially overlapping EPZs. In some instances the precise boundary of an EPZ may be difficult to determine. It is recommended that a strategic approach be used, rather than a cadastral base for defining EPZs

P4, s1, a7 ***Action 7*** ***Identify management issues arising from the core's impact on its surroundings***

This action is a significant aspect of the IBP process setting it apart from other buffer planning processes. The IBP methodology recognises that the core, being part of a much larger system, interacts in many ways with its surroundings and may itself result in adverse impacts on adjacent land and communities. The following planning actions are needed:

- *Identify all existing and potential negative threatening processes that originate in the core and understand their causes and mode of operation*
  The contextual analysis undertaken in P1, s2 provides data on which to identify existing and potential threats. Examples may include native wildlife dispersing from the core and causing damage to crops and property, or injury and death to people; and fires escaping from the core.

- *Identify the impacts of the existing and potential threats on the surrounding land and its people*
  A tabular matrix may assist in the identification of threatening processes originating from the core and enable a better understanding of their impacts on the surrounding land and its people. A similar process to that used in identifying the impacts from external threats should be undertaken (i.e. P4, s1, a1 and a2).

- *Identify 'hot spots'*
  For many of the identified threats the entire core may be the source of the threat. For example, conflict species may be distributed throughout the core and hence the whole area would be identified as the source of the threat. However, the area of impact may be specific and related to the particular nature of the threat. For example, the conflict species may have a regular daily or seasonal movement pattern thus concentrating the impact of the threat on the surrounding community. This migration route may be identified as a hot spot and management may be directed to minimising the threat along the movement corridors. Alternatively, the source of the conflict may be confined to a particular part of the core. For example, local communities may require access to specific resources of the core. In developing countries this may include areas of grassland needed for thatching, or particular sites that have cultural significance.

The identification of these threatening processes lies with those responsible for plan development, and should occur in close conjunction with representative community stakeholders.

Example:

Freshwater Creek Integrated Buffer (Peterson, 2002) The application of the IBP to an endangered regional ecosystem in Freshwater Creek (Australia) included an interaction matrix that identified how the threatening processes within the core (i.e. internal threats) affected the surrounding terrestrial and marine environment. Consideration of both external and internal threatening processes in this plan provided a comprehensive approach to developing an integrated buffer for this core. For example, water pollution was identified as a significant internal threat. The core was situated upstream of Hays Inlet, an area of mangrove, marine clay pans and estuaries. These areas were subject to tidal inundation and were a significant habitat for commercial fish species and other marine bird life, particularly resident and trans-global migrant bird species, many of which were protected under the Ramsar treaty. Some of the coastal waters were included in a marine park (Moreton Bay Marine Park). In the upper catchment of Freshwater Creek much of the vegetation had been cleared, while the riparian zone was entirely cleared, the creek being piped underground or trained by concrete channels. These structures deprived the creek of its wildlife, biodiversity and ecological processes and consequently of its biological water-purifying capacity. Water quality monitoring in the core and upper reaches of the creek indicated that water quality was poor and organically polluted as a result of several point and non-point sources of pollution. Retention of vegetation in the core, particularly along Freshwater Creek was an important mechanism for filtering pollutants and several policies were identified for implementation within the core, to maintain and improve the quality of water leaving the core and entering the adjacent, internationally significant coastal waters.

P4. s2 Step 2: Develop draft management policies and land use measures

In the IBP process, active management both within the core and buffer areas is essential. Step 2 addresses the development of policies, outcomes or strategies to achieve the stated goals and objectives of the IBP. All policies should be carefully negotiated with major stakeholders to obtain consensus and may include a statement of:

- the purposes or goal to be achieved through the prescription;
- the rules of conduct to serve those purposes;

- contingencies in which the rules apply;
- sanctions to induce compliance and/or penalise non-compliance; and
- assets to cover enforcement and administrative costs (Brunner, 1995).

This step consists of three actions:

P4. s2. a1 ***Action 1*** ***Develop draft management policies for the core and each EPZ***

- *Identify the desired outcomes*
  State the broad desired outcomes or intents for the core and each EPZ. These should be compatible with the aims and objectives (P2,s3) established early in the planning process.

- *Devise draft performance based management policies for all threatening processes*
  Develop specific policies to eliminate or minimise the adverse effects of each threatening process. The key to developing constructive mitigation strategies is determining which of the cause-and-effect pathways results in the greatest effect and focusing mitigation and rehabilitation strategies on these pathways. Where possible, avoidance or minimisation should be the aim, rather than mitigating significant effects after they have occurred. For example, attempting to remove contaminants from water may be less effective than preventing pollution discharges entering a watershed.
  Proactive approaches will require extensive policy coordination at several levels. The choice of policies will depend on the local circumstances and the consensus view of the major stakeholders. In general, the policies should ensure that benefits accrue to the local communities, in terms of enhanced employment and economic opportunities, compensation for lost use rights, or stewardship payments for the management of natural areas. In particular, where a national benefit is achieved through the IBP process, it may be inequitable for local communities to pay the full cost of conservation. Where possible, sustainable use strategies should be implemented within the core and buffer, as these may assist in raising the level of support and acceptance of the core and its role in nature conservation within the community. The tenure arrangements within the core and buffer will influence the procedure for implementation of the necessary policies.
  A recommended approach for drafting the policies is to identify relevant performance criteria and acceptable solutions. The performance

criteria will relate to the underlying outcomes that are desired and may state particular standards to be achieved, while leaving the means flexible and open to innovative approaches. Hence, a particular resource use or development would be permitted to occur if it is able to protect the resource values, irrespective of the manner in which it is to be carried out. Thus, performance criteria should emphasise the purpose of the controls rather than outline rules and regulations. An approach, which specifies permitted and prohibited uses and activities, may antagonise local communities and be counter-productive in the long-term. By adopting a performance based approach, local communities can utilise traditional methods and institutions to achieve the required performance outcomes. Hence, 'all is possible' within the confines of the performance criteria, as a range of acceptable solutions may be identified to show how the criteria could be implemented. The agency responsible for implementing the IBP should consider developing a 'guideline document' to explain in detail the performance criteria and give more specific indications of the likely direction of future planning decision making within the buffer.

- *Determine the need for temporal policies*
  Temporal considerations may necessitate the establishment of performance standards specific to a particular time period or season. For example, grass collection is permitted in Royal Chitwan National Park on a few specified days of the year (Sharma, 1990). Within the Koala Coast area of southeast Queensland, temporal policies were applied to motorists, who were required to reduce their vehicle speed at specific times of the day (dusk to dawn) and year (October – February) to help minimise injury and death to koalas during periods when the koalas were more mobile (Queensland Government, 1997).

Policy development should be based on wide consultation with all stakeholders. Policies ultimately must be acceptable to the community, otherwise compliance is likely to be low, especially where enforcement provisions are limited. This phase of policy development may require extensive consultation, perhaps even over a number of years. It may need to be accompanied by an education strategy that targets those affected by the proposed policies.

Examples:

Coastal planning (Graham and Pitts, 1996) The draft 'Good Practice Guidelines for Integrated Coastal Planning' (Graham and Pitts, 1996) illustrates the relationship between objectives, deemed to comply standards and performance criteria (Table 7).

Table 7 Coastal planning guidelines

Objectives	Deemed to comply standards	Performance criteria
The natural role of wetlands and riparian areas in filtering nutrients and absorbing soluble pollutants in water shall be maintained. Use and development of land adjoining wetlands should be consistent with ecological processes that contribute to the values of wetlands.	Within 30m of a boundary of a wetland there shall be no: • clearance of vegetation; • erection of structures; • construction of any roads tracks or accesses; or • use of herbicides or artificial fertilisers. Water table level not to deviate from mean seasonal level by more than 10 per cent.	Roads, tracks and engineering works in and near wetlands and waterways must allow for the maintenance of ecological processes. Buildings in and near wetlands shall be sited so as to minimize loss of natural vegetation cover and so as not to interfere with any natural flow regimes of the wetland.

(*Source*: Graham and Pitts, 1996)

Guidelines for Biodiversity Conservation: A guide to development and sustainable living within significant natural areas and buffers (Peterson, 2002) The guide was developed to assist in the implementation of an IBP for an endangered regional ecosystem that provided important habitat for the koala in Pine Rivers Shire (Australia). The document provides detailed and practical guidance to assist local and state government officers, decision makers, planners, developers, construction workers and residents, who are involved in making decisions that affect the conservation of significant natural areas. The guide includes regional and local/site scale recommendations to address the threats identified in the IBP process and to better conserve these habitat fragments, while at the same time allowing sustainable development within the important habitat and surrounding areas.

The guide adopts a performance based approach (Table 8), stating broad outcomes to be achieved in relation to the identified threats. Acceptable solutions are described and illustrated to indicate how the outcomes may be achieved. For example, Figure 12 provides a suggested solution for the placement of a road within a road reserve to help maintain vegetation and movement corridors for wildlife and to reduce the impacts of edge effects on fauna and flora. The solutions are not prescriptive as a range of acceptable solutions may be possible in particular circumstances.

P4. s2. a2 *Action 2* ***Develop an 'Issues Paper' and analyse the submissions***

An 'Issues Paper' should be developed to obtain feedback on the reliability of the data and to provide an avenue for comment on possible future directions. The following actions are required:

- *Prepare an issues paper*
  The issues paper could take a variety of forms: detailed report; brochure; information sheet; a series of maps and text. The paper should include a statement of the problem, the proposed aims and objectives of the plan, the proposed core and important habitat, EPZ boundaries and the draft management policies.

- *Ensure wide distribution of the issues paper*
  Wide publicity is required to ensure that all the stakeholders are informed of the draft proposals. The mechanisms could include direct mail, community meetings in local areas, advertising in local community centres, the media (print and electronic) and provision of hard copies in relevant government offices and public libraries. Provide clear information on the time frame for submissions and the place for their lodgment (e.g. postal and/or e-mail address).

- *Consider the submissions presented concerning the issues paper*
  All submissions should be considered by the planning team and a document produced that provides an analysis of the submissions. This document should be made available to all stakeholders and other contributors. The 'analysis of submissions' paper should include:

  - the names of the respondents;

Table 8 Guidelines for Biodiversity Conservation: Sample of performance criteria and acceptable solutions

Performance Criteria The intent may be achieved where:	Acceptable Solutions The Acceptable Solutions illustrate various ways of meeting the associated Performance Criteria.
The street network design reflects the importance of maintaining habitat connectivity and minimising disruption to wildlife movement. The street network avoids duplication and unnecessary destruction of habitat.	The carriageway: • is offset to one side of the reserve to enable a wider strip of vegetation to be retained on the other side, thus reducing edge effects and ongoing management needs (Figure 12) or • meanders through important stands of vegetation to minimise the loss of important trees or ecosystems; or • is divided to minimise overall loss of habitat and to decrease fragmentation, so long as this does not result in increased wildlife mortality.

(*Source*: Peterson, 2002)

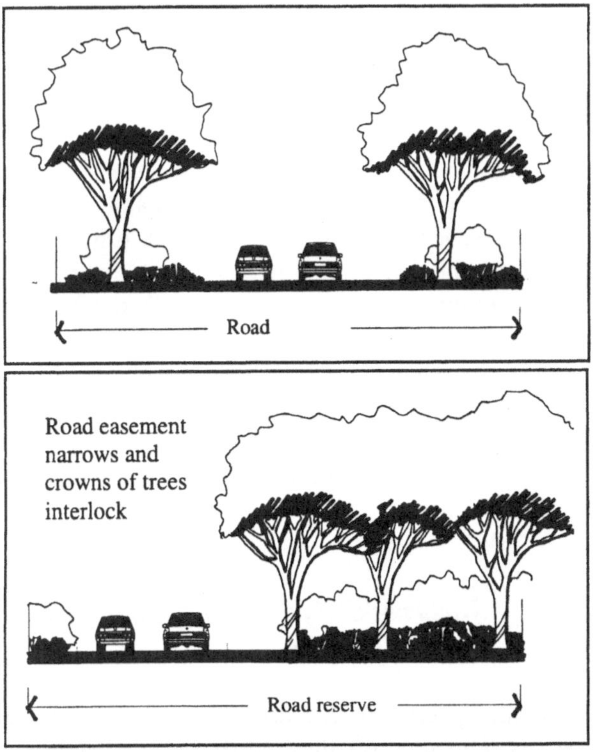

A. Road centred within the road reserve

B. Road located to one side of the road reserve

Figure 12 Location of the carriageway on one side of the road reserve (B) to retain vegetation within the road reserve

- a statement of the content of the submission made by each respondent (this may be produced on a thematic basis);
- an analysis of the submissions (on a thematic basis); and
- recommendations in relation to each theme.

P4. s3 **Step 3:** **Identify the integrated buffer zone and related policies**

The final IBP boundaries and associated policies are developed in this step, which consists of three actions.

P4. s3. a1 ***Action 1*** ***Synthesise EPZ boundaries to delineate the draft integrated buffer zone***

The final buffer boundary will be a composite of the EPZ boundaries and associated policies. The following planning action is required:

- *Synthesize all EPZ boundaries*
  Overlay or sieve all EPZ boundaries (P4,s1) to produce a 'Composite EPZ'. Group areas having similar threats into sub-zones of the integrated buffer to reduce the complexity of the final map. The IBP may contain areas in which several threats have been identified and which require appropriate use controls. In other areas of the buffer, fewer restrictions may be required, due to the smaller number of identified threatening processes. Hence within the buffer there may be a great variety in the uses and activities that may be permitted. Final buffer boundary delimitation is a difficult task and may require minor adjustment of the synthesised EPZ boundaries. It may be necessary to develop guidelines or principles as a basis for establishing this boundary (Table 9).

P4, s3, a2 ***Action 2*** ***Synthesise EPZ policies and integrate with the core's policies***

The buffer should have policies specific to the area of operation of the identified threatening processes. This requires the synthesis of EPZ policies

Table 9 Buffer boundary guidelines

The buffer boundary should:
⇒ encompass ecosystems or landscape units
⇒ be easy to implement, enforce and allow for dynamic management
⇒ be practical for landowners and others affected by the buffer plan – the buffer boundary should facilitate the existing lifestyles and cultural practices of those within and adjacent to the buffer
⇒ be easy to identify – once the ecological base for the boundaries has been determined, minor alternations may be introduced to extend buffer boundaries to recognisable features, either natural (catchment boundary, ridge line) or cultural (e.g. property boundaries, roads)
⇒ be recorded on maps (e.g. topographic and cadastral) and be readily accessible to the public
⇒ be adequately marked on the ground (e.g. with signs at roads entering the buffer area or critical habitat)

to reflect the sub-zones of the draft buffer (P4, s3, a1). In this way a heterogeneous buffer zone is produced, one that responds to the source of particular threats. It is important to ensure that the policies developed for particular EPZs are compatible with the management objectives of the core.

P4. s3. a3 *Action 3* ***Produce preliminary draft plan for limited distribution***

As wide consultation and involvement with all stakeholders has been an integral part of the process, the draft document must now be circulated to key individuals, groups and agencies for further comment and modification before a more extensive consultation processes is embarked upon. The following specific actions are required:

- *Prepare a preliminary draft document*
  The draft document, should contain the following:

- background or context of the plan and the reason for its development;
- the agency/group responsible for the plan's preparation;
- how the plan links with other strategies and policies in the region;
- a vision statement;
- a statement of the goal(s) and objectives;
- an endorsement process;
- a monitoring and review process;
- the physical buffer plan consisting of a description of the methodology, the buffer boundaries and policies, and performance indicators; and
- an implementation strategy.

• *Edit and print the preliminary draft plan*
Where necessary, use an editor to prepare the final draft and format the document for printing.

- Prepare for limited distribution of the preliminary draft plan
- Ensure critical evaluation of the draft by selected groups, including:
- internal review by key people or units within the organisation;
- key stakeholders; and
- experts e.g. individuals/organisations with relevant expertise.

A document based on the review of these submissions may be useful. It could identify the major issues raised by the respondents, the reviewer's comments and proposed recommendations.

Example:

Cooloola National Park (Peterson, 1991) Cooloola's buffer is a partial application of the IBP process as it did not examine and map threats from the core to the surrounding landscape. A synthesis of the 16 EPZs resulted in an all-encompassing buffer zone (Figure 13), comprising several sub-zones, each with land use policies applying to the area of operation of the particular threat they were designed to control. The buffer zone identified the specific areas where threats were known to occur or had the potential to occur, and targeted management strategies to these areas. The eastern beach zone was a 'hot spot' with eleven broad policies being required, while the

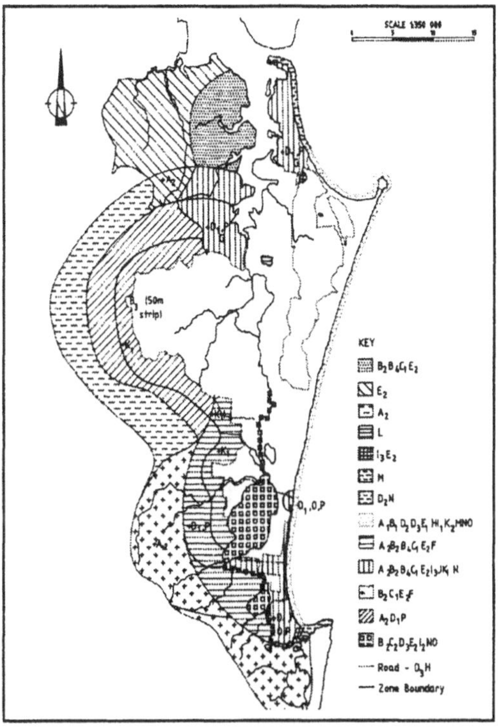

Figure 13 **Cooloola National Park buffer zone** (Peterson, 1999: 278)

(Key indicates sub-zones where specific policies apply. Letters indicate the policy type and the numbers refer to specific actions)

western boundary of the park required only one policy. Thus a specific set of control measures and land use policies were devised, responding to the heterogeneous environment surrounding the park, and the varied nature of threats that emanated from this environment.

Freshwater Creek IBP (Peterson, 2002) A partial application of the IBP was undertaken for this small habitat fragment and corridor system along Freshwater Creek (Australia), the purpose being to develop a physical buffer plan and associated biodiversity guidelines and an implementation strategy. Thirteen individual threats (external and internal) to the environmental components of the core and its surrounds were identified. Several threats were amalgamated, resulting in the delimitation of six EPZs (vegetation loss, degradation and fragmentation; barriers to movement;

predators; altered surface hydrology; fire; and other minor impacts). The synthesis of the EPZs resulted in the identification of 17 management areas. However, as several of these areas were very small, particularly near the creek corridor, they were synthesised with neighbouring lands, where these areas contained more stringent land use policies. The final synthesis (Figure 14) resulted in the definition of 14 management areas within the IBP, each having specific management requirements. Land adjacent to the southern bank of the Freshwater Creek corridor contained all six threatening processes, while the northern and western areas contained only one major threat. An intermediate level of threat existed in the eastern coastal areas, this resulting primarily from processes within the core and upper catchment that impacted on the values of these areas. The final IBP reflected the heterogeneous nature of the environment and was based on an examination of underlying ecological processes, with the EPZ boundaries encompassing the area of influence of the major threatening processes. It took into account the impact that the core had on its surrounding landscape and the policies were relevant to the source of the threatening processes and to the existing planning and management framework at the local, regional and wider levels.

In association with the physical buffer plan a comprehensive guideline document was developed with identified performance criteria and acceptable solutions to be implemented in the EPZ areas. The guide provided practical recommendations on how to better conserve the values of the core and buffer, while at the same time allowing sustainable development of these areas. The guide took a performance-based approach to provide land managers with the scope to implement innovative strategies within the core and buffer. As remnant patches of vegetation were scattered throughout the landscape, the guide provided recommendations at two major levels of planning, the strategic/regional level and the site level. Consideration of the regional spatial structure of vegetation was an important to ensure the inter-connection of major habitat patches and the maintenance and/or restoration of genetic diversity. At the site level the guide provided recommendations, which would help to ensure individual developments were consistent with the overall regional and sub-regional framework. In a practical sense, the illustrations within the guide provided a very useful visual stimulus for planners and land managers.

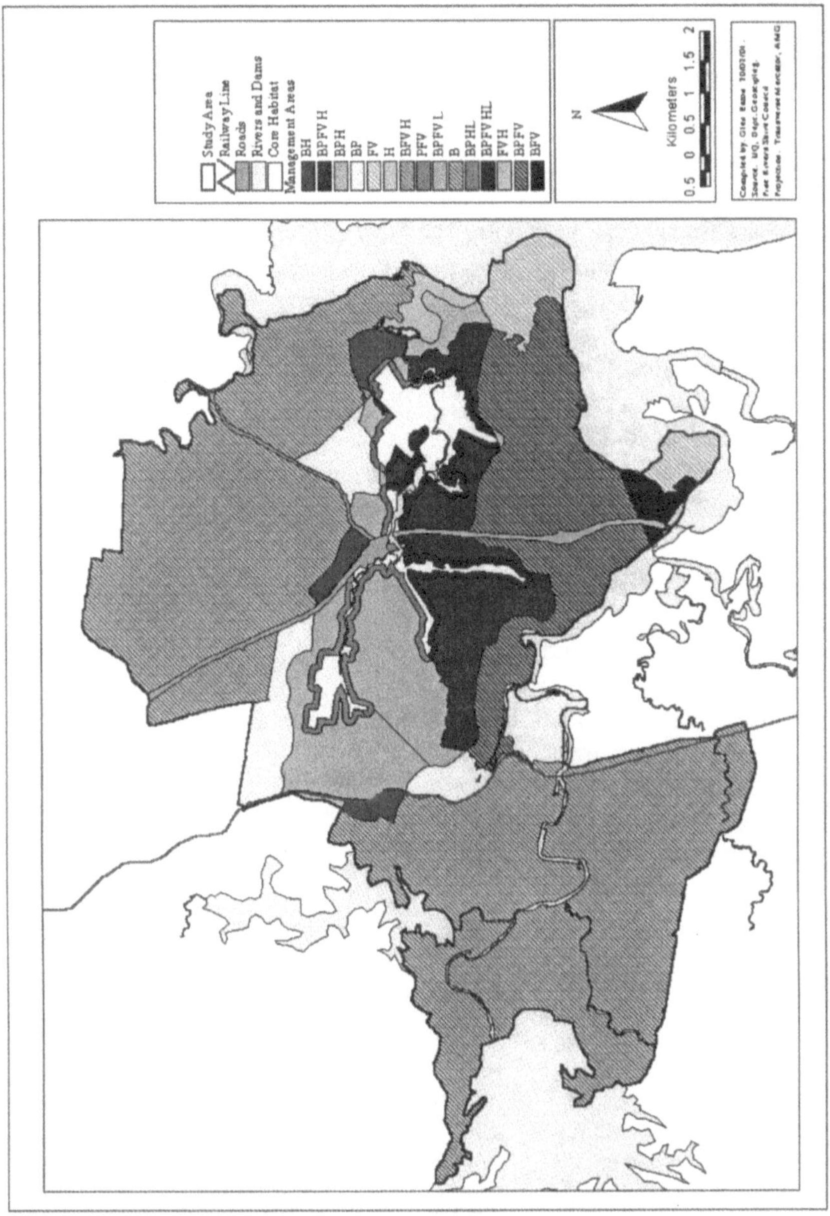

Figure 14 Freshwater Creek IBP (Peterson, 2002:277)
(Where management areas represent relevant EPZs: V – Vegetation loss, fragmentation and degradation; B – Barriers to Movement; P – Predators; H – Hydrology; F – Fire; L – Light and noise)

PHASE 5

FINAL INTEGRATED BUFFER PLAN

Brief Description

This phase focuses on developing the final IBP and the institutional plan to ensure effective monitoring, plan review, enforcement and implementation (Figure 15).

Information Input

- preliminary draft IBP plan;
- map of core areas and related management policies; and
- map of IBP zone and management policies.

Expected Outcomes

- institutional plan including:

  - performance evaluation and monitoring program;
  - research program;
  - education program;
  - implementation program;
  - enforcement program; and
  - review program.

P5. s1 **Step 1:** Develop a performance evaluation and monitoring program

Good quality decision making and implementation require that the decision makers, planners, managers and the public have regular, accurate and accessible information, at an appropriate scale, particularly concerning the condition of the environment (e.g. how ecosystems function and respond to disturbance). The broad objectives or outcomes of the IBP plan's monitoring program may include the following:

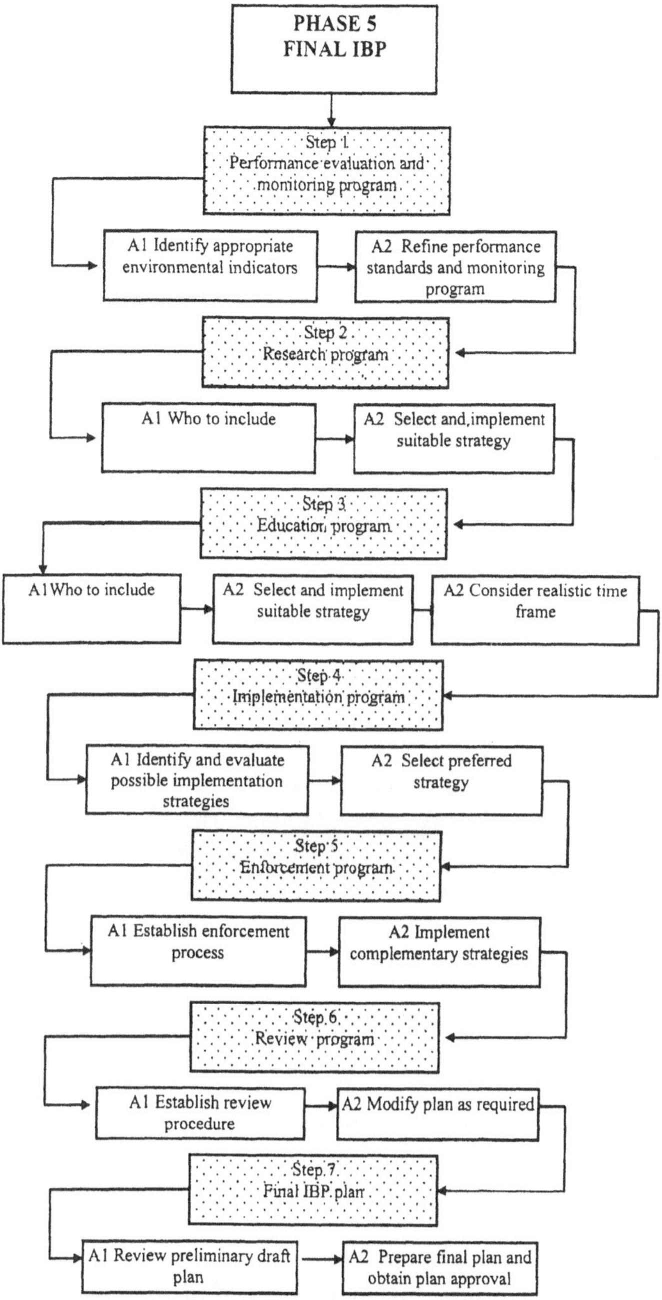

Figure 15 Phase 5 – Final IBP

- to develop an iterative process that establishes feedback linking management planning and its consequences;
- to provide an early warning of potential problems, the data being used to ensure that specified critical thresholds are not exceeded and to improve management;
- to report on the plan's effectiveness in achieving stated aims, objectives and policy outcomes;
- to contribute to an assessment of the region's progress towards achieving sustainable development, including conserving biodiversity;
- to help decision makers to make informed judgements about the environmental, social and economic (e.g. cost effectiveness) consequences of the IBP's policies and to develop alternative strategies for achieving the plan's aims, where necessary;
- to help ensure the effective allocation of resources/funds to appropriate areas; and
- to increase public understanding of the IBP and environmental issues generally.

P5. s1. a1 ***Action 1*** ***Identify appropriate environmental indicators***

To assess the overall effectiveness of the IBP it is necessary to establish indicators against which environmental performance may be reviewed. Indicators are variables (e.g. social, economic, biological and physical) which, when quantified, provide information about a particular attribute of the environment, to help explain how things change over time. As the concept of environmental indicators is a recent one, there is limited research to draw upon to help develop effective indicators for the IBP. The collection of data for indicators is often achieved through monitoring. The monitoring program for that indicator consists of 'repeated measurements of the variables that make up the indicator in various places and times, and in a defined way' (Saunders et al., 1998: 5).

- *Identify appropriate selection criteria for key indicators*

  The task of selecting the key indicators is critical to the development of a monitoring program and should be based on valid criteria (Table 10), with the chosen indicators meeting as many of the selection criteria as possible.

Table 10 What is a good indicator?

Selection Criteria	Meaning
Relevant	⇒ to the policy and management needs of the IBP plan
Measurable	⇒ quantitative/qualitative measures of the component can be obtained and recorded over a suitable time period
Important	⇒ reflects an important aspect of the environment
Scientifically valid	⇒ based on scientifically valid collection methods and is able to be validated
Understandable	⇒ are easy to understand (e.g. by most members of the community)
Cost effective	⇒ data can be obtained at 'reasonable cost' in relation to the information obtained
Temporal	⇒ shows trends over time and are linked to targets or thresholds
Spatial	⇒ shows areal changes in extent
Sensitive	⇒ can give early warning of potential problems
Cumulative	⇒ responds to the cumulative impacts of several threatening processes
Comparable	⇒ has a baseline or guideline against which to compare it
Participatory	⇒ facilitates community involvement, where possible and appropriate
Consistent	⇒ with other indicators used at the local, state, regional and national level
Integrative	⇒ portrays linkages between the social, economic and environmental dimensions of sustainability
Specific	⇒ able to distinguish between local and non-local sources of environmental degradation and effects

(*Source*: adapted from Saunders et al., 1998)

- *Identify the organisational level or spatial scale at which the indicators are to be applied*
  Choosing the appropriate spatial scale for expressing indicators of biodiversity is critical (Saunders et al., 1998). Monitoring of ecosystem change and condition, as well as the status of the target species, should occur at a scale at which management agencies can implement effective strategies. This scale should also be closely linked to natural ecosystem boundaries, rather than administrative boundaries, or the site boundaries in which a proposed development is to take place.

  Noss's (1990) hierarchical approach to monitoring is a possible strategy. The hierarchy recognises three primary attributes of ecosystems, namely composition, structure and function and examines these attributes at four levels of organisation: regional landscape; community-ecosystem; population-species; and genetic. Not all levels of the organizational structure may need to be dealt with at the same level of detail. Cost factors may also preclude some types of monitoring, particularly genetic monitoring. Indicators could be selected from the following levels:

  - regional (e.g. dispersal corridors) - measured using air photo surveys and satellite images;
  - population (e.g. population size, fecundity, survivorship, age and sex ratios, movement) - measured using sightings, groupings, numbers and movement patterns;
  - individual (e.g. physiological parameters); and
  - genetic (e.g. heterozygosity) (Noss, 1990).

- *Identify the temporal scales at which the indicators are to be applied*
  Choosing an appropriate temporal scale for expressing indicators of biodiversity is critical (Saunders et al., 1998). Temporal scales will vary depending on the indicator and such scales should be established separately for each monitoring program. The issues and elements being reported on will have different natural dynamics. Monitoring programs designed to detect change need to employ temporal scales appropriate to the natural scales of change, but modified, according to the management needs for information on rates of change. Elements that change slowly may be measured infrequently, but if a small change is important, then measurements need to be more frequent. The timing of monitoring may need to reflect seasonal breeding trends. It may also be important to monitor during and after periods of rapid change e.g. after a major event such as a fire, or during an extensive drought.

- *Select relevant key indicators for use in the IBP*
  The IBP is based on identifying the existing values of the core, examining processes that threaten it and implementing planning measures to ensure sustainable outcomes. This strategy mirrors the framework in which State of the Environment Reporting indicators are often developed. Hence, one approach for developing appropriate indicators is to identify:

  - indicators of the main components or elements of the 'condition' of the core area and its target wildlife (Table 11A);
  - indicators of the main 'pressures' or threatening processes (Table 11B); and
  - indicators of the 'responses' implemented to ameliorate the threatening processes (Table 11C). Relevant indicators need to be identified to suit local conditions.

  A monitoring program may be needed to measure some indicators or to test specific hypotheses that are relevant to the IBP's objectives and performance criteria. Blind data gathering will be of little value. Monitoring should continue throughout the life of the plan and beyond and its long-term objective should be to establish a methodology that ensures consistent data are collected so that valid comparisons, particularly in terms of direction and rate of change can be made over a long time frame. Monitoring must also enable the buffer manager to distinguish changes that are attributable to use and management from those due to other factors.

- *Identify 'hot spots' and areas/ecosystems at high risk*
  'Hot spots' (refer to P4, s2) may be areas of concentrated biodiversity (e.g. high levels of species richness and endemism) or areas where a range of threatening processes are significant. They may warrant more intensive monitoring.

- *Establish 'baseline' conditions for strategic monitoring*
  The baseline conditions can be used to identify and evaluate change. Where possible, monitoring should begin in Phase 3 with a description of baseline conditions in core areas, corridors and the buffer. It may be advantageous also to survey the local community before the project begins to establish baseline data against which to measure subsequent project effects. This will provide the basis for an 'environmental audit' to be made of the IBP's impacts.

Table 11A Suggested 'Condition' Indicators

Category	Possible Environmental Indicators
Genetic diversity	• number of distinct entities (e.g. subspecies, ecotypes, and geographic, morphological, physiological, behavioural or chromosomal races) readily recognisable within a species • number of individuals within each population • number of discrete populations • degree of physical isolation between populations • population amplitude – extent to which a species maintains occupancy of the full range of habitats in which it naturally occurs (both current and historic) • genetic diversity at marker loci within individuals and populations
Species diversity	• estimated number of species • conservation status of species on a local/regional basis (presumed extinct, endangered, vulnerable, rare or common) • percentage of species known to be changing in distribution (+/-) • number, distribution and abundance or migratory species by taxon per biogeographic region • demographic characteristics of target taxa, including common species (population size and breeding success)
Ecosystem diversity	• number, identity, condition and area of native vegetation types and marine habitat types at the bioregional scale • number and extent of ecological communities of high conservation potential

(*Source*: adapted from Saunders et al., 1998)

Table 11B Suggested 'Pressure' Indicators

Category	Possible Environmental Indicators
Human population	• distribution and density of population (numbers per unit area) • change in human population density • number of building permits granted • number and extent of new developments (e.g. houses, commercial, tourist, agricultural, as appropriate)
Clearing/ fragmentation of habitat	• extent and rate of clearing or major modification, as classified by agent or sector (hectares per annum of native vegetation type) • geographic location of remnants of native vegetation, by type, and the ratio of total length of edges of the remnants to area • the degree of fragmentation of native vegetation
Exotic species	• rate of extension of exotic species into the IBP area • pests – type and numbers
Pollution	• pollution point sources
Altered fire regimes	• extent, frequency and seasonality of burning by vegetation type

(*Source*: adapted from Saunders et al., 1998) (contd/ Part C)

Table 11C Suggested 'Response' Indicators

Category	Possible Environmental Indicators
Protected areas	• the extent of each vegetation type/marine habitat within protected areas • percentage of regional ecosystems within protected areas
Off-reserve conservation	• number of properties/area of land included in off-reserve conservation programs (e.g. voluntary conservation agreements, Land for Wildlife, local government reserves etc)
Knowledge base	• proportion of IBP area covered by biological surveys
Management	• the amount of funding provided for the implementation of the IBP
Management of native biota	• area of land with controls on clearing of native vegetation • area of clearing officially permitted • area cleared in relation to the area revegetated • the number of lending institutions which considering biological diversity as a component of their lending policies • area of land under 'best practice' management
Control of exotic species	• the number of management plans for exotic species • the number of research programs in relation to the impact of, and control of, exotic organisms
Pollution control	• amount of funding for, and number of research programs into the effects of pollution on biological diversity and how to alleviate them • the number of local laws or regulations that deal with the impacts of pollution on biological diversity
Altered fire regimes	• the number of management plans which take account of the impacts of fire on biological diversity, by vegetation type
Planning	• the percentage of local governments with management plans for conservation and maintenance of biological diversity • number of recovery plans, conservation plans • the percentage of companies with management plans for conservation and maintenance of biological diversity • the percentage of farm plans and catchment management plans completed
Knowledge	• number of research programs and the levels of funding for research aimed at indicating the role of biological diversity in ecological processes • number and location of long-term ecological monitoring sites by vegetation type within bioregions • percentage of state/local government budgets spent on conservation of biological diversity
Community involvement	• number of local governments which employ an officer responsible for the conservation and maintenance of biological diversity • the number and size of community groups involved in conservation and maintenance of biological diversity (e.g. Landcare)

(*Source*: adapted from Saunders et al., 1998)

- *Identify monitoring sites and treatments*
  In selecting monitoring sites it is important to consider the following:

  - what is to be measured;
  - what are the best locations;
  - are these sites representative;
  - are the sites suitable for the possible monitoring techniques; and
  - will the sites be available over a long time frame.

- *Design a sampling scheme*
  Monitoring should be conducted through flora/fauna studies based on permanent transect lines or quadrats within ESAs. Long-term monitoring projects will invariably result in many different consultants being contacted over time to undertake the survey work. It is important that the monitoring methodology is designed and standardised according to current best practice and recorded accurately. Once the original monitoring methodology is endorsed, the method should be followed by all future consultants involved in the monitoring program. The results will be of limited value if the survey methodologies used by the different consultants are inconsistent and this will prevent the direct comparison of results. Repeatability is important to ensure that trends are real and not due to differences in monitoring activity.

- *Establish the required frequency and timing of data collection*
  The frequency and timing of monitoring should be responsive to the indicator being measured. The duration of monitoring may vary according to the nature of the buffer area and its threatening processes, the ecosystems being monitored and financial considerations. A minimum, or regular monitoring duration should be established, and where possible, long term monitoring should be the target.

- *Establish a recording system*
  Recording of collected data is important for updating the plan and providing effective management strategies. Data must be presented at the appropriate technical level and be easy to understand. For long-term effectiveness of the IBP, the maintenance of a reliable database is crucial.

- *Undertake a feasibility study*
  It is necessary to determine whether the preferred monitoring strategies are feasible. They may be assessed against indicators of cost, time and staff requirements.

- *Conduct a pilot project*
  A trial monitoring program should be conducted before long-term funding is dedicated to the program. The program should be assessed and modified, where necessary.

- *Implement monitoring program and evaluate the data*
  Baseline data should be collected for the indicators to provide a basis for future evaluation. This should be used to assess the congruence between the plan's objectives and its impacts.

- *Determine the frequency of monitoring reports*
  Regular progress reports (e.g. annually at first) will help assess the changes occurring in the plan area and provide an evaluation of the impact of the plan's initiatives on the maintenance of the nature conservation values and allow action to be taken should results indicate a negative trend in one or more of the environmental indicators. Monitoring reports may provide a useful stimulus to help maintain the community's enthusiasm for the project. Annual reports may be a useful starting point and will allow management to adapt to changing situations.

- *Determine a reporting mechanism and procedure for modification of the plan*
  Monitoring is of limited use unless the monitoring results reach the manager of the site/area and remedial action is initiated when a failure to achieve the plan's objectives, or performance indicators is detected. This requires an effective reporting mechanism, which may be achieved through a community network, a public information access line or a permanent organizational structure (e.g. an 'implementation unit'), which responds to current needs. The results of the monitoring assessment should then be translated into positive changes in the IBP, particularly in relation to management policies and performance criteria.

- *Identify possible barriers to the monitoring strategy and mechanisms to overcome them*
  Barriers may include: time constraints; lack of funding; and a lack of skilled human resources. Solutions to these will be ongoing over the life of the IBP. The planning team should develop the monitoring program in conjunction with stakeholders with expertise in conservation biology e.g. local research and educational institutions, government agencies, community groups and individual landowners (e.g. they may monitor the threat of weed invasion by taking repeat, standardised photos). Information obtained from monitoring should be included in the local government's or area's State of the Environment Report and there should be an expectation that the results of monitoring are acted upon.

P5. s1. a2 *Action 2* ***Refine performance standards and monitoring program***

Following monitoring it may be necessary to refine the performance indicators and aspects of the monitoring program. This includes:

- *Evaluate results of monitoring program and modify the monitoring program*
  If the aims/objectives of the IBP are not being met the performance standards may need to be adjusted to ensure a higher level of compliance. The monitoring program should be modified, as necessary.

P5. s2 **Step 2: Develop a research program**

Ongoing research may be necessary to provide data on the biodiversity of the plan area, the operation of identified threatening processes, and the social and cultural aspects of the local community.

P5. s2. a1 *Action 1* ***Who to include?***

Aim for a cooperative research program that may include the agency responsible for developing the IBP, universities, NGOs and individual members of the community. This action should be initiated by the planning team, in conjunction with research institutions. It may be necessary to establish a committee or working group to oversee the continuing development of the research program.

P5. s2. a2 *Action 2* *Select and implement a suitable strategy*

Where a research committee is established, it should be responsible for evaluating alternative options, costing these options, prioritising actions over a selected time frame and selecting a preferred strategy.

P5. s3 **Step 3: Develop an education strategy**

To minimise the continuing loss and damage to sensitive ecosystems it may be necessary to change the way people perceive their environment. An education strategy should develop a view amongst all stakeholders that people are a part of their environment, not separate from it, and that their survival depends on maintaining a fully functional environment. The community must also understand that there are limits to the degree of alteration that natural systems can tolerate before these systems will degrade. Although simply stated, this is difficult to achieve. This step consists of three actions:

P5. s3. a1 *Action 1* *Who to include?*

The community of the planning region should be targeted. Specific groups include students, landholders, local government officers, developers and builders, conservation groups and politicians.

P5. s3. a2 *Action 2* *Select and implement a suitable strategy*

Implementation of the buffer policies requires the selection of an appropriate strategy. Thus it may be necessary to identify the range of possible implementation tools and to establish criteria to evaluate them. These may include criteria such as cost effectiveness, political and social acceptability, and administrative efficiency (refer to Peterson [2002], where the document, 'Implementation Strategies for Biodiversity Conservation' evaluates a range of education and awareness raising strategies).

P5. s3. a3 *Action 3* *Consider a realistic time frame*

Educational strategies should consider long time frames, as it may take decades to change community attitudes. However, specific time frames will be relevant to the particular mix of strategies identified in P5, s5, a2. Appropriate avenues for adequate financing should be considered.

P5. s4 **Step 4: Develop an implementation program**

Implementation is an important part of the IBP process, for those who prepare the plan may be different from those who implement it and both groups may have different values and knowledge of the IBP. The identification of implementation strategies should occur throughout the plan's development. This implementation step consists of two actions.

P5. s4. a1 *Action 1* *Identify and evaluate possible implementation strategies*

Those responsible for implementation were identified as important stakeholders in P2, s1, a1 and implementation issues and problems have been considered in all phases of the developing plan. It is the role of those who implement the IBP to ensure that its aims and objectives, policy outcomes and performance criteria are met. Clark (1992: 427) stresses that 'policy is effectively made by the people who implement it' and hence these people/organisations should play a major role in the policy process. The following planning actions are required:

- *Identify possible implementation strategies*
  Specific policies may apply in the core and buffer. However, the way in which these are implemented may vary. Where possible, a range of implementation strategies should be developed and evaluated and an effective mix of strategies identified.

- *Identify possible implementation barriers and develop contingency plans to overcome them*
  Barriers to implementation, which may limit the effectiveness of the plan's intent, may include:

- a lack of commitment to implementation by government, other organizations and/or the community;
- organizational inertia, or resistance to change;
- lack of specialist/technical knowledge for implementation;
- inappropriate delegation of responsibilities to organizations or agencies which are ill-suited or even opposed to the plan's aims;
- lack of clear understanding of the aims and objectives of the plan;
- lack of specific plans for implementation;
- shortage of resources, including money to accomplish the plan's aims;
- the use of discretion which alters the meaning of the plan to be implemented; and
- limitations of power.

To overcome these barriers it may be necessary to consider:

- a clearer definition of the aims and specification of operational objectives (refer to P2, s3);
- more detailed specification of the actions to be undertaken;
- briefing sessions for those involved in implementation, including training programs to explain the IBP process and policies;
- wide distribution of guideline documents explaining the IBP planning process;
- community education programs through a range of media e.g. community newspapers, industry journals, school programs, extension programs and community information sessions etc. As local people may need to be motivated to care for their local area, the IBP should be allied with a program of education about conservation of biodiversity to raise support for the plan;
- incentive schemes, both financial and non-material;
- developing communication networks between the community and the administrators of the buffer to ensure a two-way flow of information and continuing community involvement in the evolution of the buffer strategies;
- deploying adequate field staff to manage the core and buffer;
- a single implementing agency, or a dominant one;
- adequate resources to support the plan;
- allowing sufficient time to elapse for implementation to become effective;
- early project evaluation to assess implementation problems; and
- developing enabling legislation/action programs and assign responsibility for their enactment.

- *Consider the timing of the implementation strategy*
  The implementation of particular strategies may need to be well timed to enhance community support and to direct attention to the most crucial areas. The implementation strategy should consider the degree of threat, or 'the value of an area combined with some assessment of the urgency with which it should be protected' (Pressy and Logan 1997: 411). Hence the degree of threat may affect the sequencing of policy implementation within the IBP. High priority may need to go to areas that are subject to imminent clearing, land in future urban designations that is subject to development pressure, or prime sites such as low elevation, gently sloping land that may be subject to development pressure. More secure areas, in the short to mid-term, may be able to have their IBP policies implemented at a later date. Consideration of the timing of the implementation strategy may thus optimise the conservation benefits to be derived from the IBP process.

P5. s4. a2 *Action 2* *Select the preferred implementation program*

- *Select the preferred strategy and implement the plan*
  All stakeholders should be involved in selecting the preferred implementation strategy. The implementation of the plan should be well publicised. How this occurs will depend on the nature of the area in which the IBP is to be established.

P5. s5 Step 5: Develop an enforcement program

The long term effectiveness of the IBP in controlling the loss of species and their habitats will be limited without adequate enforcement, and a change in the attitudes and behaviours that have contributed to the identified biodiversity problem(s). Violation of regulations (and policies) may be caused by attitudes that promote non-compliance and hence effective enforcement requires clear, realistic guidelines for environmental management, a sound legislative framework and informed, cooperative attitudes towards compliance on the part of the regulators and regulated (Tai, 1996). This step consists of two actions.

P5. s5. a1 ***Action 1*** ***Establish an enforcement process***

The IBP strategy may rely on enforcement for conservation effectiveness. Although buffer plan benefits may help to compensate individuals and communities for lost use rights they may not remove the motive for exploitation. Enforcement may be necessary to ensure the buffer does not become an open access resource. The following are required:

- *Ensure enforcement staff have a thorough knowledge of the local area and the plan provisions*
  Local people are likely to understanding their community and the issues surrounding possible poor compliance and may be effective enforcement personnel. To enhance enforcement capabilities consideration should be given to using staff from other enforcement agencies, who may only need to be given delegated powers to enforce legislation outside their usual powers.

- *Target enforcement at voluntary processes within the community*
  Target enforcement from within the community (e.g. utilise honorary officers or conservation groups to improve compliance) and through local customs. Strengthen enforcement along with community development projects and build on traditional practices, where possible (Tai 1996).

- *Develop or utilise an existing enforcement agency within the implementation organisation*
  Enforcement may need to be implemented through the local/state government management agencies and must be adequately funded. Where human resources are unavailable or expertise is lacking, a new enforcement arm may be required, thus necessitating adequate training of the new staff.

P5. s5. a2 ***Action 2*** ***Implement complementary strategies to enhance enforcement***

- *Assess the need for a compliance enforcement system*
  A compliance system aims to secure conformity by means of direct compliance with set guidelines, or by taking action to prevent law violation without the necessity of detecting or penalising violators (Tai,

1996). The law achieves compliance through negotiation, the use of incentives (social, economic or symbolic), or a support strategy.

- *Assess the need for a deterrence enforcement system*
  A deterrence system aims to secure conformity with the law by detecting and penalising violators. For example, where new developments occur in the core and buffer, monitoring or inspection should take place in the construction and maintenance phase of a development to ensure that all conditions, codes of practice or relevant guidelines are adhered to. This may include spot checks, as well as more regular inspections by staff who know the area and the relevant provisions that have to be complied with. A well informed public can play an important and effective role in the abatement of environmental damage if they cooperate with regulatory agencies in detecting non-compliant behaviour (Tai 1996).

- *Develop an appropriate strategy*
  The selected strategy may combine elements of both a compliance and deterrence system and should be specific to the needs of the particular area.

P5. s6 Step 6: Develop a plan review process

This step establishes the review procedure for the buffer plan. It consists of two actions.

P5. s6. a1 *Action 1* *Establish a review procedure*

Although ongoing monitoring and evaluation will enable periodic changes to be made to the IBP, a formal review enables the plan's performance to be evaluated to enable changes to be introduced, as and when required. The evaluation should consider the entire strategy and not merely the biological aspects. This will include social, political, economic, organisational and other major variables. The following actions are required:

- *Consider the need for both external and internal reviews*
  Whether an external and/or internal review is undertaken will largely depend on any legislative requirements, cost and political factors. Consideration should be given also to comparing pre-questionnaire

responses with similar questions administered after 'x' years of the buffer plan.

- *Establish checklist for review*
  Establish what circumstances would initiate a review of the plan (King Cullen, 1989). These may include:

  - poor performance (e.g. critical thresholds, for example, relating to the extent of remnant habitat or the population size of the target species, are exceeded) (Note: it may be necessary to identify threshold values for a number of important elements);
  - legislative requirements;
  - reprinting of the original plan;
  - changes to relevant legislation and other documentation;
  - content revision;
  - new issues;
  - changes to names;
  - organisational redesign; and
  - political change.

P5. s6. a2 ***Action 2*** ***Modify the plan as required***

Final plans should not be regarded as permanent, static documents. They should be modified as circumstances change and new data are obtained. Thus following the review and monitoring stages the plan may need to be amended. Where the plan is produced under specific legislation, the process for amendment will usually be specified. In other circumstances, a partial planning process should be initiated through feedback loops to relevant stages in the IBP methodology, beginning with a re-examination of the plan's aims, strategies and objectives (P2, s3).

P5. s7 ***Step 7:*** ***Prepare a final IBP***

The specific requirements of this step may be stated in relevant legislation. However, in general the following three actions should be considered.

P5. s7. a1 ***Action 1*** ***Review the preliminary draft plan based on the submissions received***

- *Analyse submissions on the preliminary draft*
  All comments on the preliminary draft (refer P4, s3, a3) should be considered by the reviewers and where significant issues have been raised, it may be necessary to hold further stakeholder meetings or workshops to refine aspects of the plan. Produce an analysis of the submissions document (refer to the format in P4, s5, a2).

- *Amend the preliminary draft plan, where necessary*
  The recommendations proposed by the reviewers should be incorporated into the revised draft plan.

- *Prepare final draft plan for distribution to wide stakeholder and community groups*
  The final draft document should contain the following:

  - the preliminary draft document and any recommended changes;
  - performance evaluation and monitoring program plan;
  - research program;
  - education strategy;
  - plan review process;
  - implementation program; and
  - enforcement program.

- *Distribute the draft plan widely*

Prepare a mailing list and distribute the draft plan widely. It is important to obtain comments from a broad range of stakeholders and the general public. Copies should be available at the offices of public agencies (e.g. libraries, government departments, relevant NGOs).

P5, s7, a2 ***Action 2*** ***Prepare the final plan and obtain plan approval***

- *Analyse submissions*
  Analyse all submissions, produce an analysis of submissions document.

- *Prepare final IBP*
  Modify the draft plan, as required and produce a final IBP.

- *Prepare for plan approval*
  The planning team should consult relevant legislation or organizational policies to prepare the accompanying documentation for plan approval. This may include a cabinet submission on the plan, ministerial briefing notes, the analysis of submissions document and the like. Each organisation or government department will have a particular approval process that must be complied with.

- *Prepare for printing and distribution of the final plan*
  Undertake required procedures to have the document printed. Distribute the document to relevant outlets.

Conclusion

The IBP model presented in this Guide is based on several important principles of good planning and ecologically sustainable development and incorporates elements of good buffer design from a range of existing buffer strategies. The IBP model incorporates a 'science-based' approach in combination with effective 'practice-based' approaches, resulting in an 'innovative' model to plan for the conservation of biodiversity. Although many individual elements of the IBP model are not new, it is the step-by-step logical structure, multi-disciplinary/cross-sectoral focus, ecological base and participatory processes that are the main innovations and contributions of the model to sustainable land use planning and management in relation to buffers.

The IBP model enables the delineation of integrated buffers for ESAs in a consistent manner using a number of criteria, thus assisting in the creation of scientifically defensible buffers and a more transparent planning approach. The stepped nature of the methodology aims to facilitate consensus building amongst stakeholders at various critical points in the process. The methodology is systematic and comprehensive in its consideration of the major components of the ecosystem and the nature of interrelationships with existing and proposed human activities and land uses. It aims to provide a logical structure for the development of a physical buffer plan and an administrative and procedural guide for the plan's preparation within a planning framework that incorporates consideration of wider planning issues. The model thus provides links between the physical buffer plan and the planning environment. It also incorporates aspects of timing and emphasises the need for an educational and research strategy, as well as enforcement, monitoring and review processes. Importantly the model is a flexible approach allowing individual solutions to identified problems, based on the cooperative involvement of stakeholders. The model simplifies a potentially complex planning task, minimises data gathering and can be wholly or partially applied, dependent on local circumstances and needs.

References

Australian and New Zealand Enironment and Conservation Council (ANZECC) (1998), National Koala Conservation Strategy, AGPS, Canberra.

Brewer, G.D and deLeon, P. (1983), *The Foundations of Policy Analysis*, The Dorsey Press, Illinois.

Callaghan, J., Leathley, S. and Lunney, D. (1994), 'Port Stephens Koala Management Plan. Draft for Public Discussion', NSW NPWS, Pt. Stephens Council and Hunter Koala Preservation Society, np.

Chenoweth EPLA (2000), 'Common Conservation Classification System', available at: http://www. wesroc.qld. gov.au. [14 December 2000].

Clark, T.W. (1992), 'Practicing Natural Resource Management with a Policy Orientation', Environmental Management, 16(4), pp. 423-33.

Clark, T.W. and Kellert, S.R. (1998), 'Toward a policy paradigm of the wildlife sciences', *Renewable Resources Journal*, 6(1), pp. 7-16.

Canadian Environmental Assessment Agency (CEAA) (2000), 'Cumulative Effects Assessment Practitioners Guide', prepared by G. Hegmann, C. Cocklin, R. Creasy, S. Dupuis, A. Kennedy, L. Kingsley, W. Ross, H. Spaling and D. Stalker, available at: http://www.cea.gc.ca/publicationse.cumul/ 3.0_e.htm [26 February 2001].

Council on Environmental Quality (1997), 'Considering Cumulative Effects Under the National Environmental Policy Act', available at: http://ceq.eh.doe. gov/nepa/ccenepa/ccenepa.htm [7 March 1999].

Curtin, A. and Lunney, D. (1995), 'A comparison of community-based survey and field-based survey for koalas in a large reserve system on the outskirts of Sydney, New South Wales', in Australian Koala Foundation (ed.), 'Proceedings on a Conference on the Status of the Koala in 1995 Incorporating the Fourth National Carers Conference', Brisbane, pp. 186-7.

Dietz, J.M., Deitz, L.A. and Nagagata, E.Y. (1994), 'The effective use of flagship species for conservation of biodiversity: the example of lion tamarins in Brazil', in P.J.S. Olney, G.M. Mace and A.T.C. Feistner (eds), *Creative Conservation. Interactive management of wild and captive animals*, Chapman and Hall, London, pp. 32-49.

Environmental Protection Agency (EPA) (2003), 'Regional Nature Conservation Strategy for South East Queensland', EPA, Brisbane.

Graham, B. and Pitts, D. (1996), 'Draft Good Practice Guidelines for Integrated Coastal Planning', prepared for RAPI with assistance from the Federal Department of Environment, Sport and Territories, Canberra.

Hegmann, G. (1995), *A Cumulative Effects Assessment of Proposed Projects in Kluane National Park Reserve*, Yukon, Parks Canada, Yukon.

Hruza, K.A. (1993), 'Buffer Zone Planning. A Possible Management Tool for Fraser Island', Bachelor of Regional and Town Planning thesis, Department of Geographical Sciences and Planning, University of Queensland, Brisbane.

Imperial Oil Resources Ltd (IORL) (1997), 'Cold Lake Expansion Project. Biophysical and Resource Use Assessment', 2, part 1, AXYS Environmental Consulting Ltd, Calgary.

International Union for the Conservation of Nature (IUCN) (2003), 'Recommendations of the Vth IUCN World Parks Congress', World Parks Congress, Benefits Beyond Boundaries, Durban, 9-17 Sept., available at http://www.iucn.org/themes/wcpa/wpc2003/english/outputs/recommendations.htm.

King Cullen, R. (1993), 'A Model for Preparing Planning Guidelines', PhD Thesis, University of Queensland, Brisbane.

Lasswell, H.D. (1971), *A Preview of the Policy Sciences*, American Elsevier, New York.

Lunney, D., Callaghan, J. and Leathley, S. (1997), 'Port Stephens Koala Management Plan', NSW NPWS, Pt. Stephens Council and Hunter Koala Preservation Society, np.

Lynch, K. (1989), *Good City Form*, The MIT Press, London.

Noss, R.F. (1990), 'Indicators for Monitoring Biodiversity: A Hierarchical Approach', *Conservation Biology*, 4(4), pp. 355-64.

Peterson, A. (2002), 'Integrated Landscape Buffer Planning Model', PhD thesis, Department of Geographical Sciences and Planning, The University of Queensland, Brisbane.

Peterson, A. (1991), 'Buffer Zone Planning for Protected Areas: Cooloola National Park', Master of Urban and Regional Planning thesis, Department of Geographical Sciences and Planning, University of Queensland, Brisbane.

Peterson, A. and Kozlowski, J. (2005), *Integrated Buffer Planning: Towards Sustainable Development*, Ashgate, Aldershot.

Pressey, R.L. and Logan, V.S. (1997), 'Inside looking out: findings of research on reserve selection relevant to "off-reserve" nature conservation', in P. Hale and D. Lamb (eds), *Conservation Outside Nature Reserves*, Centre for Conservation Biology, The University of Queensland, Brisbane, pp. 407-18.

Queensland Government (1997), 'State Planning Policy 1/97' (Conservation of Koalas in the Koala Coast), Queensland Government [Brisbane].

Reading, R., Clark, T.W. and Kellert, S.R. (1991), 'Towards an endangered species reintroduction paradigm', *Endangered Species UPDATE*, 8, pp. 1-4.

Roughan, J. (1986), 'Planning for Buffer Zones – An Application of Protection Zone Planning to Nicoll Rainforest', Bachelor of Regional and Town Planning Thesis, The University of Queensland, Brisbane.

Samedi, I. (1994), 'Analysis of threats to koala habitat: Application of a threshold approach to planning issues in the Shire of Pine Rivers, Queensland', in Koala Management Strategy for Pine Rivers Shire, F. Carrick (ed.), prepared for Pine Rivers Shire Council, pp. A.1-A.27.

Saunders, D., Margules, C. and Hill, B. (1998), 'Environmental indicators for national state of the environment reporting – Biodiversity, Australia: State of the Environment (Environmental Indicator Reports)', Department of the Environment, Canberra.

Sayer, J. (1991), *Rainforest Buffer Zones: Guidelines for Protected Area Managers*, IUCN, Gland, Switzerland.

Sharma, R. (1990), 'An Overview of Park-People Interactions in Royal Chitwan National Park, Nepal', *Landscape and Urban Planning*, 19, pp. 134-144.

Tai (1996), 'Principles of Compliance Enforcement', paper presented at Conference on Enforcement and Monitoring, Brisbane.

Torgerson, D. (1985), 'Contextual orientation in policy analysis: The contribution of Harold D. Lasswell', *Policy Sciences*, 18, pp. 241-61.

Watson, J.R. (1995), 'Community Involvement in the Fitzgerald Biosphere Reserve Success Story', paper presented at the International Biosphere Conference, Seville.